Scalable and Secure Internet Services and Architecture

CHAPMAN & HALL/CRC
COMPUTER and INFORMATION SCIENCE SERIES

Series Editor: Sartaj Sahni

PUBLISHED TITLES

HANDBOOK OF SCHEDULING: ALGORITHMS, MODELS, AND PERFORMANCE ANALYSIS
Joseph Y-T. Leung

THE PRACTICAL HANDBOOK OF INTERNET COMPUTING
Munindar P. Singh

HANDBOOK OF DATA STRUCTURES AND APPLICATIONS
Dinesh P. Mehta and Sartaj Sahni

DISTRIBUTED SENSOR NETWORKS
S. Sitharama Iyengar and Richard R. Brooks

SPECULATIVE EXECUTION IN HIGH PERFORMANCE COMPUTER ARCHITECTURES
David Kaeli and Pen-Chung Yew

SCALABLE AND SECURE INTERNET SERVICES AND ARCHITECTURE
Cheng-Zhong Xu

CHAPMAN & HALL/CRC COMPUTER and INFORMATION SCIENCE SERIES

Scalable and Secure Internet Services and Architecture

Cheng-Zhong Xu
Wayne State University
Detroit, MI

CRC Press is an imprint of the
Taylor & Francis Group, an **informa** business
A CHAPMAN & HALL BOOK

The Ambassador Bridge in Detroit, the busiest international border crossing in North America, carries more than 25% of all merchandise trade between the U.S. and Canada. Since the tragedy of 9/11, its tightened security has caused trucks to wait an average of 4 hours to cross the Detroit River. The mission of the bridge is to build a secure border without hindering the free flow of goods. This corresponds with the theme of this book in the control of information flow on the Internet. Photography by Zhiping Liu.

CRC Press
Taylor & Francis Group
6000 Broken Sound Parkway NW, Suite 300
Boca Raton, FL 33487-2742

First issued in paperback 2019

© 2006 by Taylor & Francis Group, LLC
CRC Press is an imprint of Taylor & Francis Group, an Informa business

No claim to original U.S. Government works

ISBN-13: 978-1-58488-377-7 (hbk)
ISBN-13: 978-0-367-39266-6 (pbk)

This book contains information obtained from authentic and highly regarded sources. Reasonable efforts have been made to publish reliable data and information, but the author and publisher cannot assume responsibility for the validity of all materials or the consequences of their use. The authors and publishers have attempted to trace the copyright holders of all material reproduced in this publication and apologize to copyright holders if permission to publish in this form has not been obtained. If any copyright material has not been acknowledged please write and let us know so we may rectify in any future reprint.

Except as permitted under U.S. Copyright Law, no part of this book may be reprinted, reproduced, transmitted, or utilized in any form by any electronic, mechanical, or other means, now known or hereafter invented, including photocopying, microfilming, and recording, or in any information storage or retrieval system, without written permission from the publishers.

For permission to photocopy or use material electronically from this work, please access www.copyright.com (http://www.copyright.com/) or contact the Copyright Clearance Center, Inc. (CCC), 222 Rosewood Drive, Danvers, MA 01923, 978-750-8400. CCC is a not-for-profit organization that provides licenses and registration for a variety of users. For organizations that have been granted a photocopy license by the CCC, a separate system of payment has been arranged.

Trademark Notice: Product or corporate names may be trademarks or registered trademarks, and are used only for identification and explanation without intent to infringe.

Library of Congress Cataloging-in-Publication Data

Catalog record is available from the Library of Congress

Visit the Taylor & Francis Web site at
http://www.taylorandfrancis.com

and the CRC Press Web site at
http://www.crcpress.com

To Jiwen, Jessica, and Julia

Preface

In the past decade we have witnessed explosive growth of Internet services. With the popularity of Web surfing, instant messaging, on-line shopping, and other Internet services, people rely more and more on the Web to share information, communicate to each other, and even conduct business. There is little doubt that the ever-expanding Internet services will continue to change the way we live.

Due to the unprecedented scale of the Internet, popular Internet services must be scalable so as to respond to thousands or even hundreds of thousands of concurrent requests in a timely manner. The requests can be from clients in different geographical places and with different types of access networks and devices. They may also have different expectations for quality of services (QoSs) for various reasons. This scalability requirement goes beyond the capability of any of today's uniprocessor servers. Large-scale Internet services are becoming an increasingly important class of driving applications for scalable computer systems. In particular, server clusters, which are locally or globally distributed, become commonplace behind popular mission-critical Internet services; Web caches, proxies, and content delivery networks are widely deployed at network edges to reduce access latency and save network bandwidth. For example, Google search engine used more than 6,000 Linux/Intel PCs to serve an average of more than 200 million queries per day as of 2003; Akamai formed an Internet overlay network based on over 14,000 servers deployed in more than 1,100 networks to help reap the performance and reliability of media-rich content delivery services; peer-to-peer (P2P) computing model on the horizon pushes the envelope even further by allowing Internet users to serve each other directly. More research projects are ongoing for scalable and highly available Internet services.

Scalability aside, security is another primary concern in Internet services because client and server are often in different administrative domains and the underlying TCP/IP communication protocols are insecure by design. Many technologies have been developed over the years for secure Internet services. A notable is secure socket layer (SSL) protocol for secure electronic transactions. It lays a security foundation for more than $5.5 billion e-commerce business in the 2003 U.S. retail section. Digital signature technology and its legitimation have paved a way to paperless business transactions on the Web. Java Virtual Machine (JVM) provides an agglomeration of security technologies toward general-purpose trusted computing on the Internet.

On the other hand, with the penetration of Internet services into our daily lives, people beome more concerned than ever about security and privacy of cyber activities. People are suspicious of service scalability technologies such as mobile code and prefetching, and more reluctant to accept those without proof of security. Security has a strong interplay with scalability in Internet services.

This book is intended to provide readers with an overview of scalable and secure Internet services and architecture and in-depth analysis of a number of key scaling technologies on the horizon. The topics include

- Server clusters and load balancing
- QoS-aware resource management
- Server capacity planning
- Web caching and prefetching
- P2P overlay network
- Mobile code and security
- Mobility support for adaptive grid computing

The coverage of each topic starts with a problem definition, a comprehensive review of current representative approaches for the problem. It is then followed by a detailed description of novel technologies that we recently developed at Wayne State University. The book stresses the underlying principles of the technologies and the role of these principles in practice with a balanced coverage of concepts and engineering trade-offs. It demonstrates the effectiveness of the technologies by rigorous mathematical modeling/analyses, simulation, and practical implementations. Most of the technologies were originally presented in peer-reviewed technical conferences and journals. This book is not a collection of these published works. It blends concepts, principles, design, analysis, and engineering implementations of a wide array of technologies in a unified framework for scalable and secure Internet services. It represents a systematic treatment of the subject, based on our own cutting-edge research experience over the years.

This book features a balanced coverage in breadth and depth of advanced scaling technologies in support of media streaming, e-commerce, grid computing, personalized content delivery services, distributed file sharing, network management, and other timely Internet applications. A number of software packages that we have recently developed as companions of our research publications are also covered.

Intended Audience

This book is meant for researchers, students, and practicing engineers in the fields of distributed computing systems and Internet applications. The relevance for these fields is obvious, given the increasing scalability and security concerns in Internet services. This book should provide researchers and students with a thorough understanding of major issues, current practices, and remaining problems in the area. For engineers who are designing and developing Internet services, this book provides insights into the design space and engineering trade-offs of scalability and security issues. They may also find the technologies described in this book applicable to real world problems. This book assumes that readers have general knowledge about computer systems and the Internet. It can be used as a text for senior and graduate students, as well as a reference for information technology (IT) professionals.

Acknowledgments

I wish to thank all those who contributed to this book. I thank Ramzi Basharahil, Aharon Brodie, Guihai Chen, Wanli Chen, Ngo-Tai Fong, Song Fu, Daniel Grosu, Tamer Ibrahim, Young-Sik Jeong, Manoj Kona, Shiyong Lu, Jayashree Ravi, Haiying Shen, Weisong Shi, Philip Sokolowski, Ravindra Sudhindra, Le Yi Wang, Jianbin Wei, Brain Wims, Minghua Xu, George Yin, Xiliang Zhong, Xiaobo Zhou, and other former and current CIC group members and colleagues of Wayne State University, without whom this book would not be possible. Part of this book draws on the papers that we have written together in the area of scalable and secure Internet services in the past several years.

A number of other colleagues have contributed to this book in various ways. I would especially thank Dharma P. Agrawal, Laxmi N. Bhuyan, Monica Brockmeyer, Jiannong Cao, George Cybenko, Vipin Chaudhary, Sajal K. Das, Minyi Guo, Kai Hwang, Weijia Jia, Hai Jin, Vijay Karamcheti, Francis C. M. Lau, Minglu Li, Philip McKinley, Burkard Monien, Lionel M. Ni, David Nicol, Yi Pan, Alexander Reinefeld, Nabil J. Sarhan, Loren Schwiebert, Mukesh Singhal, Pradip K. Srimani, Xian-He Sun, Xiaodong Zhang, Cho-Li Wang, Jie Wu, Li Xiao, Zhiwei Xu, Tao Yang, Yuanyuan Yang, and Wei Zhao for reading the manuscript or offering comments and suggestions. I am grateful to Professor Sartaj K. Sahni, the Editor-in-Chief of *Computer and Information Science Series* of Chapman & Hall/CRC, for his encouragment and suggestions during the course of this project. He reviewed the very first version of this book during IPDPS'04 in Santa Fe, New Mexico. I appreciated very much his insightful comments and suggestions.

I also thank acquiring editor Bob Stern, project coordinator Theresa Del Forn, and project editor Sarah E. Blackmon for essential support thoughout the arduous process of getting this book into print. Sarah scrupulously read the final book draft for presentation flaws. A thanks also goes to Zhiping Liu for the cover photo of the Ambassador Bridge in Detroit.

This research was supported in part by U.S. NSF CDA-9619900, EIA-9729828, CCR-9988266, ACI-0203592, NASA 03-OBPR-01-0049, and Wayne State Univerisity's Research Enhancement Program and Career Development Chair Award. Any opinions, findings, and conclusions expressed in this material are those of the author and do not necessarily reflect the views of the funding agencies.

After all, I would like to thank my wife for all she has done and put up with over the years. Without her constant love and support, this book would be an impossible mission. My two adorable daughters also deserve a big thank-you for the joy they bring. I dedicate this book to them in gratitude and love.

Cheng-Zhong Xu
Detroit, Michigan U.S.

Contents

Chapter 1

Internet Services

1.1 Introduction

Internet services (or network services) are a form of distributed applications in which software components running on networked hosts coordinate via ubiquitous communication protocols on the Internet. The components that accept and handle requests are often referred to as server and the components that issue requests to and receive responses from servers are clients. Both client and server are essentially user-level processes, which may be run in the same host or different hosts on the Internet. Most of the network services are organized in a server/client model, in which servers and clients are running on different hosts. There are recent file sharing services in a peer-to-peer model, in which each host assumes both server and client roles. The communication protocols in network services must be open and published. They can either be standardized or industry de facto standards. The Hypertext Transfer Protocol (HTTP) is the most popular standard protocol endorsed by the WWW Consortium (W3C) for communication between Web servers and clients like Microsoft IE and Mozilla Firefox.

Network service emerged with the inception of ARPAnet, an early predecessor of Internet in the late 1960s and early 1970s. E-mail programs were the first type of services demonstrated publicly on the network. Network services took off in the 1980s, in the form of Internet services, with the deployment of TCP/IP and a proliferation of local area networks. Services like FTP, telnet, and e-mail became well received in academy as new national networks like BITnet and NSFnet were phased in. Today's popularity of Internet services was mainly due to the birth of HTTP-based killer Web applications in the mid-1990s. The Web commercialization wave in late 1990s triggered an explosive growth of Internet services. People talk to each other on the Internet using more than e-mail. Instant messages and audio/video-based multimedia communication become more and more popular. At a higher application level, we are relying on the Internet for more than information sharing. In the past decade we have witnessed the creation of a whole new sector of e-business (on-line shopping, e-banking, e-trading, distance learning, etc.). There is little doubt that the ever-expanding Internet services will continue to change the way we live.

This chapter presents an overview of Internet services, key characteristics of example services, and a road map to this book on the technologies for scalable and secure Internet services.

1.2 Requirements and Key Challenges

A defining characteristic of Internet services is openness. An open system means its components can be extended or reimplemented without affecting its original system functions. It is in contrast to closed systems in which developers must have complete system-level knowledge and full control over the disposition of all system components. A key requirement for the openness is well-defined and stable interfaces between the system components. Due to the use of open and well-published communication protocols in Internet services, clients are decoupled from servers so that they can be developed independently by different vendors and deployed separately on their own discretion.

This raises other four key challenges in the design of Internet services: transparency, scalability, heterogeneity, and trustiness.

1.2.1 Transparency

Transparency in Internet services refers to a feature that permits the free passage of data and information flow between the components, without needing to know whether the other components are remote or local (location transparency), if there were possible failures and recoveries in provisioning or delivery (failure transparency), if the service is provided by a group of servers or not (replication transparency), if the service requesting component is stationary or mobile (mobility transparency).

Sun Microsystems' NFS is an example of location transparent network services. It provides access to shared files through a Virtual File Systems (VFSs) interface that runs on top of TCP/IP. Users can manipulate shared files as if they were stored locally. The Google search engine is another example featuring replication transparency. Although it is running on clusters of servers in different data centers, it provides users a single interface.

Indirection is the primary approach to ensure transparency. The golden rule in computer science is, "A layer of indirection can solve every problem," except the problem of having too many layers of indirection. Domain Name System (DNS) is a layer of indirection for location transparency. It supports a hierarchical naming scheme for the use of symbolic domain names for hosts and networks. The naming hierarchy reflects the social organizational structures of the hosts and networks and is independent of their physical locations. Domain names are often the only published interfaces to most Internet services. The DNS translates the domain name of a service to its actual IP address on the Internet and hence hides the physical location of the service.

Web proxy is another example of indirection in access. It serves as a middleman between client and server so as to avoid the need for direct communication between them. This form of indirection decouples client from server and makes their locations and configurations transparent to each other.

Indirection ensures network transparency in function. It cannot mask the difference between local access and remote access in performance. To tolerate latency and hide the access delay due to geographical distance, techniques such as caching and prefetching, content delivery network (CDN) are widely used. They will be discussed in Section 1.2.2 in more detail.

1.2.2 Scalability

The openness of Internet services allows their clients to be deployed independently. Due to the unprecedented scale of the Internet, popular Internet services must be scalable so as to support up to millions of concurrent requests reliably, responsively, and economically as well. These scalability requirements go beyond the capability of any single machine server and the capacity of its networking channels. Internet services have become an important class of driving applications for scalable computing technologies.

An Internet service is scalable if it remains "effective" in performance when there is a significant increase of requests at the same time. Two primary performance metrics are throughput (or bandwidth) and response time. The former is a server-centric metric, referring to the number of concurrent requests that can be handled without causing significant queueing delay. The latter is a client-centric metric, referring to the client-perceived delay from the time when a request is issued to the time when its response is received. These two performance metrics are related. The response time of a request is the request processing time, plus its possible queueing delay in the server and the request/response transmission time in the network. A high-capacity server would handle requests instantly without causing too much queueing delay. However, such servers do not necessarily guarantee low response time. In streaming services that involve the transmission of a large volume of data in the network, their response time also critically depends on the network bandwidth.

The scalability of Internet services can be scaled up by upgrading systems with faster machines and higher bandwidth networks. However, Internet services made by a costly supercomputer and a dedicated super-bandwidth network are neither economical nor necessarily scalable. It is because in an open and dynamic environment, it is uncertain how fast is fast. Internet traffic is often highly variable. This makes capacity planning even harder. Cost concern opens up a third dimension of scalability: cost-effectiveness.

A primary measure for achieving scalability is *replication*. Replicating a server in multiple hosts to form a server cluster provides a scalable solution to keeping up with ever-increasing request load. The cluster of servers can be placed in the same physical location or distributed geographically. *Load balancing* is a key integration that distributes client requests between the servers for the objective of balancing the servers' workload and preserving request locality. It also implements a layer of indirection and supports replication transparency — hiding the fact that more than one server are up and run behind. On another track altogether, data replication implemented by caching proxies tends to place data in the physical proximity of their users so as to reduce user-perceived service access time and meanwhile to cut the requirement for bandwidth in network core. Scalability aside, replication greatly improves service availability and fault tolerance.

Server replication and data caching are two broad approaches for improving the performance of Internet services. But these two solutions share much similarity in technology and mechanism. For example, a surrogate server can serve as a caching proxy, as well; a group of cooperative caching proxies in different physical locations also forms a base of CDN over shared surrogate servers. Over the past few years, we have seen the convergence of these two replication approaches for scalable Internet services. This trend is best demonstrated by recent peer-to-peer (P2P) and grid computing technologies. In a P2P system, each host serves as both a client and a server of data resources, with various degrees of replication, in the system. Grid computing technologies go beyond the objectives of data sharing. It enables the visualization of geographically distributed (and replicated) resources, such as processing, network bandwidth, and storage capacity, and allows many-to-many sharing of these resources. More information about P2P file sharing and grid computing services will be provided in Section 1.3

Replication aside, another important scaling technique is the use of mobile codes. A mobile code, as its name implies, refers to programs that function as they are transferred from one host to the other. It is in contrast to the client/server programming model. Instead of sending requests associated with input data to a server for processing, the mobile code approach uploads codes to the server for execution.

Code mobility opens up vast opportunities for the design and implementation of scalable systems. For example, script programs in ActiveX or Javascript are mobile codes that are widely used to realize dynamic and interactive Web pages. One of the most practical uses of such mobile codes is validating on-line forms. That is, a Javascript code embedded in HTML form pages can help check what a user enters into a form, intercept a form being submitted, and ask the user to retype or fill in the required entries of a form before resubmission. The use of the mobile codes not only avoids the transmission of intermediate results back and forth from client to server and reduces the consumption of network bandwidth and server processing power, but also enhances the responsiveness of the user inputs.

Java applets are another form of mobile codes that empower Web browsers to run general-purpose executables that are embedded in HTML pages. Java applets are limited to one-hop migration from Web server to client. More flexible forms of migration are mobile agents that have as their defining trait the ability to travel from machine to machine autonomously, carrying their codes as well as running states. In Chapters 11 to 18, we give the details of the mobile code approach and related security issues.

1.2.3 Heterogeneity

The Internet is essentially a heterogeneous collection of hosts and networks. Internet services mask the heterogeneity in data representation and communication by using standard or industry de facto standard protocols, such as HTTP, Simple Mail Transfer Protocol (SMTP), Real Time Streaming Protocol (RTSP), and Extensible

Markup Language (XML). However, the heterogeneity between the hosts in processing capacity, access network bandwidth, and other performance aspects is rarely transparent. In particular, recent proliferation of wireless networks has changed the landscape of Internet services. More diversified client hosts, including resource-constrained devices like PDA, cellular phone, and set-top box, have come into the picture. Due to their different resource constraints, users tend to have very diverse service expectations. For example, users who are accessing to a streaming service on desktops with a high bandwidth network in their offices would expect to enjoy the highest quality audio/videos. But if they transfer the streaming channel to their PDAs with a WiFi network when they leave their offices, they would be happy to receive a downgraded quality of streams. There is no doubt that a "same-service-to-all" model for such Internet services with media-rich contents is inadequate and limiting. To deal with the heterogeneity of access devices and networks, promising approaches are *content adaptation* and *adaptive scheduling* that are intended to provide different levels of quality of serivde (QoS) to different clients.

Generally, provisioning of a high-quality service is likely to consume more resources at both server and network sides. By adjusting the level of QoS, service differentiation techniques are able to postpone the occurrence of request rejection as the server load increases. They achieve the scalability in terms of cost-effectiveness. In addition, service differentiation can also provide an incentive for different charging policies. Due to its significance in scalable Internet services, Chapters 4 to 7 are dedicated to the topic of service differentiation.

1.2.4 Security

The openness of Internet services makes it possible for client and server to be developed and deployed independently by different parties. Because of the decoupling, a fundamental identity assumption in existing closed systems no longer holds. In distributed systems within an administrative domain but isolated from the Internet, whenever a program attempts some action, we can easily identify a person to whom that action can be attributed and it is safe to assume that person intends the action to be taken. Since finding responsible identities for malicious or misbehaved actions related to an Internet service is a challenge in open distributed systems, neither its server nor clients can be designed in an optimistic way, by assuming the good will of each other. Security becomes a primary concern in Internet services.

In general, security in open distributed systems involves three issues: confidentiality (or privacy), data integrity, and availability. Confidentiality requires protection of information from disclosure to unauthorized parties; data integrity is to protect information from unauthorized change or tampering; availability ensures legitimate users have access anytime. Most of the security issues in Internet services can be resolved by the use of existing cryptographic key-based distributed security protocols. Because of the distrust between the components in an open system, its security measures rely on a third party for key management.

Due to the dynamic nature and unprecedented scale of the Internet, Internet services pose new security challenges.

Denial-of-service (DoS) attacks. DoS attacks intend to disrupt a service by bombarding the server with such a large number of pointless requests that serious users are unable to access it. They are often carried out in a distributed manner by the use of many hosts in different locations. MyDoom was one of the fastest-growing notorious e-mail worms for distributed DoS (DDoS) attacks. Its variants spread widely in multiple waves in 2004 and brought down its targeted SCO Web site (`www.sco.com`) on February 1, 2004 by using hundreds of thousands infected machines around the world to bombard the server at the same time, according to CNET News [244]. It also slowed down many search engines including Google, Yahoo's AltaVista, and Lycos in July.

Browser-based attacks. Although computer viruses and worm attacks remained viewed as the most threatening security risk, there is growing concern over browser-based attacks by adbots and spywares. Adbots and spywares are programs hidden in a Web page, which are intended to show ads, sabotage visiting computers, or compromise their privacy by secretly monitoring user activity and performing backdoor-type functions to relay confidential information to other machines. Their attacks make clients feel plagued more often than ever and may even pose the next big security nightmare, according to a recent survey by CompTIA [69]. Microsoft has strived to counter browser-based attacks by fixing Internet Explorer security holes. It announced 20 vulnerabilities on a single day of April 13th, 2004. The growing concerns over Microsoft IE security has given an early boost to the open-source Firefox browser.

Content tampering attacks. Existing cryptographic key-based secure channel techniques ensure data integrity in transmission. However, network services often rely on content caching and adaptation technologies in support of scalability and heterogeneity. These technologies raise new challenges for service integrity. For example, to deliver a media-rich content to resource-constrained device like PDA, content transcoding is often used to downgrade the image resolution and size. But such an adaptation causes problems in the reasoning of content authenticity.

Security in mobile codes. Adbot and spyware are mobile codes that spread by visiting Web sites, e-mail messages, instant messengers, etc. Mobile codes are not necessarily hostile. Mobile codes like script programs in ActiveX and Javascript, Java applets, and mobile agents provide promising approaches for achieving the scalability of Internet services.

In an open environment, mobile codes can be written by anyone and execute on any machine that provides remote executable hosting capability. In general, the system needs to guarantee the mobile codes to be executed under a controlled manner. Their behaviors should not violate any security policy of the system. For mobile agents that are migrated from one system to another, they may carry private information and perform sensitive operations in the execution environments of the hosts they visited. An additional security concern is the agents' privacy and integrity. Security measures are needed to protect the agent image from being discovered and tempered by malicious hosts.

1.3 Examples of Scalable Internet Services

1.3.1 Search Engine

With the explosive growth of WWW, one of the grand challenges facing users is information search. The Internet without search is like a cruise missile without a guidance system. Search engine is such a service that enables users to locate information on the Web. The Google search engine [44] developed by Brin and Page has turned Web searching into a profitable industry. In addition to Google, other major search engines in the Web, as listed in Search Engine Watch (`www.searchenginewatch.com`), include AllTheWeb.com, Yahoo's Inktomi, and Ask Jeeves' Teoma.

At the core of each search engine is a crawler (or spider) that visits on a regular basis all the Web sites it knows, reads the pages, and follows the links to other pages site by site. The collected pages are then indexed to make them searchable by a search engine software. It sifts through the indexed Web pages to find matches to a search query and rank them in order of their relevance. All crawler-based search engines share a similar baseline architecture. These engines differ in how these components are turned. For example, the Google search engine ranks the importance of a Web page or site by its citations [44]. It interprets a link from page A to page B as a vote for page B by page A and assesses the importance of a page by the number of votes it receives. Each vote is also given a weight based on the importance of its casting page. Additional hypertext-matching analysis determines the pages that are most relevant to a specific query.

Search engines are often run on Web clusters to achieve scalability and high availability. According to Hennessy and Patterson [152], the Google search engine used more than 6,000 Linux/Intel PCs and 12,000 disks in December 2000 to serve an average of almost 1 thousand queries per second as well as index search for more than 1 billion pages. A recent report in *Newsweek* revealed that the Google query load increased to over 200 million searches in the course of a day and its index was probed more than 138,000 times every minute worldwide in 90 languages [192].

Search engines are characteristic of dynamic Web pages. In contrast to static Web pages prestored on a server, dynamic pages are generated by programs on the server every time a request is processed. For a static Web page, the processing time of its requests is mainly determined by the page file size. Past Web traffic characterization studies pointed out that the file popularity distributions were Zipf-like and the file size distributions were heavy-tailed and that the transferred Web pages were typically smaller than 50 kbytes of size [13]. Recent Web traffic studies also showed Web pages are shifting from static pages to dynamic contents [12, 238]. Dynamic contents provide a far richer experience for users than static pages, but generating dynamic pages on the fly incurs additional overhead on server resources, especially on CPU and disk I/O. Moreover, dynamic pages exhibit a low degree of reference locality and benefit little from caching. All these factors pose challenges for scalable and highly available search engines.

1.3.2 On-line Shopping and E-Commerce

The Web initially debuted as an agent for information dissemination and sharing. With the explosive growth of Internet services in the past decade, the Web is becoming a major venue for people to conduct bidding, shopping, and other on-line e-business. According to quarterly reports of U.S. Census Bureau (`www.census.gov/mrts/wwwi`), the U.S. retail e-commerce sales in 2003 went up 26.2% from year 2002, reaching $5.5 billion. There is little doubt that the ever-expanding Internet services will change the way we live.

Amazon.com is a leading on-line bookseller. According to a recent report in eWeek.com [128], over the 2003 holiday season, Amazon.com handled the shipment of more than a million packages per day, and processed more than 20 million inventory updates daily. In October 2003, for the first time in its 8-year history, Amazon.com posted a profitable quarter that wasn't driven by holiday shopping.

Amazon database contains more than 14 terabyte of data and continues to double each year in both size and query volume. Its data warehouse is under migration from current Sun servers to clusters of Linux servers, running Oracle's Real Application Clusters (RAC) software.

E-commerce workloads are composed of sessions. A session is a sequence of requests of different types made by the same customer during a single visit to a site. During a session, a customer can issue consecutive requests of various e-commerce functions such as browse, search, add-to-the-shopping-cart, register, and pay. Most of the requests are for dynamic pages. Moreover, different customers often have different navigational patterns and hence may invoke different functions in different ways and with different frequencies. Menascẽ and Almeida [218] presented analytical techniques of workload and performance modeling as well as capacity planning for e-commerce sites.

Three key QoS issues on e-commerce sites are responsiveness, availability, and security. As in search engines, server clusters are popular approaches for achieving responsiveness and availability. But load balancing strategies must be session-oriented. That is, they must direct all consecutive and related requests of a session from one customer to the same server. This is known as session integrity (also referred to as sticky connections). For example, a customer may add an item to the shopping cart over a TCP connection that goes to server A. If the next connection goes to server B, which does not have the shopping-cart information, the application breaks. This required coordination in serving of related requests is referred to as client affinity (affinity scheduling). Hence, simple round robin load balancing strategies are not applicable in the stateful e-commerce services while locality-aware request routing mechanisms provide a convenient way to support session integrity.

Security is another major concern with e-commerce. Secure socket layer (SSL) protocol is a baseline technology used by most e-commerce sites to provide confidentiality and authentication between Web client and server. Processing SSL transactions puts an extra load on server resources. This means customer transaction requests and browsing requests have different requirements for server-side resources. They also have different impacts on the revenue of the sites. On the other hand, an e-commerce merchant tends to share servers handling transaction requests with other merchants in a Web cluster. Load balancing strategies must deal effectively with these issues.

1.3.3 Media Streaming Services

On the Internet we are seeing the gradual deployment of media streaming applications, such as audio over IP, video conferences, and video-on-demand (VoD). These media streaming applications generate traffic with characteristics and requirements that differ significantly from conventional Web application traffic.

Audio and video streams are continuous and time-based. The term "continuous" refers to the user's view of the data. Internally, continuous streams are represented as a sequence of discrete data elements (audio sample, video frame) that replace each other over time. They are said to be time-based (or isochronous) because timed data elements in audio and video streams define the semantics or "content" of the streams. Timely delivery of the elements is essential to the integrity of the applications. In other words, streaming applications have stringent timing and loss requirements. However, unlike typical Web pages, streaming media do not necessarily require the integrity of the data elements transferred. There can be some packet loss, which may simply result in reduced quality perceived by users.

Continuous streams are often bulky in transmission. This is especially so for reasonably high-quality videos. Streaming media servers need to move data with greater throughput than conventional Web servers do. Typical audio stream rates remain on the order of tens of kilobytes per second, regardless of the encoding scheme. Video streams span a wide range of data rates, from tens of kilobytes per second to tens of megabytes per second. Server disk I/O and network I/O are prone to traffic bottleneck. The characteristics of encoded videos (frame size and frequency) vary tremendously according to the content, video compression scheme, and video-encoding scheme. In addition, streaming traffic usually consists of a control part and a data part, potentially using different protocols like RTSP and Voice over IP (VoIP), while conventional Web traffic is homogeneous in this case. Hence, a traffic manager must parse the control channels to extract the dynamic socket numbers for the data channels so that related control and data channels can be processed as a single, logical session.

These distinct characteristics of streaming applications impose a great impact on caching and load balancing. For example, statistics of video-browsing patterns suggest that caching the first several minutes of video data should be effective. However, this partial caching notion is not valid for Web object requests. The large video size and the typical skews in video popularity suggest that several requests for the same object coming within a short timescale be batched and a single stream be delivered to multiple users by multicast so that both the server disk I/O and network bandwidth requirements can be reduced.

Due to the media popularity distributions, data placement methods can also be crucial to load balancing in distributed streaming servers. We investigate the effect of data placement on the performance of streaming services in Chapter 5.

1.3.4 Peer-to-Peer File Sharing

The ubiquity of the Internet has made possible a universal storage space that is distributed among the participating end computers (peers) across the Internet. All peers assume an equal role and there is no centralized server in the space. Such architecture is collectively referred to as the peer-to-peer (P2P) system. Since the concept was first used by Napster [140] for music sharing in the late of 1990s, P2P file sharing has rapidly become one of the most popular Internet services, with millions of users online exchanging files daily. P2P designs harness huge amounts of resources — the content advertised through Napster has been observed to exceed 7 TB of storage on a single day, without requiring centralized planning or huge investments in hardware, bandwidth, or rack space. As such, P2P file sharing may lead to new content distribution models for applications such as software distribution, distributed file systems, searching network, and static Web content delivery.

The most prominent initial designs of P2P systems include Napster and distributed Gnutella [133] and Freenet [65]. Unfortunately, all of them have significant scaling problems. For example, in Napster a central server stores the index of all the files available within the Napster user community. To retrieve a file, a user queries this central server using the desired files well-known name and obtains the IP address of a user machine storing the requested file. The file is then downloaded directly from this user machine. Thus, although Napster uses a P2P communication model for the actual file transfer, the process of locating a file is still very much centralized. This makes it both expensive (to scale the central directory) and vulnerable (since there is a single point of failure). Gnutella decentralizes the file location process as well. Users in a Gnutella network self-organize into an application-level network on which requests for a file are flooded with a certain scope. Flooding on every request is clearly not scalable. Freenet goes a step further by addressing the security issues and privacy issues, but it still employs a flooding-based depth-first search mechanism that incurs heavy traffic for large systems. Both Gnutella and Freenet are not guaranteed to find an existing object because the flooding has to be curtailed at some point.

People started the investigation with the question: could one make a scalable P2P file sharing system? It is known that central to any P2P system is an indexing scheme that maps file names (whether well known or discovered through some external mechanism) to their locations in the system. That is, the P2P file transfer process is inherently scalable, but the hard part is finding the peer from whom to retrieve files. Thus, a scalable P2P system requires, at the very least, a scalable indexing mechanism. In response to these scaling problems, several research groups have independently proposed a second generation of P2P systems that support scalable routing and location schemes. Unlike the earlier work, they guarantee a definite answer to a query in a bounded number of network hops while maintaining a limited amount of routing information; among them are Tapestry [351], Pastry [266], Chord [301], and CAN [256].

1.3.5 Open Grid Service

A grid is a layer of network services that support the sharing and coordinated use of diverse resources in a dynamic, distributed virtual organization. A VO is defined as as a group of geographically and organizationally distributed individuals and institutes that appear to function as a single unified organization with a common mission. The network services in the grid, often referred to as grid service, can be aggregated in various ways toward the virtual organization mission.

A cornerstone of grid computing is an open set of standards and protocols, e.g., Open Grid Service Architecture (OGSA, `www.globus.org/ogsa`), which enable communication across participants in a virtual organization. Grid technologies were first developed to optimize computing and data resources for scientific and engineering applications. A grid connects diverse computing resources in a VO, from supercomputers down to desktop PCs and give users unprecedented access to massive computing power and storage capacity to tackle the most challenging computational problems. In addition to computing and storage resources, the grid could also integrate access to various databases and scientific instruments such as microscopes and telescopes that produce volume data in real time. By virtualizing such diverse resources, the grid would enable significant scientific advances.

TeraGrid (`www.teragrid.org`) is an ongoing project with the aim of building and deploying the world's largest, most comprehensive, distributed infrastructure for open scientific research. By 2004, the TeraGrid will include 20 teraflops of computing power distributed at five sites, facilities capable of managing and storing nearly 1 petabytes of data. All the TeraGrid components will be tightly integrated and connected through a 40 Gigabit network.

Early research on grid technologies focused on resources and resource sharing. The recent focus is to align grid technologies with Web services via OGSA so as to capitalize on desirable Web service properties including service description, discovery, etc.

1.4 Road Map to the Book

This book is organized in two parts. Part I is on cluster-based approaches and related server-side resource management for scalabliblity and high availability in Internet services. Part II is on network-side scaling techniques with an emphasis on mobile code approaches for network-centric Internet services. Figure 1.1 shows a

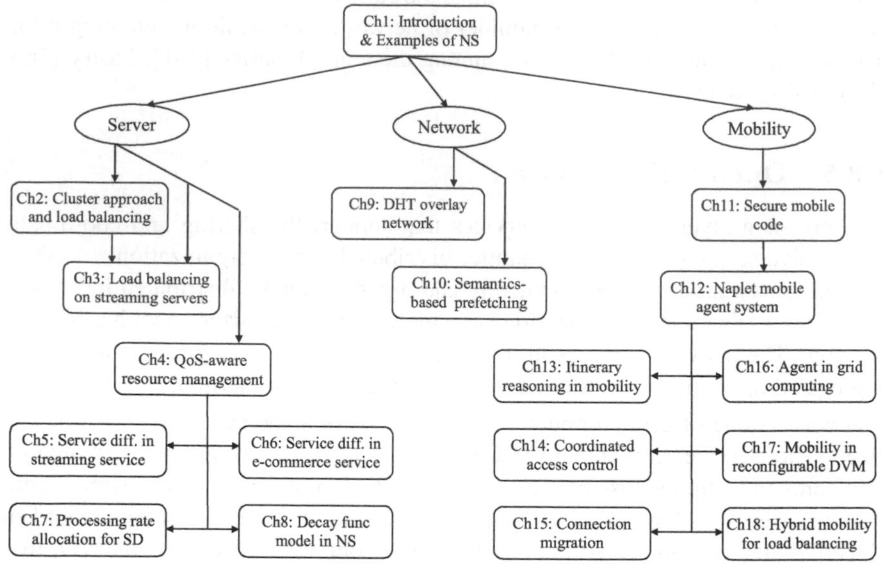

Figure 1.1: Road map to the book.

road map to the book.

Chapter 2 gives an overview of the cluster-based approach for scalable Internet services. The focus is on load balancing technologies for locally distributed servers and geographically distributed content delivery networks. It concludes with a unified W5 model of the load balancing problem for understanding its key issues, in particular, the rationale behind load balancing strategies in parallel computers, in distributed systems, and on the Internet.

On server clusters with no shared storage, data placement is crucial for streaming services where the data volume is too large to be replicated on each server. In Chapter 3, we study the stream replication and placement problem on VoD servers. Since the quality of VoD services is largely constrained by the storage capacity and the server network I/O bandwidth, we present an optimal strategy that maximizes the encoding bit rate and the number of replicas of each video, and meanwhile maintains balanced utilization of network I/O bandwidth of different servers.

Chapter 4 presents an overview of QoS control technologies for scalability and high availability of Internet servers. Service differentiation technologies not only can provide degraded levels of QoS to client requests when a server is heavily loaded, but also adapt the service quality to meet the variety of client preferences and device diversity.

To provide ubiquitous access to the proliferating rich media on the Internet, scalable streaming servers must be able to provide different levels of quality of services to various client requests. Recent advance of transcoding technologies makes it possible for streaming servers to control the requests' network-I/O bandwidth usages on the fly. In Chapter 5, we study the service differentiation problem in streaming servers and present a harmonic bandwidth allocation strategy that maximizes the system utilization and meanwhile provides a guarantee of proportional sharing of constrained network I/O bandwidth between different classes of requests, according to their predefined differentiation weights.

The service differentiation model can also be applied to e-commerce servers to maximize profit. Chapter 6 presents a two-dimensional proportional slowdown differentiation (PSD) model to treat requests in different states within a session and between sessions differently according to their profitability. It gives a simple rate allocation scheme based on M/M/1 queueing model for PSD provisioning. In Chapter 7, we revisit the problem in a general M/G/1 model and develop an integrated rate allocation strategy with feedback control to correct deviations caused by bursty input traffic patterns.

In Chapter 8, we study the impact of the second moments of input traffic process on server load in a general context. The chapter presents a decay function model of request scheduling algorithms that facilitates the study of statistical correlations between the request traffic, server load, and QoS of requests.

The second part of this book is on network-side scaling techniques. Chapter 9 presents a P2P technology for distributed caching and robust information sharing. It starts with a generic architectural model that characterizes the fundamental requirements of a variety of distributed hashing tables (DHT). It presents a constant-degree lookup efficient DHT that we have recently developed.

Chapter 10 is on prefetching technologies to reduce the average Internet access latency. It presents a semantics-based prefetching architecture that improves upon existing prefetching strategies by exploiting semantic locality in requests. Its effectiveness is demonstrated in a personalized network news service.

The rest of the book is dedicated to mobile code approaches and their applications. Chapter 11 surveys existing mobile code technologies and approaches to enhance the security of mobile code systems. Focus is on mobile agents, a general form of mobile codes.

Chapter 12 presents the design and implementation of a Naplet mobile agent system under development at Wayne State University. It features a structured itinerary language, coordinated tempo-spatial access control, and connection migration. Details of these features are described in Chapter 13, Chapter 14, Chapter 15, respectively.

Chapter 13 presents an itinerary language, MAIL, for safety reasoning and assurance of agent mobility. It is an extension of the core itinerary constructs of the Naplet system. The language is structured and compositional so that an itinerary can be constructed recursively from primitive constructs. We prove MAIL is regular-completeness in the sense that it can specify any itinerary of a regular trace model.

Chapter 14 focuses on server protection in a coalition environment where the servers are generally cooperative and trustworthy. It presents the design and implementation of an agent-oriented access control model in the Naplet system and a coordinated spatial-temporal access control model in the coalition environment.

Chapter 15 presents a reliable connection migration mechanism that supports transparent migration of transient connections between mobile agents. It also guarantees exactly-once delivery for all transmitted data.

Chapter 16 presents mobility support for adaptive grid computing. Due to the dynamic nature of the Internet, grid computing services on the Internet ought to rely on run-time support for being adaptive to the change of the environment. Mobility is a primary function to support such adaptation. Chapter 16 focuses on a resource brokerage system that matches resource demands with supplies, based on mobile agents. The system contains an integral distributed shared array (DSA) component in support of distributed virtual machines (DVMs) on clusters.

In addition to agent migration, the virtual machine services ought to be mobile for the construction of reconfigurable DVMs in adaptive grid computing. Chapter 17 applies code mobility for service migration and presents the design and implementation of a mobile DSA for reconfigurable DVMs. Mobile agents in reconfigurable DVMs with support for service migration define a hybrid mobility problem. Chapter 18 addresses the hybrid agent/service migration problem for load balancing in DVMs.

While reading the chapters in order is recommended, this is not necessary. After Chapter 1, readers can move on to Chapters 2 and 3 for load balancing on server clusters, Chapters 4 through 7 for QoS differentiation provisioning, Chapter 8 for server capacity planning, Chapter 9 for P2P overlay network, Chapter 10 for prefetching, and Chapters 11 through 18 for mobility codes and mobility support for adaptive open grid services.

Chapter 2

Network Load Balancing

Due to the unprecedented scale and dynamic nature of the Internet, popular Internet services are often run on Web clusters so as to guarantee the scalability and availability of the services. Web clusters rely on load balancing technologies to distribute the service traffic between the cluster servers transparently to clients. In this chapter, we define the load balancing problem on Internet servers and discuss the requirements of service traffic for load balancing strategies. We present a survey of hardware and software approaches in both server and network sides. The concept and principle of load balancing have long been studied in a general context of parallel and distributed systems. We conclude this chapter by presenting a unified W5 model of the load balancing problem for understanding its key issues, in particular, the rationale behind load balancing strategies in different types of the systems.

2.1 The Load Balancing Problem

The explosive growth of Web technologies in the past decade has resulted in great opportunities for the deployment of a wide variety of Internet services and making them accessible to vast on-line users. Due to the unprecedented scale and dynamic nature of the Internet, popular Internet services must be scalable to support a large number of concurrent client requests reliably, responsively, and economically. These scalability and availability requirements pose great challenge on both processing power and networking capacity. Internet services have become an important class of driving applications for scalable computer systems. In particular, server cluster architecture is gaining momentum due to its salient feature of cost-effectiveness. For example, the Google search engine used more than 6,000 Linux/Intel PCs to service an average of 1 thousand queries per second as well as index search for more than 1 billion pages [152], A recent survey of large-scale Internet services also revealed that all the representative services, including search engines, online shopping, and bidding, were running on a cluster of hundreds or even thousands of servers at more than one data center [234].

A Web server cluster is defined as a group of servers that work together as a single coherent system for the provisioning of scalable and highly available Internet services. It relies on load balancing technologies to distribute service traffic between its back-end servers efficiently and transparently to the clients. Scalability is a measure of the system ability to meeting the increasing performance demands as the service traffic, or load, grows. The system capacity can be measured by the number of concurrent connections that the server can support without causing significant queueing delay, the number of connections per second, or throughput in bit rate of traffic through the server internal infrastructure.

A client-centric performance metric is responsiveness, or client-perceived service time. It is the request processing time in the server, including its possible queueing delay, plus the request/response transmission time in the network. The system capacity and responsiveness metrics are related. A high-capacity server would cause less queueing delay and consequently lead to quick responses to requests. Load balancing strategies enable a Web cluster to scale out with more participating servers to keep up with ever-increasing request load.

In addition, load balancing technologies improve the system availability by taking advantage of the server redundancy. Availability refers to the server's ability to provide uninterrupted services over time. It is often measured as the percentage of uptime. For example, an availability level of 99.9% translates to 8 hours 45 minutes of downtime per year, whether planned or unplanned. If one of the cluster servers fails or goes off-line, its load is automatically redistributed among the other available servers with little or no impact on the service as a whole.

In a Web cluster, the servers are not necessarily located in the same site. They can be dispersed in different geographical locations. The content of the origin server can also be cached in proxy servers in different places. Due to the scale of the Internet, service transmission time in the network is a significant performance factor, in particular, in the delivery of media-rich contents. Deployment of servers in different sites tends to move services closer to their clients so as to reduce the network transmission delay. They can also reduce the downtime due to terror attacks, earthquake, fires, power outage, and other environmental factors. The globally distributed servers, including origins and cache proxies, are connected directly or indirectly by a service-oriented logical network on the Internet. Such a network is more often referred to as a *service overlay network.* The network introduces an additional layer for load balancing among different sites to guarantee the scalability and high availability of the Internet services.

A simple form of service overlay networks is *server mirroring*. It is a process of replicating the entire document tree of an origin server on several mirrors. This process is widely used for on-line distribution of new releases of large software packages, such as RedHat Linux (`www.redhat.com`) and Mozilla browser (`www.mozilla.org`). Although server mirroring shares the same objective of service overlay networks, it provides no support for transparent load balancing. Instead, it advertises the servers' addresses and leaves the selection of servers to the clients.

Load balancing on server clusters involves a number of key issues. The foremost is workload measurement. As an abstraction of the work to be performed, workload has different meanings in different applications. In Internet services, a basic scheduling unit of load balancing is the client request and a simple server load index is the number of active connections in response to client requests. The requests are nonuniform in terms of their resource demands because they may need to access different data and cause different types of operations. Requests for dynamic Web pages impose an extra overhead on server resources, especially on CPU and disk I/O. Load balancing based on the metric of active connections may not necessarily suffice.

Moreover, requests of the same e-commerce transaction from a client often have a strong data-dependent relationship. Requests from different clients may also exhibit rich data locality, if they are about to access the same data. Effective load balancing on a server cluster must take into account all these workload factors. Chapter 1 characterizes the workload of popular Internet services and presents their implications on load balancing and cluster technologies.

Two other key issues in load balancing are traffic switching policy and mechanism. A load balancing policy determines the target server for each incoming request, while a mechanism enforces the traffic distribution policy efficiently and transparently to clients. There are many load balancing policies and mechanisms with different characteristics. We present a survey of the representative strategies for server clusters and service overlay networks in Section 2.2 and Section 2.3, respectively.

2.2 Server Load Balancing

Server load balancing assumes a special system, called load balancer, to be deployed at the front end of a server cluster. All the back-end servers can be connected to the balancer directly or indirectly via a switch. The load balancer and the servers form a virtual server to clients. The virtual server has an open IP address for clients to access it. The IP address is often referred to as a virtual IP (VIP) address, configured on the load balancer as a representation of the entire cluster. The load balancer takes as input all requests destined for its VIP and forwards them to selected back-end servers based on its built-in load balancing policies. To avoid creating a single point of failure, two or more load balancers can be deployed and configured as an active–standby or active–active redundancy relationship. Since this chapter focuses on load balancing principles and policies, we assume a single load balancer in a cluster. Readers are referred to [40] for details of redundant balancer configurations.

There are two types of load balancers according to the layer information of the packets that the balancers take into account. A layer-4 balancer switches packets based on the transport layer's protocol header information contained. TCP and UDP are two most important transport layer protocols for Internet services. The port numbers contained in their headers can be used to distinguish between the popular HTTP, SSL, FTP, and streaming applications. The IP source address in the network protocol header makes it possible for client-aware load balancing algorithms. Another type of load balancers is the layer-7 balancer. It switches packets based on application-level payload information contained in the packets. Information like URLs of access objects and cookies help the balancer to make more intelligent switching decisions.

2.2.1 Layer-4 Load Balancer

A load balancer is essentially a traffic switch that diverts requests to different backend servers according to various load balancing policies. The load balancer also determines the return path of server responses. A load balancer can be organized in a two-way or one-way manner according to the return path. A one-way (or one-arm) organization allows servers to return their responses directly to the clients, bypassing the load balancer. In contrast, a two-way (or two-arm) organization requires all the responses to pass the balancer. The one-way organization is often called *direct server return*. *Packet double rewriting* and *packet forwarding* are two representatives of the organization, respectively.

Packet double rewriting. Packet double rewriting is a two-way approach for load balancing, based on IP Network Address Translation (NAT) [182]. NAT provides a way to translate a packet IP address in one network to another IP address in a different network. In packet double rewriting, the servers are often deployed on a separate network from the VIP of the load balancer. Each server is configured with a private (nonrouted) IP. The load balancer is configured as a layer-3 default gateway between the public network and server cluster.

On the receipt of a client packet, the load balancer selects a target server according to a load balancing policy and then rewrites the packet's destination IP address accordingly. The server's response packets need to go through the load balancer because it acts as the server's router. When the load balancer receives the response packets, it replaces their source IP addresses with the VIP address, making them look as though they are coming from the virtual server. Due to the modification of IP head, the checksum needs to be calculated twice.

NAT-based load balancing has another limitation in streaming servers. It is known that a streaming protocol typically involves a TCP-based control connection and a UDP-based data connection. To access a media server, a client first opens a control connection to the server and then negotiates with the server for an IP address and a port number for subsequent data delivery. For packet rewriting in data connections, the load balancer must watch the negotiation process and translate any private IP address and port number to public ones. Since the agreement is contained in the control packet payload, the load balancer must be enhanced with a capability of taking into account application-level information. We present the details of such layer-7 load balancers in Section 2.2.2.

Load balancers with a two-way configuration can be implemented easily in hardware and deployed with no need for modification of existing environments. Examples of systems that are based on packet double rewriting include Cisco LocalDistributor, F5 Networks Big/ip, Foundry Networks ServerIron, Linux Virtual Server; see [40, 178] for details. However, such a load balancer is prone to traffic bottleneck because it has to process both incoming requests and outgoing responses [276]. This problem becomes even more severe when responses are bandwidth-intensive, particularly in streaming services. In contrast, the one-way configuration would be more efficient in this respect.

Packet forwarding. Packet forwarding assumes that the load balancer is located in the same subnet as the servers, connected by a layer-2 switch. They all share one virtual IP address. The load balancer configures the VIP as its primary address and the servers use the VIP as their secondary address configured on their loopback interfaces. The servers disable Address Resolution Protocol (ARP) functions so that the incoming packets can reach the load balancer without conflict.

On the receipt of a client packet, the balancer forwards it to a selected server on the same subnet by changing its destination MAC address to the server. The receiving server processes the packet and sends back the response packets to the client directly, without intervening the balancer. Since all the response packets use the VIP as their source, service transparency is retained. Packet forwarding is used in systems such as IBM Network Dispatcher, ONE-IP, Intel NetStruture Traffic Director, Nortel Networks Alteon, and Foundry Networks ServerIron; see [40, 178] for the details of these products.

Packet tunneling. Packet tunneling (IP tunneling or IP encapsulation) was initially designed for constructing a secure network segment on the Internet [182]. It is to wrap each IP datagram within another IP datagram as payload, thus providing a method of redirecting the IP datagram to another destination. In this approach, the IP addresses of a load balancer and the back-end servers should be configured in the same way as in the packet forwarding approach. After deciding on a server for an incoming datagram, the balancer encapsulates the datagram via IP encapsulation; the source and destination of the outer header are set to the VIP and a selected server IP address, respectively. On the receipt of such a wrapped datagram, the selected server unpacks the datagram and gets the original datagram. Because the server shares the VIP with the load balancer, the original IP datagram is delivered to the server through its private (secondary) address. The packet tunneling approach requires the IP encapsulation support in the load balancer and all the servers.

Note that the load balancer in a server cluster is not necessarily a specialized hardware device. It can be a general-purpose computer with an installation of cluster-aware operating system. Linux kernel releases (2.0 and later) contain an IP Virtual Server (IPVS) module that acts as a load balancer for TCP and UDP traffic [349]. Although this software-based approach delivers a limited scaling performance, in comparison with hardware-based layer-4 switches, it is flexible in support of various system configurations. For example, the Linux/IPVS module provides built-in support for all three one-way and two-way cluster organizations.

2.2.2 Layer-7 Load Balancer

TCP uses a three-way handshake protocol to open a connection. In the protocol, an initiator, say host A, sends a TCP SYN packet to a recipient host B. Host B responds with a TCP SYN ACK packet, if it accepts the connection request. The connection is established after host A acknowledges the receipt of the packet from B with a new TCP SYN ACK packet. Layer-4 load balancer starts to select a server on the receipt of a TCP SYN packet from a client. The TCP SYN packet is forwarded to a target server right after it is determined. The target server completes the connection setup procedure with the client under the assistance of the load balancer as a middleman.

The client information available to layer-4 load balancers is limited to the IP source address and TCP port number in the TCP SYN packet. To take advantage of application-level information, such as cookies and URL addresses, in subsequent TCP data packets for smarter decisions, a layer-7 load balancer must be able to serve as a *TCP gateway*, establishing a connection on behalf of the target server before it is determined. That is, the balancer delays the binding of a TCP connection to a server until the application request data are received. Once the balancer determines the target server, it opens a TCP connection with the server and then forwards application request and response data between the two connections. In TCP communication protocol, each packet is assigned with a sequence number to help ensure reliable data delivery of the packet. Since the two TCP connections associated with the load balancer have different initial sequence numbers, the balancer must also translate the sequence number for all request and response packets to keep the TCP protocol semantics. It is prone to bottleneck in traffic flow. In the following, we present two efficient layer-7 load balancers that improve over the TCP gateway organization by avoiding sequence number translations.

TCP splicing. TCP splicing relies upon a special communication component in between the network and the MAC layers [67]. When a load balancer (or dispatcher) is requested by a client to establish a TCP connection, it accepts the request instead of redirecting it to a server. The TCP splicing component inspects the contents of the client's subsequent data requests and then establishes another TCP connection with a selected back-end server. When both connections are set up, the TCP splicing component transfers all the requests and responses between the two connections by altering their headers (source and destination IP addresses, IP and TCP header checksums, and other fields) and makes them appear to be transferred in a single TCP connection. Figure 2.1 presents an illustration of the TCP splicing principle: (1) when a client with IP address 200.0.0.1 wants to access `http://www.xyz.com`, (2) the load balancer accepts this request and establishes a connection with the client. (3) After getting the client's data request, the load balancer selects a server at 10.0.0.2 to handle the request and then splices the two connections. (4) On the receipt of a client request with a source address of 200.0.0.1 and a destination of 141.200.10.1, the TCP splicing component modifies its source and destination addresses to 200.0.0.1 and 10.0.0.2, respectively. (5) The modified request is sent to the server directly. (6) Similarly, when it receives a response

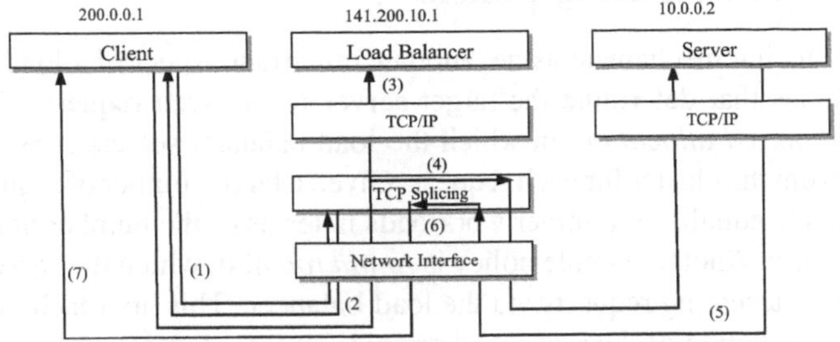

Figure 2.1: TCP splicing architecture.

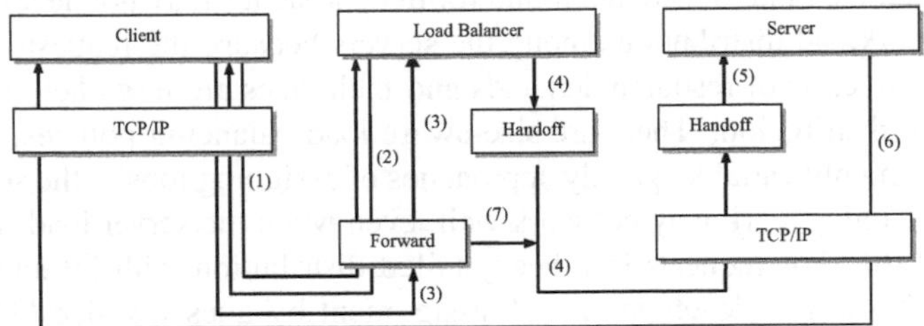

Figure 2.2: TCP hand-off architecture.

from a server with a source address of 10.0.0.2 and a destination of 200.0.0.1, the splicing component changes them to 141.200.10.1 and 200.0.0.1, respectively. (7) The modified response is forwarded to the client without intervening the load balancer. TCP connection splicing avoids many protocol overheads at high-level network layers.

TCP handoff. TCP handoff requires an implementation of a TCP hand-off protocol on top of the TCP/IP stack in the load balancer as well as the back-end servers [239]. Figure 2.2 depicts the traffic flow of this approach as follows: (1) a client issues a request to establish a connection with the load balancer; (2) the load balancer accepts the request and sets up the connection; (3) the client sends a data request to the load balancer; (4) the load balancer receives the request, inspects its contents, and hands the connection to a selected server using the hand-off protocol; (5) the back-end server takes over the established connection; (6) the server's responses are returned to the client directly as if they were from the virtual server; (7) a forwarding module located between the network interface driver and the TCP/IP stack is used for acknowledgment packets. The module first checks whether a packet should be forwarded. If so, it is modified and sent to the appropriate server without traversing the TCP/IP stack.

2.2.3 Load Balancing Policies

Load balancing mechanism aside, another important aspect of a load balancer includes policies that determine the target server for a given request. The simplest one is *randomized allocation*, in which the load balancer selects a back-end server randomly from the cluster for each request. Given a large number of requests, this approach tends to equalize the server workloads in terms of the number of requests in a statistical sense. Another simple policy is *round robin*, in which the servers in a cluster take turns accepting requests via the load balancer. This approach guarantees an equalized distribution of the requests deterministically. A variant of the round robin policy is weighted round robin. It assigns different numbers of requests to different servers according to their predefined weights. The weights are often set based on the servers' static configuration like CPU speed, memory size, and network bandwidth. Albeit simple, the randomized and round robin approaches may not necessarily lead to a good workload distribution among the servers because the requests are of different sizes in terms of resource demands and their sizes are less likely to conform to a uniform distribution. There are size-aware load balancing policies. But when job sizes are highly variable, greedy approaches of assigning jobs to the servers with the least remaining work may not work well, even when the server loads are known precisely [146]. For requests in a heavy-tailed distribution with infinite variance, Harchol-Balter *et al.* developed a task assignment by guessing size (TAGS) policy [146, 148] with better performance. It improves the user-perceived performance, as measured by the requests' waiting time relative to their service times.

The randomized and round robin policies are not adaptive to servers' changing conditions. A widely used adaptive policy is *least connection*, in which the load balancer assigns requests to the server with the least number of active connections. There are other adaptive load balancing policies based on response time, backlog queue length, etc. Adaptive load balancing policies make decisions based on current server workload information. The information can be collected either by the balancer periodically, or reported by the servers whenever there is a state change like exceeding a predefined overload threshold. Based on the past workload information, the load balancer can estimate the servers' processing rates and assign requests in a weighted round robin way accordingly.

Adaptive load balancing policies can be enhanced further by exploiting the data locality in client requests. It is noticed that a significant portion of the processing time is spent in disk access. Locality-aware request distribution (LARD) [239] makes server selection based on client request contents for load balancing as well as improving cache hit rate. Consider a request for objects of a certain type. LARD suggests the request be handled by the same server of previous requests for the same type of objects, if the server is not overloaded. Otherwise, the request is dispatched to the server with least active connections. To reduce the overhead of the load balancer, the request content can be inspected in a distributed manner. In scalable content-aware request distribution [15], the client requests are first diverted to different back-end servers by a simple load balancing scheme like random or round robin. The servers inspect the request contents independently and forward the requests to their most appropriate servers for actual processing. All the servers use the same policy as LARD to determine the target server for each request.

Data locality aside, client requests may have other possible logical relationships. Requests of the same session in an e-commerce transaction, such as on-line shopping, are related. They should be assigned to the same server so that the shared state can be carried over between the requests. This feature is called *session persistence*. Session persistence is a crucial performance factor in server clusters. We define client-aware load balancing as a server resource allocation process that optimizes the objective of load balancing under the constraint of session persistence. We review the policies and enforcement mechanisms in detail in the following section.

2.2.4 Load Balancing with Session Persistence

The HTTP protocol was initially designed in support of stateless applications. A stateless application means there is no state information necessary to be carried over between Web interactions. Requests from the same client or different clients are independent of each other, and idempotent in the sense that they can be executed multiple times in any order without affecting their semantics. The stateless protocol decouples servers from clients in an open environment and ensures the scalability of Web services.

However, most Internet applications cannot be completed in a single request. In fact, many applications involve a sequence of requests that share an execution context in the server. The requests form a stateful session by the use of cookies on a stateless server. Session persistency refers to the ability of the load balancer to divert all requests of the same session to the same server. In a cluster of e-commerce servers, session persistence also requires assigning all the sessions of a client to the same server for the duration of an application transaction. Session persistence is also referred to as *sticky connection*.

Like the objective of load balancing, session persistence can be achieved based on layer-4 and layer-7 protocol header information. A layer-4 load balancer has access to IP source addresses. The balancer can switch packets among the servers according their source IP addresses. It maintains a session table containing the current mappings from IP addresses to servers. The IP address mappings ensure that all requests, except the first one in a session, follow the assignment of the session head request.

Source IP-based session persistence is simple. It imposes few extra overheads to identify a client in a layer-4 load balancer. However, this identification approach may cause an *originator problem* in the situations that clients access the services through forward proxy servers provided by Internet Service Providers (ISPs), such as America Online and Comcast, and enterprises. Since the proxies make requests to Internet services on behalf of their clients, the IP sources of the requests represent a group of clients, instead of an individual client. Source IP-based session persistence methods lead to severe load imbalance between the servers of a cluster if the proxies account for a large percentage of the total traffic.

To overcome the originator problem in the presence of large proxies, a layer-7 load balancer uses application level information contained in the requests to distinguish sessions from different clients. The most common approach is based on cookies. A cookie is defined as an HTTP header in the form of "keyword:value" for the purpose of passing state information between Web interactions while retaining the stateless nature of HTTP servers. Cookies are set by Web servers and passed onto clients as piggybacks of responses. The cookies are stored in client browsers for resuming connections with respective servers in the future.

A layer-7 load balancer can use a cookie like "server=1" to divert traffic containing the cookie to server 1 so as to preserve session persistence. The cookie can be set by a back-end server when it is handling a session header request. On receiving a request, the load balancer checks the server cookie contained in the packet and assigns the request to the server specified by the cookie. This approach is simple to implement. However, it requires the server applications to be aware of server identifiers of a cluster configuration. This requirement would cause much administrative burden whenever the cluster is reconfigured.

To make the Web application independent of cluster configuration, a remedy is to allow the servers to assign a server cookie to a default value. It leaves to the load balancer the updating of the cookie to a specific server identifier before it forwards the response containing the cookie to clients. We note that the cookie initialization step cannot be skipped. Without an initial cookie setting, the load balancer has to insert a cookie, which causes a change of packet size. Due to the packet size limit, any packet size change would make a load balancer to be too complex to implement.

In addition to cookies, a layer-7 load balancer can also use the URL address of each access object to preserve stick connections and exploit data locality between the requests for the same object. URL-based switching has a number of advantages. It is known that the document size follows a heavy-tailed distribution, for which a reservation policy for big documents is much better than single waiting queue [72]. In a server cluster with URL-based switching, one or a few servers can be reserved for serving large documents. URL-based switches can direct requests for different contents to different servers.

URL-based switching can also be implemented in layer-7 content-aware request distribution policies [15, 239] to exploit data locality between requests. It takes advantage of data locality in requests for the same content and saves disk access time. In particular, for dynamic Web pages, the static components, such as background and styles, can be reused.

URL-based switching is not compatible with HTTP/1.1 protocol. HTTP/1.1 supports persistent connection. It means multiple HTTP request/response exchanges can be combined within one TCP connection so as to save the time for TCP connection setup and teardown. The multiple interactions can even be carried out in a pipeline fashion. URL-based switching cannot distinguish between the Web interactions within the same connection.

Cookie and URL-based switching work for content-aware load balancing among Web servers. They can also be applied for HTTPS e-commerce servers. HTTPS is an HTTP protocol running on top of SSL, for exchanging sensitive information between client and server. It is an industry de facto standard protocol for e-commerce applications. SSL is a stateful protocol that uses a session identifier to uniquely identify each session. To protect communication privacy and integrity, all application data are encrypted before transmission. The encrypted data can be decrypted only by client and server. Untill SSL Ver 3.0, session identifier was encrypted and there was no chance for a layer-7 load balancer to implement session persistence methods for HTTPS servers. In the latest SSL version, session identifier is no longer encrypted. Session identifiers in plain texts make it possible to implement a layer-7 load balancer for session persistence.

2.2.5 Distributed Approaches for Load Balancing

The load balancers reviewed in preceding sections assume a centralized hardware or software component for load balancing. An alternative to this centralized approach is a decentralized organization, in which all servers share a VIP and cooperate in the process of load balancing. When a server detects a client request, it decides whether it should accept this request or not based on a predefined formula or load balancing policy. These servers can be locally distributed or geographically distributed. Since each server can decide the most appropriate recipient, there is no centralized point to become the potential performance bottleneck. Normally, these schemes assume that client requests can reach different servers, which can be achieved through broadcasting or DNS-based round robin. Furthermore, different load-balancing policies can be applied to these implementations.

The redirection mechanism can be implemented in various ways such as IP tunneling, IP filtering, HTTP redirection, URL rewriting, and packet rewriting. IP filtering is mostly used for distributed load balancing on server clusters. HTTP redirection and URL rewriting are often used for global load balancing. They will be discussed in Section 2.3.1.

In IP filtering, all servers of a cluster are deployed in the same subnet, sharing a VIP address and running an IP filtering component. Each component is essentially a network driver that acts as a filter between the cluster adapter driver and the TCP/IP stack of the server. The servers are enabled to concurrently detect incoming TCP and UDP traffic for the VIP address and determine whether they should accept packets based on a distributed load balancing policy. If so, the packets are forwarded up to the TCP/IP stack for further processing.

Microsoft Network Load Balancing (NLB) [222] is an example of such a software service, provided in the Windows Server 2003 family. On each server, the cluster adapter driver determines the mapping of packets in a fully distributed way based on their source IP addresses and port numbers. Although the servers make load balancing decisions independently, they need to coordinate in response to cluster changes like server failure, join, or leave. The coordination is realized by periodically exchanging "heartbeats" within the cluster via IP multicasting. Each heartbeat signal contains information of a server status. Whenever a cluster state change is detected, the NLB software at each server starts a stabilization process to reach a new consistent cluster state.

2.3 Load Balancing in Service Overlay Networks

For scalability and availability, popular Internet services are often run on more than one geographical site on the Internet. Service overlay network is to distribute requests between different sites for balancing the workload of the sites and reducing the service transmission delay on the network. The geographically distributed servers or proxies can be organized in different ways. This section focuses on popular content delivery networks (CDNs) and related load balancing strategies.

2.3.1 HTTP Request Redirection and URL Rewriting

The simplest approach for workload distribution between different Web sites is HTTP request redirection. In this approach, clients send requests to an advertised server IP address. The server then replies to them with other servers' addresses according to a load balancing policy, together with a status code, such as 301 (Moved Permanently) and 302 (Moved Temporarily). By checking the status code, the clients send new requests to the selected servers, if needed. The client requests can also be redirected automatically.

URL rewriting is based on dynamic URL generations. When a client requests a Web page, the URLs contained in this page are generated dynamically according to a load balancing policy. These URLs point to different servers so that future requests for these URLs are distributed accordingly. For example, `http://141.200.10.1/document.pdf` is a link inside `http://www.xyz.com/index.html`. When a client accesses this index page, the server can replace it with `http://10.0.0.2/document.pdf`. As a result, any future requests to the site 141.200.10.1 will be redirected to the new location.

2.3.2 DNS-Based Request Redirection

The most common approach for global load balancing is DNS-based request redirection. It integrates load balancing functionalities into the authoritative DNS for a particular service and advertises the VIP on the load balancer as the service address.

Figure 2.3 shows a typical information flow architecture of the Internet services. To access an Internet service via its URL address like `http://www.yahoo.com` a client first contacts its local DNS to get the IP address of the domain name, `www.yahoo.com`. If its local DNS does not contain the IP address, the domain name resolution request goes upward through a number of intermediate proxies to root name server until the service's authoritative DNS is reached. The resolved IP address either refers to a real server or a server cluster. Since the authoritative DNS will receive all DNS queries, it can distribute requests to different servers during the process of URL address resolution. Like the load balancer in a server cluster, the DNS can make load balancing decisions based on information about client identifiers, locations, request contents, as well as site health conditions. For example, F5's 3-DNS Controller

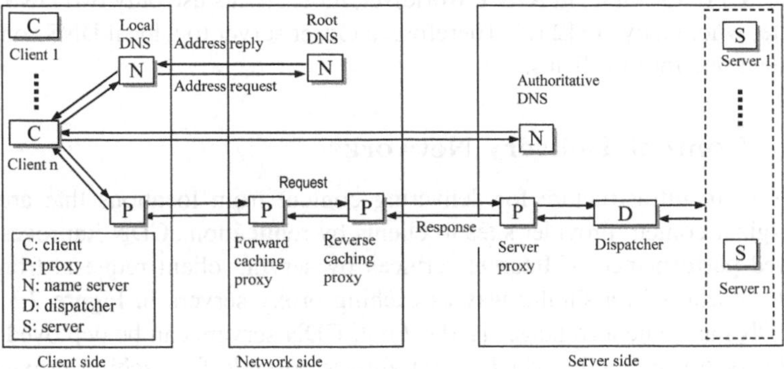

Figure 2.3: A typical information flow infrastructure of Internet services.

(www.f5.com) implemented a strategy based on proximity-based address mapping for geographically distributed servers. By comparing the client's IP address with the IP addresses of the available servers, the controller can determine the proximity of the servers and assign the client's requests to the nearest server.

To reduce the response time for future DNS name resolution requests, a resolved IP address of a client query is often buffered in the client's local DNS and intermediate proxies for reuse in the future. The name resolution caching alleviates a possible bottleneck problem with the authoritative DNS. It, however, comes with several limitations for the objective of load balancing. Due to the presence of name resolution caching, the authoritative DNS can have control over only a small fraction of address mapping queries that are missed in the caches. Although the authoritative DNS can set a time-to-live (TTL) value to limit the duration of each resolution in the caches, it will not take effect without the cooperation of the clients' local DNS and other intermediate caches. In addition, a small value of TTL would lead to a bottleneck in the authoritative DNS. Moreover, although the authoritative DNS server can detect failed servers and exclude them from the list of available servers, clients may continue to send them requests because of their cached DNS resolutions.

Akamai (www.akamai.com) deployed a two-level DNS redirection. On the top level, some DNS servers are used to resolve requests for higher level domain names such as .com or .org. On the bottom level, the DNS database is replicated or partitioned among a number of geographically distributed DNS servers. When receiving a resolution request, a top-level DNS server responds with the address of a bottom-level DNS server that is close to the client. Therefore, it is possible to set a small TTL value without making the DNS servers overloaded. Another approach is to place fully replicated DNS servers on the edges of the Internet. Resolution requests are directed to the closest DNS server to the client.

Another issue with DNS-based request redirection is the originator problem. It is known that an authoritative DNS receives a request from a client's local DNS server instead of individual clients. The clients often locate in different places from their local DNS. For example, in AT&T WorldNet, most clients use only two DNS servers no matter where they are [251]. Therefore, a closer server to a local DNS sometimes is a distant one for the clients.

2.3.3 Content Delivery Networks

CDN is an infrastructure for delivering content from locations that are closer than original content provider sites to clients by replication. CDN improves client-perceived performance of Internet services by serving client requests from CDN servers. It works in a similar way as caching proxy servers in Figure 2.3. However, CDN has some advantages [251]. First, CDN servers can be deployed around the network transparently, which is not always possible for caching proxies. The second one is that service providers have full content control, such as content replication. For caching proxy servers, it is hard to control which content should be placed on which proxy servers. One more advantage of CDN is that it can improve performance for uncachable content, such as dynamic content and streaming content. Finally, the content of Internet services can be placed on CDN servers before being accessed by clients.

There are two kinds of replication modes for CDN: full replication and partial replication. In the full replication mode, the origin servers are modified to use the authoritative DNS provided by CDN companies. Client requests can be delivered to CDN's servers or forwarded to the origin servers. Partial replication needs modifications of the hyperlinks inside web pages. For example, a hyperlink `http://www.xyz.com/image.gif` from xyz company's homepage can be changed to `http://www.cdn.com/www.xyz.com/image.gif`. Therefore, when a client sends a request for this object, the host name needs to be resolved by CDN's name servers. Examples of such DNS redirection systems include WebSphere from IBM (`www.ibm.com/websphere`) and EdgeSuite from Akamai (`www.akamai.com/en/html/services/edgesuite.html`).

Akamai's CDN fought for scalability and high availability of Internet services by using more than 12,000 servers in over 1,000 networks, as of year 2002 [85]. On receiving a client's resolution request, Akamai's CDN name server determines the target server using both client-side and server-side information and taking metrics like network topology, bandwidth, and server's load into account. Moreover, the target server should have the requested content. For example, a media stream request should not be directed to a server that handles only HTTP. To prevent a server from overloading, a CDN load balancing system must monitor the state of services, and their servers and network. Akamai uses a threshold method for this purpose. All sites periodically report their service states and site load information to a monitoring application. When the load of a site exceeds a predefined threshold, more servers are allocated to process client requests. If the load exceeds another threshold, the site is excluded from the available site list until it is back to the threshold.

Large-scale Internet service providers often deploy and administer their own *enterprise CDNs*. They use multiple levels of load balancing to distribute client requests for a single giant Internet service. At the top level, the ISPs distribute their servers geographically based on various request redirection strategies to offer higher availability and better performance. The site selection can take into account the site's load and network's traffic information. After receiving a client request, the front-end load balancer (or dispatcher) selects an appropriate server from the local pool of back-end servers, according to the server load balancing approaches as we reviewed in preceding sections.

The Web site of the 1998 Olympic Winter Games is a good example of enterprise CDNs that IBM developed [157]. It supported nearly 635 million requests over the 16 days of the game and a peak rate of 114,414 hits in a single minute around the time of the women's freestyle figure skating. For scalability and availability, Web servers were placed in four different global wide sites. Client requests to `http://www.nagano.olympic.org` were routed from their local ISP into the IBM Global Services Network, and IGS routers forwarded the requests to the server site geographically nearest to the client. Each site had an installation of one SP-2 machine configured with ten R6000 uniprocessors and an eight-way symmetric multiprocessor and had 12 open IP addresses. Load balancing at each site was conducted in a hierarchical way. At the top level were two routers as the entry points of each location. One is the primary point and the another is a backup. Between the routers and the Web servers were four IBM Network Dispatchers (NDs) for load balancing. Each of these four NDs was the primary source of three of the 12 IP addresses and the secondary source for two other addresses. These NDs were assigned to different costs based on whether they were the primary or secondary source for an IP address. Secondary addresses for an ND were assigned a higher cost. Therefore, the routers delivered incoming requests to the ND with the lowest cost. On the receipt of the request, the ND distributed it to a server based on servers' load.

Oppenheimer and Patterson reviewed three more examples of large-scale Internet services that were run on their own CDNs [234]. They were representatives of online service/Internet portals (online), global content-hosting services (content), and high-traffic Internet services with a very high read-to-write ratio (readmostly). At the survey time, both the "Online" and "ReadMostly" services supported around a hundred million hits per day and the "Content" service supported around 7 million hits per day. The "Online" service used around 500 servers at two data centers, the "Content" 500 servers at four data centers, and the "ReadMostly" more than 2000 servers at four data centers. They generally use policies that take into account the server's load and availability to distribute client requests. The "Online" service provided its client an up-to-date list of appropriate servers for their selection. In the "Content" service, each two of its four sites worked in redundant pairs. Its clients need to install a software package that points to one primary and one backup site of the content service. The "ReadMostly" service used DNS-based load balancing strategies on the basis of its sites' load and health information. Among locally distributed servers, round robin DNS servers or layer-4 dispatchers were used to direct client requests to the least loaded server.

Finally, we note that CDN was originally proposed for fast delivery of media-rich contents. For CDN to deliver live streaming media, there come new challenges because the live stream cannot be stored in advance on CDN servers. IP multicast or application-level multicast are used to distribute streaming media to CDN servers. For IP multicast, a multicast group is created by the original server before the stream is distributed. A CDN server joins or leaves the group when it receives the first client request or finishes the last client request. Application-level multicast is based on a distribution tree of intermediate servers that make use of IP unicast to deliver stream from the root (origin server) to the leaves (CDN servers).

2.4 A Unified W5 Load Balancing Model

The concept of load balancing is now new. It has been studied for many years in parallel and distributed systems. In particular, the cluster architecture is widely used as a cost-effective solution for high-performance applications and high-throughput computing, because coupled commodity machines cost much less than conventional supercomputers of equivalent power. Load balancing is a key technology in parallel computers and computing clusters, as well.

A load balancing technology needs to answer the following W5 questions: (1) Who makes the load balancing decision? (2) What information is the decision based on? (3) Which task is selected for balancing according to the decision? (4) Where is the task to be performed? (5) When (or why) is the load balancing operation invoked? In the following, we delineate the model and show various answers to each question in different contexts.

2.4.1 Objectives of Load Balancing

Narrowly speaking, load balancing is a process that distributes or redistributes workload between different system components in such a way that their workloads are "balanced," relative to the peers' capacities, to a certain degree. The system components are processors in a parallel computer, nodes in a computing cluster, servers or proxies in a Web cluster or service overlay network. We refer to the components as load balancing peers. This is a basic requirement for all load balancing strategies, although the abstraction of their workloads has different meanings in different contexts.

Load balancing is not an isolated performance issue. There are other performance issues, such as communication cost and data locality that need to be addressed in the process of workload distribution. The load balancing problem is often defined as a constrained optimization process with a much broader objective to address these issues together. In parallel computing, a parallel application comprises a number of concurrent tasks. The tasks often need to communicate with each other for exchange data or synchronization. An efficient execution of the parallel application on a high-performance computing server relies on a load balancing technology to balance the workload in terms of the residing tasks between the cluster nodes, meanwhile minimizing the internode communication cost.

In cluster-based distributed computing, workstations in the same department are clustered for harvesting unused CPU cycles and memory and improving the utilization of the aggregated computational resources. In such high-throughput computing clusters, users generally start their jobs, in the form of processes in execution, at their host machines. The processes are independent of each other. They can be either run locally or in remote workstations. The objective of load balancing is to ensure no machine is idle while there are jobs waiting for processing in other machines, and meanwhile remote execution of a job would not be outweighed by its migration cost and thrashing the memory/caches.

In contrast, load balancing on server clusters is to distribute traffic among the servers in such a way that the server workloads are balanced, relative to their capacities and meanwhile preserving session persistence and location proximity. If a server is down, its services are failed over to other servers.

2.4.2 Who Makes Load Balancing Decisions

By the first question, load balancing technologies can be categorized into two classes: centralized and decentralized. In a centralized approach, a designated system component collects and maintains the workload information of the peers. It makes load balancing decisions, and instructs other components to adjust their workloads accordingly during the load balancing procedure. In a decentralized approach, the system components maintain the workload information of others by themselves and make load balancing decisions individually.

DNS- and switch-based schemes in Web clusters are examples of the centralized approach. In this approach, the DNS server or dispatcher takes the full responsibility of load balancing. Centralized approaches can yield good performance in small-scale systems. In large-scale systems with hundreds or thousands of processors or systems involving much work for load balancing decisions, the designated system component is prone to bottleneck traffic. For example, locality-aware load balancing algorithms like LARD [239] need to analyze the request contents. They are limited to small-scale Web clusters. A remedy for the bottleneck is hierarchical (or semi-distributed) approaches that try to combine the advantages of both centralized and fully distributed approaches. In a Web cluster, the dispatcher forwards the client requcsts to a server in a simple way and lets the servers determine collaboratively or individually the most appropriate one for handling, as implemented in [15].

2.4.3 What Information Is the Decision Based on

There are two types of information that are useful for load balancing decisions. One is task information such as task size in parallel computing, request IP address, session identifier, and contents to be accessed in Internet services. The other type of information is system running status like workload distribution.

By the information that a load balancing decision is based on, we categorized the load balancing technologies into two classes: oblivious and adaptive. Oblivious algorithms distribute tasks to load balancing peers, without taking into account any information of workload or system status. In contrast, adaptive algorithms make load balancing decisions according to the information of tasks, systems, or both. Oblivious load balancing algorithms require no run-time support for analyzing the workload information or maintaining the peer states. They are popular in parallel processing of applications whose computational characteristics are known in advance. Block, cyclic, and other static task partitioning approaches assign tasks to threads/processes (virtual processors) at compile time. Their load balancing decisions are oblivious to processors' run-time states. Oblivious load balancing algorithms are not limited to static distribution approaches. There are run-time algorithms that schedule the execution of tasks on different processors in a randomized or round robin fashion. Randomized and round robin approaches tend to balance the processors' workload in terms of the number of requests.

Oblivious load balancing algorithms are commonplace on the Internet, as well. Static load balancing algorithms are often used in a CDN to assign the client requests from different geographical locations (based on their request IP addresses) to different server sites. Randomized and round robin approaches are used to distribute client requests to different servers at each site. It is known that client requests for an Internet service start with its URL address resolution by an authoritative DNS. The DNS maps the service URL to a server IP address randomly or in a round robin way to achieve the objective of load balancing to some extent.

Randomized and round robin algorithms work well for applications whose processing tasks are independent and their resource demands follow a uniform distribution. They do not suffice for scheduling of dynamic tasks, CPU cycle harvesting jobs, and Internet requests.

In either centralized or distributed approaches, ideally, a load balancing peer should keep a record of the most up-to-date workload information of others. Practically, this is infeasible in distributed systems and message passing parallel computers because interprocessor communication necessary for the collection of workload information introduces nonnegligible delays. This communication overhead prohibits processors from exchanging their workload information frequently. Hence, another trade-off is a balance between incurring a low cost for the collection of system wide load information and maintaining an accurate view of the system state. This trade-off is captured in the following three information exchange rules:

- On demand — Load balancing peers collect others' workload information whenever a load balancing operation is about to begin or be initiated.

- Periodical — Load balancing peers periodically report their workload information to others, regardless of whether the information is useful to others or not.

- On state change — Load balancing peers disseminate their workload information whenever their state changes by a certain degree.

The on-demand information exchange rule minimizes the number of communication messages but postpones the collection of system wide load information until the time when a load balancing operation is to be initiated. Its main disadvantage is that it results in an extra delay for load balancing operations. The rule is rarely used in response time sensitive high-performance application or Internet services. Conversely, the periodic rule allows processors in need of a balancing operation to initiate the operation based on the maintained workload information without any delay. The periodic rule is mostly used with periodic initiation policies. The problem with the periodic rule is how to set the interval for information exchange. A short interval would incur heavy communication overhead, while a long interval would sacrifice the accuracy of the workload information used in load balancing decision making. The on-state-changing rule strikes a good balance between the information accuracy and communication overhead. All the information exchange rules are in practical use for network load balancing. In particular, periodic rules are used in Microsoft NLB in Window Server 2003 [222] and Akamai CDN [85].

2.4.4 Which Task Is the Best Candidate for Migration

A load balancing decision answers two questions: (1) Which task should be a candidate for migration (or remote execution)? (2) Where should a task under consideration be assigned or reassigned for execution? The task candidacy question is server-centric, dealing with the selection of a task to give off from an overloaded component or to take in to an underloaded one. To answer this question, task preemptivity and data locality must be taken into account.

By the selection rule, a load balancing operation can be performed either nonpreemptively or preemptively. A nonpreemptive rule always selects newly created threads/processes or new requests, while a preemptive rule may select a partially executed process if needed. Migration of a process preemptively has to first suspend the process and then resume it remotely. The process state needs to be transferred, which is generally more costly than a nonpreemptive transfer.

A simple load balancing strategy is to maintain a global task queue shared by all system components. A component fetches a task at the head of the queue for execution whenever it becomes idle. This dynamic strategy inherently provides effective load balancing. But a task may be scheduled onto a different component each time it is executed. This could lead to performance degradation because tasks thrash the system components' cache/memory as the computation proceeds. *Affinity scheduling* [297] counteracts this problem.

Affinity is a measure of cost of task migration. It must be determined quickly at the time of task creation or suspension based on an estimate of the amount of resident task state and where the data it requires are located. The cost of moving a task from a source to a destination processor consists of the time to transmit the state of the task and the time to deliver all of the data required by the task that already reside on the source. This cost is a measure of the task's affinity for the source processor. If the task migration cost is large compared to the execution time of the task, the task has high affinity for the processor. If the migration cost is small, the task has low affinity. An affinity scheduler can schedule tasks from the global queue in a biased way such that a task tends to stay on the same component [297].

For example, in Linux kernel of version 2.4 and above [41], the scheduler searches the *entire* set of processors at process creation or wake-up time to establish any potential processor to preempt. The scheduler determines whether any remote processor has a task waiting for execution in its local run-queue with a "goodness" value that is higher by at least "PROC_CHANGE_PENALTY" when compared to the best local schedulable task. The "PROC_CHANGE_PENALTY" was set for taking into account the data locality factor.

In addition to the transfer overhead, the selection rule also needs to take into account the extra communication overhead that is incurred in the subsequent computation due to process migration. For example, a splitting of tightly coupled processes will generate high communication demands in the future and erode the benefit of load balancing. In principle, a selection rule should break only loosely coupled processes. Data locality helps determine which tasks should be migrated from a heavily loaded processor to an idle one. Nearest neighbor load balancing algorithms [338] reduce load migration to nearest neighbor load shifting so as to preserve communication locality.

On Web clusters, content-aware load balancing and session persistence deal with the same performance problem as affinity scheduling. Although network load balancing has its own implementation challenges, it bears much resemblance to the classical processor scheduling problem in concept.

2.4.5 Where Should the Task Be Performed

In addition to the server-centric question, a load balancing decision addresses a task-centric peer selection question. Answers to the question of which system component a task should be assigned to lead to various load balancing policies.

The simplest task assignment policy is *randomized allocation*. It randomly selects a component to perform the task. The most common task assignment policy is *round robin*. Albeit simple, it neither maximizes the system utilization, nor minimizes the mean response time. If the task sizes are known and conform to an exponential distribution, the best policy for minimizing response time is *least-remaining work* [329]. It dispatches an incoming task to the peer with the least total unfinished work. A variant is *shortest queue* that assigns tasks to the component with the smallest number of tasks. In the case that the task sizes are unknown in advance, the shortest queue policy is optimal under an exponential task size distribution and Poisson arrival process; see [146] for a review of the task assignment policies.

All these scheduling policies were originally developed in the general context of parallel and distributed computing. They were applied to the problem of network load balancing, as we reviewed in preceding sections. Least connection scheduling is essentially an implementation of the shortest queue task assignment policy. Internet requests are highly variable in size. Their sizes often follow the Pareto and other heavy-tailed distributions. In this case, the *reservation* policy that is in common practical use has been shown to be superior to others [147]. The reservation policy reserves some system components for large requests and others for small requests. The reservation with a cycle stealing policy allows short tasks to be executed in components that are reserved for long tasks, as long as there are no long tasks waiting for service.

2.4.6 When or Why Is Migration Invoked

The last question is when (why) a load balancing operation should be invoked. An initiation rule dictates when to initiate a load balancing operation. The execution of a balancing operation incurs nonnegligible overhead; its invocation decision must weigh its overhead cost against its expected performance benefit. An initiation policy is thus needed to determine whether a balancing operation will be profitable. An optimal invocation policy is desirable but impractical, as its derivation could be complicated. In fact, the load indices are just an estimate of the actual workloads and the workload information they represent may be out of date due to communication delays and infrequent collections. Instead, primarily heuristic initiation policies are used in practice.

In pool-based multiqueues, processors are divided into multiple processor sets. Tasks are migrated periodically between run queues in order to equalize the length of their run queues. The interval is set to (refreshticks*1) milliseconds in Linux. The initiation rule in network load balancing is relatively simple. It is often initiated by a load balancer (or dispatcher) on the receipt of an incoming request. The requests can also be forwarded (migrated) to other servers for processing in distributed load balancing approaches for exploiting data locality as in scalable content-aware strategies [239] or taking advantage of physical proximity in CDN. For requests of highly variable sizes, TAGS [146] suggests initiating a request migration from a server for small requests to a server for large requests after a certain service time.

Chapter 3

Load Balancing on Streaming Server Clusters

Streaming applications often involve a large volume of media data in storage and transmission. Server clustering is one of the key technologies for the scalability and high availability of streaming services. Chapter 2 reviewed load balancing technologies for distributing service traffic dynamically among the servers of a cluster. Those technologies are complemented by static data placement that determines initial data layout across the servers. Since media contents have different popularity, they should be maintained with different degrees of redundancy. The contents should also be stored in different formats so as to meet the clients' diverse needs and the resource constraints of their access devices and networks. This chapter formulates the problem of data placement and gives it a systematic treatment in video-on-demand (VoD) applications.

3.1 Introduction

Recent advances in storage, compression, and communication have renewed interests in distributed multimedia applications in areas such as home entertainment, distance learning, and e-commerce. As the last mile bandwidth problem is being solved with the proliferation of broadband access, we are facing the challenge of designing and implementing scalable streaming servers that are capable of processing and delivering continuous time-based medias to a large number of clients simultaneously.

Early studies of scalable streaming servers focused on RAID-based storage subsystem designs as well as disk I/O and network bandwidth scheduling in single servers. The research on storage subsystems is directed toward: (1) data striping schemes in storage devices for disk utilization and load balancing; (2) data retrieval from storage subsystems in order to amortize seek time; (3) buffering of data segment for disk I/O bandwidth utilization; and (4) disk scheduling to avoid jitter. Readers are referred to [131] for a survey of early studies on multimedia storage servers. Video streaming consumes a large amount of network I/O bandwidth and network interfaces are prone to bottleneck traffic because of video stream isochronous requirements. User-centered batching, in combination with multicasting, allows clients to

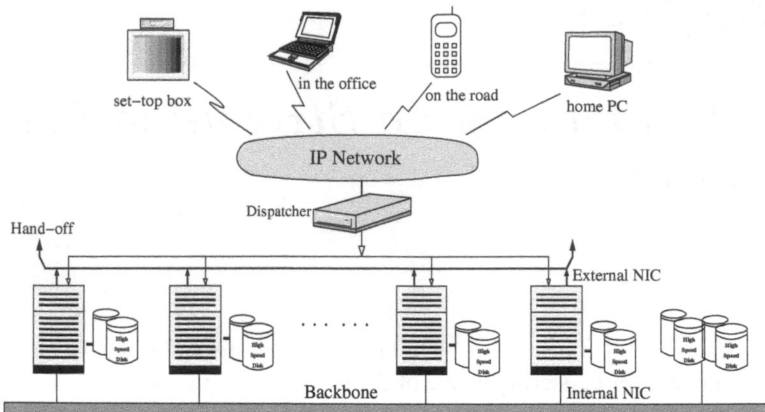

Figure 3.1: A cluster architecture for large-scale streaming applications.

share multicast data so as to reduce network traffic.

Clustering is a cost-effective approach to building up scalable and highly available media streaming servers. Figure 3.1 shows a generic architecture of a VoD server cluster. In front of the cluster is a dispatcher that makes admission decisions and determines the server for each incoming request. To avoid network traffic jams around the dispatcher, the cluster is often organized in a one-way configuration so that the servers can send responses to the clients directly.

In general, the VoD server cluster can be in either shared or distributed storage organization. A shared storage cluster is usually built on RAID systems [134]. Video data are striped into blocks and distributed over the shared disk arrays. Such systems are easy to build and administrate and the cost of storage is low. However, they have limited scalability and reliability due to disk access contention. As the number of disks increases, so do the controlling overhead and the probability of a failure.

In a distributed VoD cluster, each server has its own disk storage subsystem. The servers are linked by a backbone network. Due to the server autonomy, this cluster architecture can offer better scalability in terms of storage and streaming capacity and higher reliability. This type of clusters can also be employed in wide area networks.

A key issue in the design of distributed storage clusters is data placement. It determines where a media content (e.g., video) should be stored and in which formats. There are two complementary data placement approaches: data striping and data replication. Data striping is a data layout scheme in which videos, or multimedia objects, in general, are divided into fixed size data blocks (also called stripe units) and striped over multiple disks in round robin fashion, as illustrated in Figure 3.2(a). It decouples storage allocation from bandwidth allocation, avoiding potential load imbalances due to variations in video popularity. However, wide data striping can induce high scheduling and extension overhead. As the streaming cluster scales up with more videos and more storage spaces, expansion of the system requires a re-

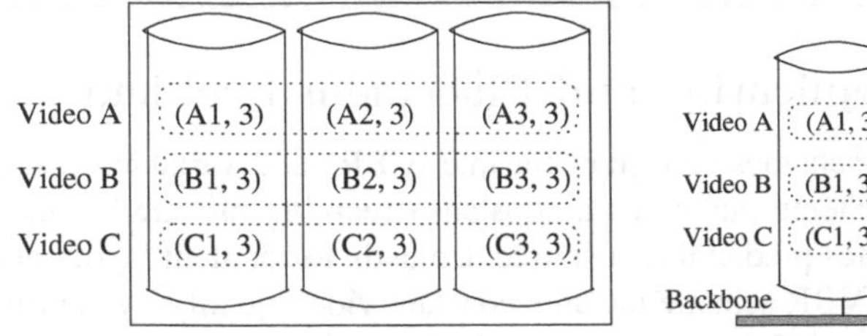

(a) Striping in a shared storage

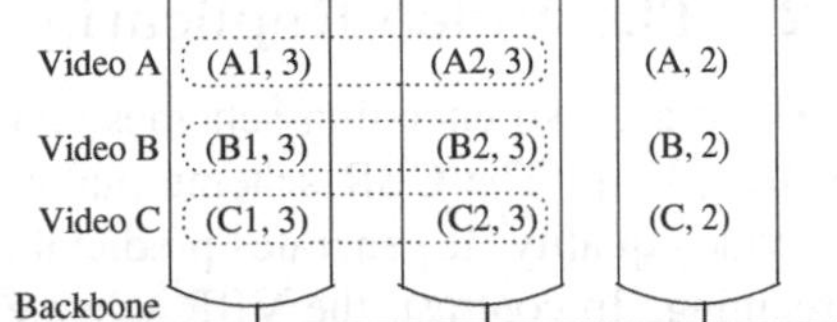

(b) Striping and replication in a distributed storage

Figure 3.2: Data layout in shared and distributed storage subsystems. The first integer label indicates the block index of a video and the second one specifies the encoding bit rate.

striping of the entire data set. Replication isolates the servers from each other at the cost of high storage overhead. It enhances the system scalability and reliability. Figure 3.2(b) illustrates a hybrid data placement strategy, in which each video has two copies with different encoding bit rates: one is striped into two blocks and the other is stored in a separate disk.

A defining characteristic with video streams is that a video can be encoded in different bit rates for different qualities at the cost of different storage and network I/O bandwidth requirements. This unique feature distinguishes video replication and placement problem from classical file allocation problems [89]. Due to huge storage requirements of videos, full replication is generally inefficient if not impossible. For example, a typical 90-minute MPEG-2 video encoded in a constant bit rate of 6 Mbps requires as much as 4 GB storage. Assume *a priori* knowledge about video popularity. The objective of this study is to to find an efficient video placement with partial replication.

The problem is how to determine the encoding bit rates and the replication degree of videos and determine the placement of the replicas on the distributed storage cluster for high quality, high availability, and load balancing under resource constraints.

High quality requires a high encoding bit rate. High availability in this context has two meanings: low rejection rate and high replication degree. The objective of load balancing is to improve system throughput in rush-hours and hence reduce the rejection rate. It is known that increasing the replication degree enhances the flexibility of a system to balance the expected load. Multiple replicas also offer the flexibility in reconfiguration and increase the dynamic load balancing ability. On the other hand, the replication degree is limited by encoding bit rates of videos and storage capacity of the cluster. Encoding bit rates also are constrained by the streaming bandwidth and the peak request arrival rate. In [358, 359], we presented a systematic treatment of the video replication and placement problem. We developed a group of replication and placement algorithms that make a good trade-off between the encoding bit rate for service quality and the replication degree for service availability and load balancing. In the following, we present the details of the algorithms.

3.2 The Video Replication and Placement Problem

There are two encoding schemes: constant bit rate (CBR) and variable bit rate (VBR) [315]. The CBR scheme maintains a constant streaming bit rate by varying video quality. It generates predictable resource demands for content layout and streaming. In contrast, the VBR scheme ensures constant video quality by varying streaming bit rate. However, VBR streams exhibit a high variability in their resource requirements, which can lead to low utilization of disk I/O and network I/O bandwidth at the server side. In this work, we consider CBR videos.

3.2.1 The Model

We consider a cluster of n homogeneous servers, $S = (s_1, s_2, \ldots, s_n)$, and a set of m different videos, $V = \{v_1, v_2, \ldots, v_m\}$. Each server has a storage capacity C and an outgoing network bandwidth B. We consider all videos in set V have the same duration d, say $d = 90$ minutes for typical movies. If a video v_i is encoded in constant bit rate b_i, the storage space for video v_i is calculated as $c_i = d \cdot b_i$.

It is known that the popularity of videos varies with a number of videos that receive most of the requests. We consider the replication and placement for the peak period of length d. This is because one of the objectives of the replication and placement is high service availability during the peak period. Load balancing is critical to improving throughput and service availability during the peak period. In this work, we focus on outgoing network I/O bandwidth, a critical type of server resource. We make the following two assumptions, regarding the video relative popularity distributions and the request arrival rates:

1. The popularity of the videos, p_i, is supposed to be known before the replication and placement. The relative popularity of videos follows Zipf-like distributions with a skew parameter of θ. Typically, $0.271 \leq \theta \leq 1$ [4, 313]. The probability of choosing the ith video is $p_i = i^{-\theta} / \sum_{j=1}^{M} j^{-\theta}$. The distribution is a pure Zipf distribution when $\theta = 1$ and is reduced to a uniform distribution when $\theta = 0$. A pure Zipf distribution represents heavy skew.

2. The peak period is the same for all videos with various arrival rates. Let $\bar{\lambda}$ be the expected arrival rate during the peak period and $p_i\bar{\lambda}$ be the expected arrival rate for video v_i. Because of the same peak period assumption, the video replication and placement are conservative, as they place videos for the peak period. This conservative model provides an insight into the key aspects of the problem.

3.2.2 Formulation of the Problem

The objective of the replication and placement is to have high service quality and high service availability. Replication enhances availability and load balancing ability by placement. Load balancing improves system throughput and hence the availability. Increasing the encoding bit rate receives high quality but decreases the replication degree due to the storage constraint. The encoding bit rate of videos is also limited by the outgoing network I/O bandwidth constraint of the servers. Thus, the objective of the replication and placement is to maximize the average encoding bit rate and the average number of replicas (replication degree) and minimize the load imbalance degree of the cluster.

Let L denote the communication load imbalance degree of the cluster. Let r_i denote the number of replicas of video v_i. Specifically, we define the optimization objective as:

$$Obj = \sum_{i=1}^{m} b_i/m + \alpha \cdot \sum_{i=1}^{m} r_i/m - \beta \cdot L, \tag{3.1}$$

where α and β are relative weighting factors. Load imbalance degree can be measured in various ways. We measure the degree in a norm-1 formula,

$$L = \max_{\forall s_i \in S} \mid l_i - \bar{l} \mid, \tag{3.2}$$

where $\bar{l}$ is the mean outgoing communication load of the servers, i.e., $\bar{l} = \sum_{i=1}^{n} l_i/n$. The objective is subject to the following constraints: (1) server storage capacity, (2) server outgoing network bandwidth, and (3) distribution of all replicas of an individual video to different servers.

Recall that b_i denotes the encoding bit rate of video v_i and the storage requirement of video v_i is $c_i = d \cdot b_i$. All r_i replicas of video v_i have the same encoding bit rate since they are replicated by the same video. Let $\pi(v_i^j)$ be the index of the server on which the jth of replicas of video v_i, v_i^j resides. Also, $\pi(v_i) = k$ means that a replica of video v_i is placed on server s_k. The communication weight of each replica of video v_i is defined as $w_i = p_i/r_i$. By the use of a static round robin scheduling policy, the number of requests for video v_i to be serviced by each replica of v_i during the peak period is $w_i \cdot \bar{\lambda} \cdot d$. Let l_k be the outgoing communication load on server s_k. Specifically, we give resource constraints from the perspective of server s_k $(1 \leq k \leq n)$ as:

$$\sum_{\pi(v_i)=k, \forall v_i \in V} b_i \cdot d \leq C, \quad \text{and} \tag{3.3}$$

$$l_k = \sum_{\pi(v_i)=k, \forall v_i \in V} w_i \cdot \bar{\lambda} \cdot d \cdot b_i \leq B. \tag{3.4}$$

The third constraint is the requirement of distribution of all replicas of an individual video to different servers. That is, all r_i replicas of video v_i must be placed on r_i servers. Specifically,

$$\pi(v_i^{j_1}) \neq \pi(v_i^{j_2}), \quad 1 \leq j_1, j_2 \leq r_i \text{ and } j_1 \neq j_2. \tag{3.5}$$

Note that placement of multiple replicas of a video in the same server implies that these replicas be merged to one replica. For this reason, we have one more replication constraint

$$1 \leq r_i \leq n \qquad \text{for all } v_i \in V. \tag{3.6}$$

In summary, we formulate the video replication and placement problem as a maximization of the objective (3.1) subject to constraints of (3.3) through (3.6). This is a combinatorial optimization because the encoding bit rate is a discrete variable and it has a finite set of possible values. In [358], we developed a heuristic algorithm based on simulated annealing to solve the problem.

We note that scalable encoding bit rates make it possible to derive optimal static replication and placement policies in a general case. However, this flexibility may lead to high dynamic scheduling overhead. The cluster dispatcher needs to know not only the video locations, but also the encoding bit rate of each replica. On streaming servers, there is a requirement for client-aware quality of service (QoS) differentiation provisioning according to the clients' access devices and networks. Video replication with different encoding bit rates supports QoS differentiation. However, a mismatch between on-line demands for videos of different encoding bit rates and the initial video replication would adversely affect the QoS. In practice, videos are often encoded in a single fixed rate. QoS differentiation can be supported by on-line transcoding that dynamically transform a video stream from its original encoding bit rate to degraded ones. We devote Chapter 5 to the topic of QoS differentiation.

3.3 Replication and Placement Algorithms

In this section, we present efficient replication and placement algorithms for videos with single fixed encoding bit rates. We consider that the video relative popularity conforms to Zipf-like distributions with a skew parameter θ, $0.271 \leq \theta \leq 1$. We assume that all videos are encoded in a fixed bit rate b.

3.3.1 Video Replication

Under the assumptions that all videos are of equal length d and encoded in the same fixed bit rate b, the storage requirement for each video becomes a constant, i.e., $c_i = d \cdot b$. We redefine the storage capacity of each server C in terms of the number of replicas. Unless otherwise specified, we use the redefinition of C in the following discussions. Note that because the encoding bit rate is fixed, the constraint (3.4) may be violated when communication load exceeds the outgoing network I/O bandwidth of the cluster.

One objective of the multiobjective function (3.1) is high replication degree. Replication can achieve fine granularity in terms of communication weight of replicas that offers more flexibility to place replicas for load balancing. We will prove in the next section that the upper bound of the load imbalance degree L generated by our placement strategy is nonincreasing as the replication degree increases. It is desirable to increase the replication degree to saturate the storage capacity of the cluster. The optimization of (3.1) is reduced to assigning the number of replicas to each video and minimizing the load imbalance degree L by placement of these replicas.

The objective of the replication is to get fine granularity of replicas in terms of communication weight for later placement. Recall the definition of the communication weight of replicas of video v_i ($w_i = p_i/r_i$) and replication constraint (3.6). We have a minmax problem as

$$\text{Minimize} \quad \max_{\forall v_i \in V} \{w_i\}, \tag{3.7}$$

$$\text{Subject to} \quad \text{constraint of (3.6), and} \tag{3.8}$$

$$\sum_{i=1}^{m} r_i \leq n \cdot C. \tag{3.9}$$

If the skew parameter of the Zipf-like distribution θ equals 0, which means a uniform popularity distribution, a simple round robin replication achieves an optimal replication scheme with respect to (3.7).

3.3.1.1 Split–Join Replication

The split–join replication algorithm proceeds recursively. It first assigns one replica to each video. For the rest replication capacity of the cluster, i.e., $N \cdot C - M$ replicas, at each iteration it gives one more replica to the video, whose replica degree is less than the number of servers and its replica(s) have the currently greatest communication weight.

Figure 3.3 illustrates the replication of five videos on three servers. Without loss of generality, suppose $p_1 \geq p_2 \geq \cdots \geq p_5$. The storage capacity of each server is three replicas. Initially, each video is given one replica and $w_i = p_i$. For the rest replication capacity of the cluster (4 replicas), at each iteration a video is chosen to be replicated. Its number of replicas is less than the number of servers and the replica(s) of this video have the greatest communication weight currently. For example, in the first iteration, video v_1 is duplicated into two replicas. The communication weight of its each replica hence is $w_1 = p_1/2$. In the second iteration, if $w_1 = p_1/2 = \max\{p_1/2, p_2, p_3, p_4, p_5\}$, the two replicas of v_1 will be duplicated into three replicas. Each replica has a communication weight of $p_1/3$. In the third iteration, if video v_1 already has three replicas, it won't be split any more. Video v_2 has the greatest communication weight and it is duplicated into two replicas with $w_2 = p_2/2$, and so on.

Theorem 3.1 *The split–join algorithm achieves an optimal replication scheme with respect to the objective of (3.7) in a time complexity of* $O(mn \log m)$.

3.3.1.2 Zipf Replication

The split-join algorithm is costly in time because it has to scan the whole replication capacity of the cluster. We present a time-efficient algorithm that utilizes

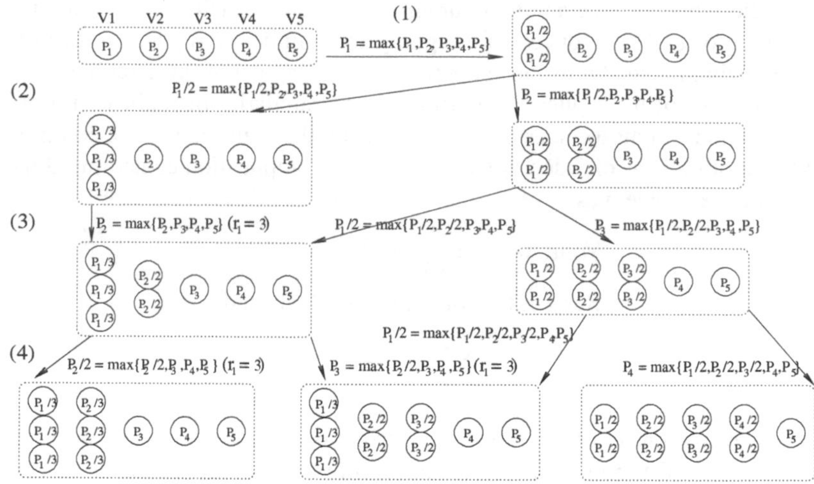

Figure 3.3: An illustration of the split–join replication.

the information about the Zipf-like video popularity distributions to approximate the optimal solution. The key idea is to classify the popularity of the videos into N intervals. The videos, whose popularity is within the same interval, are assigned the same number of replicas according to their interval indices.

The popularity partition is done in a heuristic way with another Zipf-like distribution. There are two key functions in the algorithm: $generate$ and $assignment$.

- The function $generate(\psi, u_j)$ partitions the range of $[0, p_1 + p_m]$ into n intervals according to a Zipf-like distribution with a skew parameter of ψ, i.e., $\frac{u_j}{p_1+p_m} = j^{-\psi} / \sum_{k=1}^{n} k^{-\psi}$ for $1 \leq j \leq n$. The boundaries of intervals are given by z_j. For example, if $\phi = 0$, the range of $[0, p_1 + p_m]$ is partitioned evenly into n segments.

- The $assignment(u_j, r_i)$ function assigns the number of replicas r_i to video v_i according to the interval index of its popularity. The parameter ψ determines the popularity partitions and hence the total number of replicas generated by the algorithm. We employ a binary search approach to search the parameter ψ.

Figure 3.4 illustrates a replication scenario with the setting of 20 videos, 6 servers, and popularity parameter $\theta = 0.75$. The storage capacity of the cluster is 24 replicas. The replication algorithm has $\psi = -2.33$.

As the parameter ψ increases, the boundary z_j, $(1 \leq j \leq N-1)$ decreases and the total number of replicas is nondecreasing according to function $assignment(u_j, r_i)$. We give the bounded search space of $[\psi_{\min}, \psi_{\max}]$ for the parameter ψ. If the first interval $\frac{u_1}{p_1+p_m} \geq \frac{p_1}{p_1+p_m}$, we have $r_i = n$, for all $v_i \in V$ according to the function $assignment(u_j, r_i)$. It follows that $\psi_{\max} = \theta \cdot \log m + \log n$. If the last segment

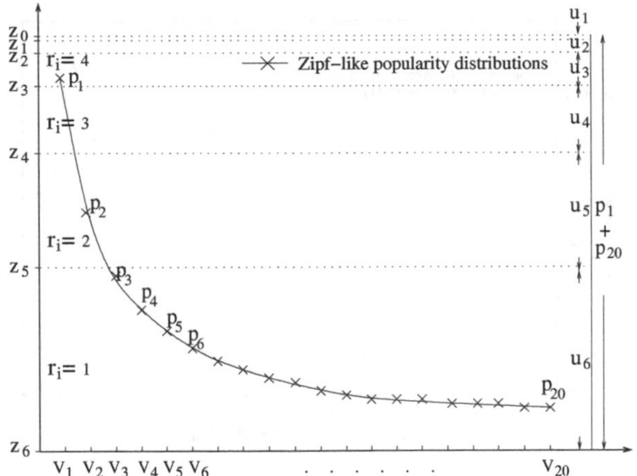

Figure 3.4: A replication scenario with the setting of 20 videos on 6 servers when $\theta = 0.75$.

$\frac{u_n}{p_1+p_m} \geq \frac{p_1}{p_1+p_m}$, we have $r_i = 1$ for all $v_i \in V$ and $\psi_{\min} = -\frac{\psi_{\max}}{\log n - \log(n-1)}$ for the same reason.

The features of Zipf-like distributions give

$$p_{m-1} - p_m = \min_{\forall v_i, v_j \in V} \{| \, p_i - p_j \, |\}.$$

In the worst case, the binary search algorithm terminates when the change of ψ, denoted by ψ_δ, becomes smaller than $\frac{p_{m-1}-p_m}{p_1+p_m}$. Hence, the bound of ψ_δ is calculated as $\frac{(m-1)^{-\theta}-m^{-\theta}}{1+m^{\theta}}$.

Note that typically $0.271 \leq \theta \leq 1$. The bound ψ_δ is minimized when $\theta = 1$ and it approximately equals m^{-2}. Due to $|\psi_{\min}| \gg \psi_{\max}$, as illustrated in Figure 3.5, the first replication interval converges to 1 much faster than the last interval does. It implies that the binary search algorithm would execute more iterations when the storage capacity is close to nonreplication than when it is close to full replication. In contrast, the split–join algorithm converges faster when the storage capacity of the cluster is close to nonreplication. A hybrid strategy can be employed and we can set $\psi_{\min} = 0$. With the search space and the termination condition for the binary search approach, the complexity of Zipf replication hence is $O(m \log m)$.

3.3.2 Smallest Load First Placement

Video placement is to map all replicas of m videos to n servers to minimize the load imbalance degree L. If the previous replication leads to a uniform communication weight for all replicas, i.e., $w_i = w_j, \forall v_i, v_j \in V$, a round robin placement achieves an optimal solution. It supposes that the replicas are arranged in groups in an arbitrary order such as $v_1^1 \ldots v_1^{r_1}, v_2^1 \ldots v_2^{r_2}, \ldots, v_m^1 \ldots v_m^{r_m}$.

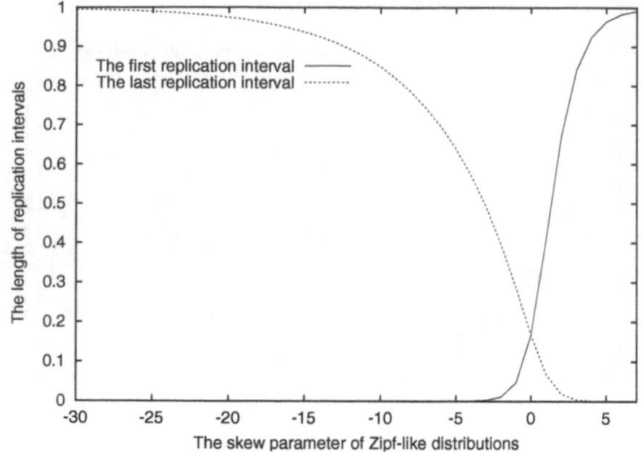

Figure 3.5: The convergence rates of the Zipf replication.

However, in most cases, the communication weight of replicas of different videos may be different. For example, the ratio of the highest popularity to the lowest popularity is m^θ. If $m^\theta > n$, $\frac{w_1}{w_m} = \frac{p_1/r_1}{p_m/r_m} \geq \frac{p_1/n}{p_m/1} > 1$ due to constraint (3.6). This placement problem is more related to load balancing problems than to bin packing problems [338]. The difference with bin packing problems is that the number of servers for placement is given, rather than minimized. The differences with load balancing problems are storage limitations for placement and replicas of one video have to be placed to different servers due to constraint (3.5). We propose a placement algorithm called *smallest load first placement* (SLFP) in Algorithm 1.

Algorithm 1 Smallest Load First Placement

```
1: arrange all replicas of each video in a corresponding group;
2: sort these groups in a nonincreasing order by the communication weight of the replicas in
   the groups;
3: for each of C iterations (C is the number of replicas that each server can contain) do
4:    select N replicas with the greatest communication weights;
5:    distribute these N replicas to the N servers — the distribution should satisfy that the
      replica with the greatest communication weight should be placed to the server with the
      smallest load and this server has not been placed with a replica of the same video;
6: end for
```

Figure 3.6 illustrates the placement strategy in a cluster of four servers. The replicas are arranged as $v_1^1, v_1^2, v_1^3, v_3^1, v_3^2, v_2^1, v_2^2, v_3^1, \ldots$ in a nonincreasing order according to their communication weights. The first iteration places the first four replicas to the four servers. In the second iteration, server 4 has the current smallest load and

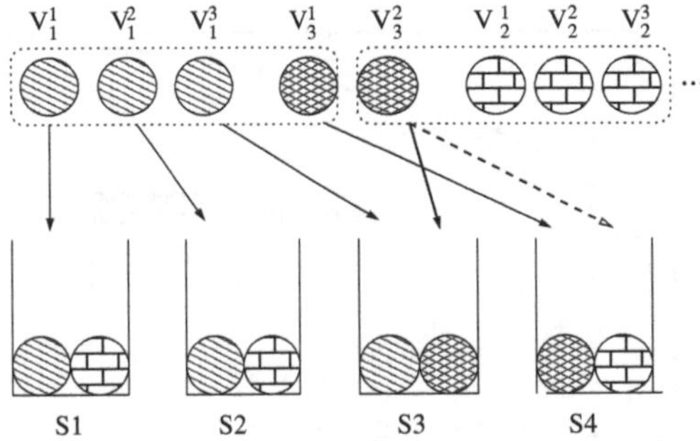

Figure 3.6: An illustration of the smallest load first placement (SLFP).

the replica v_3^2 has the greatest communication weight. However, since server 4 has already been placed with a replica of video v_3, i.e., v_3^1, the replica v_3^2 is placed to server 3 with the second smallest load, and so on.

Theorem 3.2 *In the smallest load first placement algorithm, the load imbalance degree L is bounded by the difference between the greatest communication weight and smallest communication weight of the replica(s). That is, $L \leq \max_{\forall v_i \in V}\{w_i\} - \min_{\forall v_i \in V}\{w_i\}$.*

Proof Let $w_1, w_2, \ldots, w_n, w_{n+1}, w_{n+2}, \ldots, w_{n \cdot C}$ be the communication weight of the $N \cdot C$ replicas in a nonincreasing order. It is evident that $w_1 = \max_{\forall v_i \in V}\{w_i\}$ and $w_{n \cdot C} = \min_{\forall v_i \in V}\{w_i\}$. In the worse case, the change of L at iteration $j (0 \leq j \leq C-1)$ of the placement is bounded by $w_{j \cdot n+1} - w_{j \cdot n+n}$. The summation of these changes would be the worst case for L after the placement. We have $L \leq w_1 - w_n + w_{n+1} - w_{2n} + \cdots + w_{n \cdot (C-1)+1} - w_{n \cdot C}$. It follows that $L \leq \max_{\forall v_i \in V}\{w_i\} - \min_{\forall v_i \in V}$. □

From Theorem 3.2, we know that the upper bound of the load imbalance degree is nonincreasing as the replication degree increases. This implies that high replication degree achieves fine granularity of communication weight of replicas, which in turn offers more flexibility for balanced placement.

3.4 Service Availability Evaluation

In this section, we present the simulation results by the use of different replication and placement algorithms with various replication degrees under the assumption of

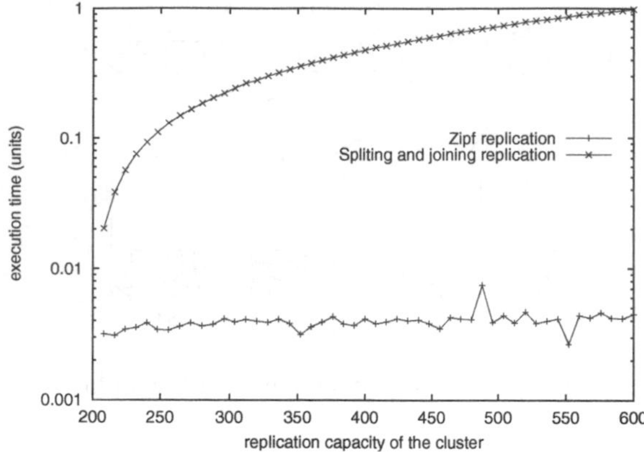

Figure 3.7: The execution time of Zipf and split–join replication algorithms.

a fixed encoding bit rate. In the simulations, the VoD cluster consisted of eight homogeneous servers. Each server had 1.8 Gbps outgoing network bandwidth. The storage capacity of each server ranged from 67.5 GB to 202.5 GB. The cluster contained 200 videos with duration of 90 minutes each. The encoding bit rate for videos was fixed with the typical one for MPEG II movies, i.e., 4 Mbps. The storage requirement of a video hence was 2.7 GB. The storage capacity of the cluster ranged from 200 to 600 replicas and the replication degree ranged from 1.0 to 3.0.

Within the peak period of 90 minutes, the request arrivals were generated by a Poisson process with arrival rate λ. Since the outgoing network bandwidth of the cluster was 3600 streams of 4 Mbps, the peak rate of λ was 40 requests per minute. The video popularity distribution was governed by a Zipf skew parameter θ ($0.271 \leq \theta \leq 1$).

The simulation employed a simple admission control that a request was rejected if required communication bandwidth was unavailable. We use the rejection rate as the performance metric. We simulated the Zipf and split–join replication algorithms and found that the Zipf replication algorithm achieved nearly the same result as the split–join algorithm, but at a much lower time complexity. Figure 3.7 shows that the cost of Zipf replication remains constant, while the execution time of split–join algorithm increases with the server capacity.

For brevity in representation, we give the results of the Zipf replication only. For placement, we compare the round-robin placement algorithm and the smallest load first placement algorithm. In the following, we give some representative results. Each result was an average of 200 runs.

In order to help understand the impact of different replication algorithms on performance, we also give a feasible and straightforward algorithm called classification-based replication. It classifies the m videos into n groups (n is the number of servers) according to the video popularity distributions. All groups contain an approximate

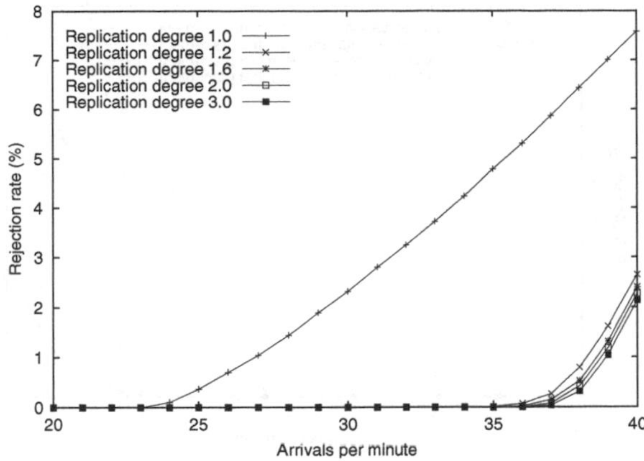

Figure 3.8: Impact of different replication degrees on rejection rate.

number of videos ($\lceil m/n \rceil$ or $\lfloor m/n \rfloor$). First, each video is given one replica. Then, if the storage capacity of the cluster allows, in each replication iteration the strategy adds one more replica to videos in corresponding groups, starting from the video with the highest popularity. Last, if there are more replicas that can be placed onto the cluster, the replication algorithm just does round robin replication.

3.4.1 Impact of Video Replication

First, we investigate the impact of replication degree (average number of replicas) on rejection rate. Figure 3.8 shows the results with a set of replication degree $\{1.0, 1.2, 1.6, 2.0, 3.0\}$ when the system is in different load conditions. Different replication and placement algorithms were employed with popularity parameter $\theta = 1.0$ (i.e., pure Zipf distribution). Figure 3.8 shows that the rejection rate drops as the replication degree increases and the rejection rate decreases dramatically from non-replication to a low replication degree of 1.2. This is because the most popular videos are given the most replicas, which significantly reduces the granularity of replicas in terms of communication weight for load balancing placement. There are no dramatic differences between the results of other replication degrees.

In [358], we also examined the impact of replication degree for videos with a different popularity parameter θ. It was found that the impact of replication degree increases as the video popularity gets more skewed. The decrease of the video popularity skew induces the fine granularity of communication weight of replicas for placement.

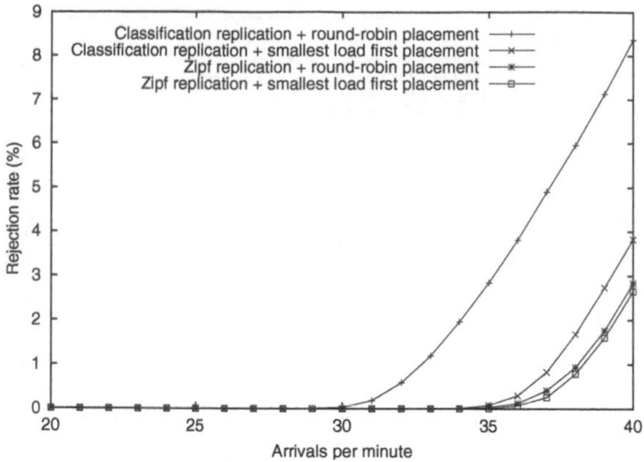

Figure 3.9: Impact of replication and placement algorithms on rejection rate.

3.4.2 Impact of Placement of Video Replicas

Figure 3.9 depicts the impact of four algorithm combinations on rejection rate when the video popularity conforms a pure Zipf distribution ($\theta = 1.0$). It was assumed that the video replication degree was 1.2. The figure shows that the Zipf replication, together with the smallest load first placement algorithm, yields the lowest rejection rate as the system becomes heavily loaded. In fact, the Zipf replication postpones the system saturation point from 30 requests per minute to 35 requests per minute, even when it is combined with a round robin placement policy. The Zipf replication produces fine-grained replicas in terms of communication weight. Given a large number of video replicas of the same size, a round robin placement generates a good load balance.

Figure 3.9 also shows the smallest load first placement algorithm can reduce the rejection rate significantly, even in the case that the video replication is approximated by a simple classification-based algorithm. The smallest load first placement algorithm outperforms the round robin slightly. In [358], we conducted experiments with different video popularity skews. It was observed that the performance gap between the smallest load first placement and the round robin algorithms enlarges with the increase of the video popularity skew.

To better understand the cause of the performance gaps due to different replication and placement algorithms, we show in Figure 3.9 the load imbalance degree L of (3.2) as the system load changes. It is expected that when the system is lightly loaded, there is no rejection rate although the system load may not be balanced between the servers. If communication traffic is perfectly balanced between the servers, rejection won't occur until the input load reaches the cluster bandwidth capacity (i.e., 40 requests per minute). The load imbalance leads to rejection in systems loaded with as light as 30 requests per minute, as shown in Figure 3.9.

One of the replication and placement objectives is to minimize the degree of load

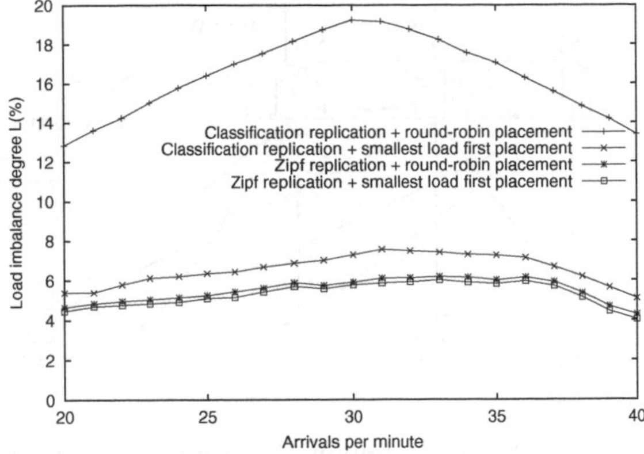

Figure 3.10: Impact of replication and placement algorithms on load imbalance degree.

imbalance. Even in the Zipf replication in combination with the smallest load first placement algorithm, the cluster load is not totally balanced. It is because the algorithms are derived based on a mean-value estimate of arrival rate. Variance and other high moment statistics of the arrival distribution yield a nonnegligible degree of load imbalance and hence rejections. When the arrival rate exceeds the bandwidth capacity, more servers become overloaded. Although the load imbalance degree starts to drop, the rejection rate rises sharply because of the cluster-wide overloaded condition.

3.4.3 Impact of Request Redirection

We note that the preceding analysis on replication is based on the assumption of *a priori* knowledge of video popularity. We have also assumed the same peak period of request rates for all videos. It might be true for videos in the same category. Nevertheless, in general, different types of videos would not have the same peak period of arrival rates. Actually, many experiments involving nonstationary traffic patterns assumed that the relative popularity of videos varied on a weekly, daily, and even hourly basis.

Balancing network traffic for servers of a VoD cluster is critical during heavy-load periods. In [359], we presented a request redirection strategy to achieve the objective of dynamic network traffic balance by taking advantage of internal backbone bandwidth. Figure 3.11 presents an illustration of the algorithm. Suppose that video v_2 is more popular than it was expected and video v_3 is less popular. Server S_2 on which video v_2 is mapped may be overloaded while server S_3 on which video v_3 is mapped becomes underloaded. Note that a request for video v_2 can be serviced not only by its hosting server S_2 when its outgoing network bandwidth is available, but also by

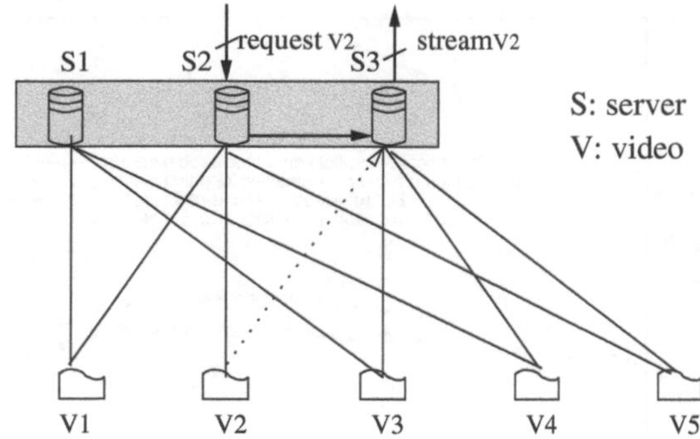

Figure 3.11: An illustration of streaming redirection.

a server with available outgoing network bandwidth, say S_3, which has no replica of v_2 but if the bandwidth of the internal link connecting these two servers is available. The dotted line shows a redirected request. Redirection avoids extra run-time overhead and excessive delays incurred by dynamic replication possibly. However, its effect is constrained by data layout methods and internal bandwidth of the backbone network. Considering the server autonomy, redirection will not be employed when there is no rejection rate. Thus, redirection does not necessarily guarantee a load balanced system, because a load balanced system is desirable but not the ultimate goal.

Figure 3.12 shows the performance of request redirection strategy combined with two different data layout methods. Each internal link between two servers was assigned 100 Mbps or 25-stream bandwidth and the average replication degree was 1.2. Obviously, request redirection reduces the rejection rate significantly by balancing the outgoing network traffic of servers of the cluster. It also postpones the occurrence of rejection. With the redirection strategy, the poor data layout method yields similar performance to the good ones. This is probably because the bandwidth of each internal link between two servers is so large that outgoing network traffic has been fully balanced by the use of redirection. Note that when the arrival rate reaches the link capacity (40 requests per minute), there is still about 1% of rejection rate. This is because of the instances of the arrival rate may exceed the average arrival rate in simulation.

3.5 Concluding Remarks

In this chapter, we have investigated the video replication and placement problem in distributed storage VoD clusters. We have formulated the problem as a combi-

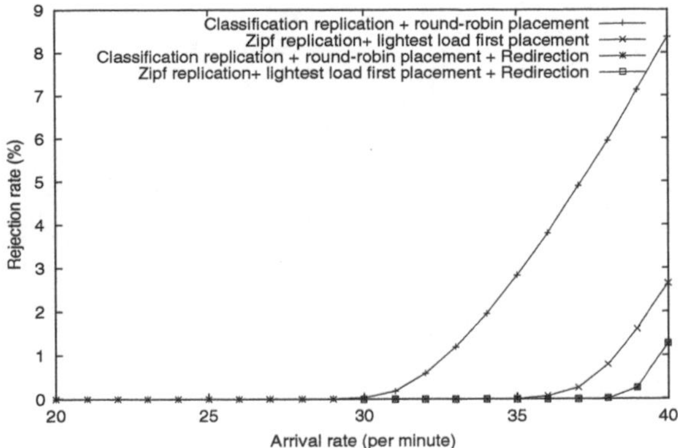

Figure 3.12: Impact of request redirection on rejection rate.

natorial optimization problem with objectives of maximizing the encoding bit rate and the number of replicas of each video and balancing the workload of the servers. The optimization process is subject to the constraints of the storage capacity and the outbound network bandwidth of the servers. Under the assumption of single fixed encoding bit rate for all videos, we have developed an optimal replication algorithm and a bounded placement algorithm for videos with different popularity. The placement algorithm achieves a tight bound of load imbalance degree. A more time-efficient algorithm utilizes the information about Zipf-like video popularity distributions to approximate the optimal solution.

We note that the algorithms are conservative because they assume *a priori* knowledge of video popularity and the same peak period of request rates for all videos. To complement the conservative strategies, we have presented a request redirection strategy that utilizes the internal backbone bandwidth to balance the outgoing network traffic between the servers during the run time.

The replication and placement framework in this chapter provides a flexible way to maintain multiple replicas of a video with different encoding bit rates. This paves a way to the provisioning of different levels of QoS to requests for various videos or to requests from various clients/devices. Adaptive QoS control on streaming and other Internet services is the theme of Chapter 4 through Chapter 7.

Chapter 4

Quality of Service-Aware Resource Management on Internet Servers

Load balancing and data placement presented in the last two chapters are to ensure scalability and high availability of Internet servers. A request accepted by a server is often handled in a best-effort manner to maximize the quality of service (QoS). There are many situations where the requests from different clients for the same or different contents should be treated differently. For example, media requests from a black/white and low resolution PDA should never be treated equivalently as those from a high-end desktop with a broadband network. QoS differentiation is a key technology for such purposes. In this chapter, we review the QoS-aware request scheduling and resource management techniques on Internet servers.

4.1 Introduction

Server cluster and load balancing rely on the principle of resource replication to improve the scalability and availability of Internet services. Internet traffic is characteristic of self-similar and bursty patterns. Server resource configuration based on mean-value analysis is prone to bottleneck. In particular, during big, often unforeseeable events such as stock market roller coaster rides, terror attacks, or Mars landing, there is a surge of Internet traffic that can quickly saturate the server capacity and hence affect the services and even lead to lawsuits due to breakage of service level agreements.

Under an overloaded condition, the server does not have sufficient resource to provide service to all clients. Instead of failing all clients, service differentiation allocates resources to client requests in a biased way so as to provide different levels of QoS, in terms of response time, bandwidth, etc., to different requests. For example, on an on-line trading site like `etrade.com` and a Web e-mail server like `hotmail.com`, a service differentiation architecture provides preferred premium customers better services and basic customers degraded or no service when the server becomes overloaded. This offers a strong incentive for a differentiated pricing model for most of today's QoS-sensitive Internet services. Similarly on a Web content hosting site, a service differentiation architecture can treat clients of various content providers differently and hence permit differentiated pricing for content hosting services. On an on-line shopping site, a service differentiation architecture can give a higher priority to sessions of loyal customers than occasional visitors so as to maximize profit. Poor performance perceived by loyal customers is a major impediment for the success of e-commerce. Moreover, the requests of a session are of different values. A checkout request is obviously more valuable than a catalog browsing request and its quality must be guaranteed.

Even on an indiscriminate Web site, a service differentiation architecture can isolate the performance of requests from different clients or proxies. By downgrading the QoS of their requests, the architecture controls the behaviors of aggressive clients and ensures fair sharing between clients, proxies, and even traffic from different domains. Fairness assurance automatically builds a firewall around aggressive clients and hence protects servers from distributed denial-of-service (DDoS) attack.

Service differentiation enhances the scalability and availability of Internet services from the server perspective. The architecture is also demanded in pervasive computing. By blending computers into the pervasive networking infrastructure, pervasive computing provides people with the most wide range of communication and information access services. The services are ensured to be accessible anytime and anywhere on any device. Provisioning of such services is a challenge because of the diversity of access devices (e.g., cellular phone, set-top box, or computer) and access networks (e.g., bluetooth, cellular, DSL, or local area network). Their capabilities to receive, process, store, and display media-rich Web content vary greatly. A user who accesses streaming services on a cellular phone must expect different service qualities from users on a high-end workstation with high bandwidth networking capacities. A service differentiation architecture supports such heterogeneous QoS requirements and preferences by adapting the stream quality in origin content servers or their proxies to various devices and access patterns.

The concept of service differentiation is not new. It was first invented for QoS-aware packet scheduling in the network core. QoS in packet scheduling is a set of service requirements that the network must meet in order to transmit data at a specified rate and to deliver them within a specified time frame. Two service architectures have been proposed: Integrated Services (IntServ) [42] and Differentiated Services (DiffServ) [38]. IntServ requires reserving routing resources along the service delivery paths using a protocol like Resource Reservation Protocol (RSVP) for QoS guarantee. Since all the routers need to maintain per-flow state information, this requirement hinders the IntServ architecture from widespread deployment. In contrast, DiffServ aims to provide differentiated services among classes of aggregated traffic flows, instead of offering absolute QoS measures to individual flows. It defines a 6-bit Differentiated Services Code Point (DSCP) field in the IP packet that enables per-hope different levels of service to be assigned to different classes. To receive different levels of QoS, each application packet is assigned to a DSCP code corresponding to the desired level of service at the network edges; see Microsoft Windows Server (`www.microsoft.com/windowsserver2003`) for an example of DSCP settings. DiffServ-compatible routers in the network core perform stateless prioritized packet forwarding, so-called "per-hop behaviors" (PHBs), to the classified packets. Due to its per-class stateless routing, the DiffServ architecture exhibits

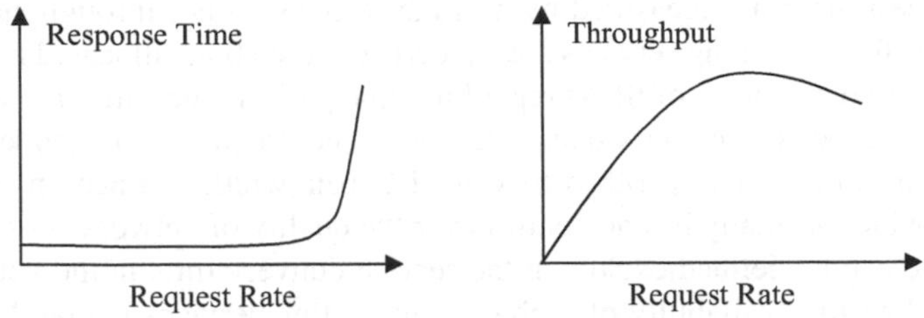

Figure 4.1: Response time and throughput versus request rate.

a good scalability.

The DiffServ architecture aims to provide a guaranteed delivery protocol for IP traffic in the network layer. Network alone is not sufficient to guarantee end-to-end QoS differentiation. An early critical path analysis of TCP transactions revealed that server-side delay constitutes over 80% of client-perceived latency for small requests when server load is high [25]. Network overprovisioning in the past decade has made QoS failure rate in the network core for even large media-rich requests. Request queueing delays and their prolonged processing time in overloaded servers have become increasing important factors of end-to-end QoS.

Figure 4.1 illustrates the response time and server throughput as functions of the request rate. A server can respond to requests quickly as long as the server is not overloaded. The server throughput normally increases with the request rate until the server is saturated. After the request rate exceeds the server capacity, the throughput drops because the server continues to take requests although most of them will be dropped. Frequent interrupts of the server by the incoming requests also attribute to the throughput drop. Application-level service differentiation is to provide different levels of QoS in terms of queueing delay and processing time to different client requests. In this chapter, we focus on the principle of application-level service differentiation. In Chapters 5 and 6, we give two service differentiation technologies on streaming and e-commerce servers.

4.2 Service Differentiation Architecture

4.2.1 The Objectives

Service differentiation is to provide a certain level of QoS guarantee to a class of aggregated requests based on predefined service level agreements (SLAs) with the clients. A service level measures the expected quality, in terms of metrics like queueing delay, allocated bandwidth, and response time, depending on the nature of QoS-sensitive applications. For example, interactive applications like on-line shopping and messenger are measured by average response time; throughput-intensive applications like streaming services are concerned more about allocated network I/O bandwidth. Two examples of SLAs regarding time performance are "$x\%$ of requests must have a response time less than y seconds" and "requests for a video must be responded in no less than y Mbps network I/O bandwidth." Since an application service provider normally has no control over the quality of network service, the response time can be defined excluding the service delivery time in the network. An SLA regarding the availability of Web content hosting services is that "the service must be available for $x\%$ of time." An availability level of 99.99%, calculated on a round-the-clock basis, would mean that the Web hosting site would experience at most 52 minutes of unscheduled downtime per year; a level of 99% equals about 3.7 days of downtime per year.

An SLA defines a contract from the perspective of clients, regarding statistical assurance of an absolute quality level for each application. We refer to this type of objectives as absolute service differentation. From the perspective of servers, there are other relative service differentiation objectives that guarantee relative service qualities between different traffic classes. That is, a traffic class with a higher desired QoS level (referred to as a higher class) should be assured to receive better (at least no worse) service than a lower class. A special form of relative architecture is proportional service differentiation. It aims to provide a guarantee of quality spacings between different request classes. In contrast to absolute service differentiation, relative service differentiation is intended to provide differentiated services to meet QoS requirements of different applications and user expectations, meanwhile maximizing the utilization of system resources and business profit.

In relative service differentiation, a service level can be specified based on the same performance metrics as in absolute service differentiation (e.g., queueing delay, bandwidth, or response time). An additional performance metric is normalized response time as defined by a term of slowdown. Slowdown measures the ratio of a request's queueing delay to its service time. It is known that clients are more likely to anticipate short delays for "small" requests and more willing to tolerate long delays for "large" requests. The slowdown metric characterizes the relative queueing delay.

4.2.2 Workload Classification

We note that the objective of service differentiation is defined on request classes, rather than on individual requests. Request classification must be done in a server front end such as the load balancer of a Web cluster or combined with an admission control component. At first, the requests need to be distinguished between applications, if they are deployed in the same IP or Virtual IP address. The Internet services on the same site are either deployed as different URL addresses via a Web hosting mechanism or on different TCP/IP port numbers. The classification based on the request target URLs or port numbers makes it possible to treat applications differently according to their QoS requirements. In the case of Web content hosting, this classification enables the server to provide different levels of QoS to requests of different content providers and ultimately to offer a strong incentive for a differentiated pricing model for Web hosting.

For the same service, requests can be further distinguished based on client information, such as packet source IP addresses and HTTP cookies. This client-centric classification makes it possible to establish preferred client classes that need premium services as opposed to clients in other service classes.

Recall that the capabilities of client access devices in a pervasive computing environment vary drastically in many aspects, such as processing power, display size and resolution, networking bandwidth, storage capacity, and battery power. HTTP/1.1 content negotiation capability and W3C Composite Capability/Preference Profile (CC/PP) [316] are mechanisms that allow a client to send its preferred version of contents and client capability information as part of HTTP requests. In HTTP/1.1 content negotiation, a server can select the best representation for a response according to the contents of particular header fields in its request message or other information pertaining to the request (such as the network address of the client). CC/PP defines a way for a client agent to specify its capabilities and user preferences as a collection of Uniform Resource Identifiers (URIs) and Resource Description Framework (RDF) text. The information is sent by the agent along with its HTTP requests. The classification based on information of client devices and preferences enables a server or its proxies to adapt media-rich contents to meet client QoS expectations.

4.2.3 QoS-Aware Server Resource Management

On the basis of request classification, the objective of service differentiation can be realized in four aspects: admission control, resource allocation, request scheduling, and content adaptation. Admission control regulates the acceptance or blocking of incoming requests to prevent the server from overloading or switches from congestion. QoS-aware admission control treats requests of different classes differently and admits high priority requests only when the system becomes highly loaded.

To provide a certain QoS level to an admitted request, the system not only needs to have sufficient computing and networking resources, but also needs to make these resources available to the request handler when they are needed. QoS-aware resource allocation and scheduling determine how much resources (e.g., CPU time, memory capacity, and network bandwidth) can be allocated and when the requests should be scheduled to consume the resources. QoS-aware resource management should not only guarantee performance isolation between different classes of service requests, but also ensure high resource utilization.

Content adaptation is to adapt delivered content quality for the adjustment of requests' resource demands. On the one hand, content adaptation is required to adapt meeting client-side variations and constraints. On the other hand, QoS negotiation and degradation can help save resource consumption and improve resource utilization on servers with constrained resources.

In the following, we present a review of the existing representative QoS-aware resource management technologies of each aspect. An application service provider does not necessarily have all these four attributes.

4.3 QoS-Aware Admission Control

Admission control of an Internet server shares the same objective of admission control in a network route. The latter employs a gateway that estimates the load by observing the length of its packet backlog queue and drops incoming packets when the queue becomes too long. A simple form of packet discard is Drop Tail, which discards arriving packets when the gateway's buffer space is exhausted. The Drop Tail algorithm is readily applicable to admission control of requests on servers. It, however, distributes rejections among different classes of traffic arbitrarily and the rejection occurrences are sensitive to bursty patterns of Internet traffic. Early Random Drop (ERD) and Random Early Detection (RED) address Drop Tail's deficiencies [112, 150]. They use randomization to ensure that all classes of traffic encounter the same loss rate. In the approaches, an incoming packet will be dropped with a probability, varying according to the backlog queue length, if the queue length exceeds a predefined threshold. By dropping packets before the gateway's buffers are completely exhausted, the ERD and RED approaches can also prevent congestion, rather than simply reacting to it.

Chen and Mohapatra [56] applied the ERD approach to QoS-aware admission control of requests in which the gateway estimates the server load based on its listen queue length. They suggested to set two thresholds on the listen queue length. Requests of lower priority classes are rejected with a higher probability after the server utilization exceeds the first threshold. All the requests are rejected when the second threshold is reached. This ERD-based approach was shown to be effective in provisioning of differentiated services in terms of queueing delays between lower and higher priority classes of traffic but with no guarantee of quality spacings. Another limiting factor of this approach is that the listen queue length is not necessarily a good indicator of server load. It is because requests of an Internet service not only are variable in size but also have strong data locality, as discussed in Chapter 2. In regard of this, the authors suggested estimating request rate based on the history of access pattern and service time of each request on its type [58]. A request is admitted if its resource demand (estimated service time) is no more than the projected available resources.

Abdelzaher *et al.* [2] treated requests as aperiodic real-time tasks with arbitrary arrival times, computation times, and relative deadlines. It is known that such a group of tasks scheduled by a deadline-monotonic policy will always meet their deadline constraints as long as the server utilization is below 0.58. They assumed a linear regression method to estimate the impact of requests on system utilization and applied linear feedback control theories to admit an appropriate number of requests so as to keep system utilization bounded.

Under the assumption of *a priori* knowledge of request arrival rate and the maximum waiting time of each traffic class, Lee *et al.* [188] proposed two admission control algorithms: maximum profit and maximum admission that either admit requests having a more stringent waiting time requirement (for maximizing the potential profit) or admit as many clients as possible into the Web server (for maximizing the popularity of the Web site). Since the admission control is based on the maximum arrival rate of each class, it is possible for a request to be serviced in a lower priority class without violation of its time requirement. The authors presented dynamic class adaptation algorithms that assign each request to a lowest possible class so as to further reduce its resource usage.

Request blocking in request-oriented admission control may lead to aborted, incomplete sessions. Since the revenue of an e-commerce site mainly comes from completed sessions, rather than requests, session integrity is a critical performance metric. Cherkasova and Phaal [60] revealed that an overloaded server can experience a server loss of throughput measured in terms of completed sessions and that the server also discriminates against longer sessions. They presented a simple session-based admission control scheme to deter the acceptance of requests of new sessions. It monitors the server load periodically and estimates the load in the near future. If the predicted load is higher than a predefined threshold, no new sessions are admitted.

4.4 QoS-Aware Resource Management

Generally there are two approaches to allocate resources for the purpose of ensuring the QoS of requests or preserving the quality spacings between two classes. One is priority-based and the other is rate-based. A priority-based approach assigns different priorities to different classes of requests. It relies on user- or kernel-level support to honor the priorities when the requests are processed. In contrast, a rate-based approach allocates a certain fraction of resources to each class. Service differentiation is achieved by rate allocation.

4.4.1 Priority-Based Request Scheduling

Priority-based request scheduling is popular in process/thread-per-request HTTP servers like Apache and Mozilla. We use the term of process to represent either process or thread hereafter in this chapter. This type of server constitutes a master process monitoring the listen queue and a pool of child processes that are spawned on demand, one for each incoming request. The master process can easily assign a priority to each child process when it is spawned or recycled for a request, according to the request's service differentiation category.

Albeit simple, a challenge with this priority-based approach is how to honor these application-level priorities in execution so as to schedule the processing of the requests in a strict priority manner. That is, low-priority requests will be blocked until no other high-priority requests exist. Requests of the same priority are serviced in a first-come-first-serve (FCFS) way. The application-level priorities can be implemented in user- or kernel-level support. A user-level implementation assumes a scheduler on top of operating systems, while a kernel-level implementation requires mapping the application-level priorities onto the priorities of execution units (processes or threads) in operating systems.

Almeida *et al.* [5] implemented a priority-based approach in both user and kernel levels for the provisioning of QoS differentiation of HTTP requests in terms of response time. All requests are assigned different priorities based on their access content. They are scheduled for services according to the strict priority policy. This is in contrast to the default FCFS scheduling in Apache server. The authors presented user- and kernel-level implementations to honor the priorities in request services. The user-level implementation adds an extra scheduler in Apache server that decides the order in which requests should be serviced. The kernel-level implementation modifies both Apache server and Linux kernel by adding new system calls to provide a mapping from request priorities into process priorities and to keep track of which processes are running at which priority level. Request priority aside, the authors also discovered that restricting the number of processes that are allocated to run concurrently for each class of application was a simple and effective strategy in obtaining differentiated performance. Similar studies on the effect of priority and degree of concurrency were conducted by Eggert and Heidemann [93]. Their studies were limited to an application-level implementation with background and foreground of two priority levels.

The user-level and kernel-level implementations of priority-based request scheduling have their own advantages and disadvantages, similar to green and native implementations of Java thread. Almeida *et al.* summarized three significant differences in the resultant process scheduling [5]. In particular, if requests are scheduled by a user-level scheduler, a low-priority request starts to be serviced and it will be executed at the same priority as those servicing high-priority requests. In contrast, a kernel-level implementation can hardly block a running process immediately, as often required by work conserving scheduling.

Priority-based scheduling with static priorities may lead to starvation for low-priority requests. There are time-dependent priority schedulers, such as PAD [88], LAD [320], and adaptive waiting-time priority [191], for proportional delay differentiation (PDD) services in network routers. They assign a time-varying priority to each packet. When a network router is ready to handle a request, it chooses the highest priority pakcet among all those at the head of backlog queues of different classes. The time-dependent priority schedulers are able to ensure predefined quality spacings between different classes of traffic. Their performance on Internet servers is yet to see.

4.4.2 Processing Rate Allocation

The processing rate of a server measures its processing power assigned to an application and determines the application quality. The processing power is, in turn, determined by allocated resources like CPU time, memory size, networking bandwidth, disk I/O bandwidth, etc. Most resources are quantifiable and their allocation can be controlled accurately to some extent. However, the dependence of processing power on allocated resources is rarely in a linear relationship. Moreover, QoS is often measured in multiple interrelated dimensions, such as timeliness, content quality, and security; each QoS metric can be dependent on more than one resource type. These complicate the problem of QoS-aware processing rate allocation.

Rajkumar *et al.* proposed an analytical Q-RAM model for multiclass resource allocation to meet multidimensional QoS requirements in a general context [253, 254]. This model allows resources to be allocated to individual applications with the goal of maximizing a system utility function under the constraint that each application can meet its minimum QoS requirement. For example, in a video-on-demand (VoD) service, the stream quality is measured by a two-dimensional QoS metric: start-up latency and stream bit rate. Stream bit rate is an integration of video quality factors like frame rate, image resolution, and color depth. It is largely constrained by the server-side network I/O bandwidth and client-side network access and device processing capacities. Start-up latency is the waiting period experienced by a user after selecting a stream to play. It is attributed to both network congestion and server load conditions. Recent advances in scalable coding and transcoding offer the potential of QoS differentiation on streaming servers. In [360], Zhou and Xu formulated the problem of network I/O bandwidth allocation as a stochastic optimization problem with a harmonic utility function of stream quality factors. They derived the optimal streaming bit rates for requests of different classes under various server load conditions and proved that the optimal allocation not only maximizes the system utility function but also guarantees proportional fair sharing of network I/O bandwidth between classes.

Another important objective function in QoS-aware resource allocation is proportional queueing delay differentiation. Recall that queueing delay is a performance metric that favors requests demanding more resources; in contrast, slowdown reflects the request's relative queueing delay to its service time. In [356], Zhou *et al.* proposed a two-dimensional proportional slowdown differentiation (PSD) service model on e-commerce servers for on-line transactions: intersession and intrasession. The intersession model aims to provide different levels of QoS to sessions from different customer classes, and the intrasession model aims to provide different levels of QoS to requests in different states of a session. The authors derived optimal processing rate allocations for given slowdown ratios between intersession and intrasession requests.

In addition to determining the amount of processing rate for each request class, another challenge with rate-based resource allocation techniques is to realize a rate allocation. There were many QoS-aware resource management mechanisms that implement a processing rate based on the maximum number of processes available for handling a class of applications [5, 93]. In a process-per-request Web server such as Apache, processes are used as both the scheduling units and the resource principals for allocation. The available number of processes limits resource usage allocated to requests of different classes. If all processes of a process pool are busy, additional requests of that class are blocked until a process becomes available. The upper bound is a parameter tunable by the administrator of the Web server based on the priority level and workload of each class. Thus, the approaches can affect QoS factors by limiting the processing rate of each class. For example, Almeida *et al.* used this tunable parameter together with the parameter of request priority implemented service differentiation in terms of response time [5]. Since the number of processes does not necessarily reflect the processing rate allocated to a traffic class, such rate-based allocation is not suitable for control of quality spacings between different traffic classes.

When an HTTP connection request arrives, it is put into a service queue to wait for an available process. HTTP/1.1 implements a persistent connection model. A server process that accepts a connection must wait for a specified amount of time for further requests from the same client. The blocking time dominates the time a process spends on serving the connection. Sha *et al.* [282] argued that the average connection service rate is determined primarily by the number of processes and not by CPU allocation and that the service rate to a traffic can be controlled approximately by adjusting the allocated processes. This processing rate allocation scheme was applied with success to implement a queueing model with feedback control for a guarantee of QoS in terms of queueing delay.

Feedback control operates by responding to measured deviations from the desired performance. When the residual errors are too big, the linear controller would lead to a poor controllability for a nonlinear system. In contrast, queueing theories provide predictive frameworks for inferring expected delays according to the input load change. Lu *et al.* applied proportional-integral (PI) control to adjust the number of processes allocated to each client (or client class) in persistent connected servers, based on periodical queueing network (QN) estimates under the assumption that each traffic class is served by a separate M/M/1 queue with *a priori* known arrival rate and service rate. The integrated approach achieved absolute and relative request delay guarantees in Web servers [201, 204, 282]. The integrated feedback control approach was recently extended for PSD provisioning between classes of traffic in a general $M/G_P/1$ FCFS queueing system with a bounded Pareto service time distribution [322]. Wei and Xu developed a self-tuning fuzzy control approach for QoS-aware resource management on Internet servers [321]. It assumes no knowledge about the input traffic. The approach was implemented in an eQoS system that assisted an Apache Web server to provide a user-perceived and end-to-end QoS guarantee.

A kernel-level implementation of the processing rate allocation scheme was due to Goyal *et al.* [136]. They proposed a hierarchical start-time fair queueing (HSFQ) CPU scheduler that allocates CPU bandwidth to application class-specific schedulers in proportion to their predefined weights, which in turn schedule the execution of their perspective applications in an appropriate way. Shenoy *et al.* applied this rate-based predictable scheduling principle to outbound network I/O bandwidth and disk I/O bandwidth allocation [286, 287]. They implemented these scheduling principles in QLinux, a Linux kernel, to meet the diverse performance requirements of multimedia and Web applications [305].

Rate-based scheduling relies on accurate accounting and predictable allocation of resources. QLinux employs a lazy receiver processing (LRP) technique [90] to ensure that network protocol processing overheads are charged to appropriate application processes. Overheads of other kernel tasks such as interrupt processing and bookkeeping operations are accounted by a CPU scheduler that provides predictable performance with a fairness guarantee.

LRP associates received packets as early as possible with the receiving process, and then performed their subsequent processing based on that process scheduling priority. This idea can be generalized by separating the concept of resource principal from that of a protection domain. Since resource allocation and scheduling primitives in today's operating systems do not extend to the execution of significant parts of kernel code, an application has no control over the consumption of resources that the kernel consumes on behalf of the application.

Resource Container is a new operating system abstraction [20]. It separates the notion of a protection domain from that of a resource principal. A resource container encompasses all system resources that the server uses to perform an independent activity, such as processing a client HTTP request. All user- and kernel-level processing for an activity is charged to the appropriate resource container and scheduled at the priority of the container. For example, a Web server wants to differentiate response time of two classes of clients based on their payment tariffs. The server can use two resource containers with different priority levels, assigning the high-priority requests to one container and the low-priority requests to another. Resource containers allow accurate accounting and scheduling of resources consumed on behalf of a single client request or a class of client requests. Thus, this new mechanism provides fine-grained resource management for differentiated services provisioning when combined with an appropriate resource scheduler. For example, lottery scheduling policy can be employed for implementing proportional share CPU scheduling among resource containers [317]. This approach can work effectively on a process-per-connection Web server, a single-process multithreaded Web server and a single-process event-driven Web server.

Xok/ExOS exokernel system gives application software more control over raw system resources [129]. It allows the network services to be implemented as independent application-level libraries for differentiated serivices provisioning. It also enables applications to employ specialized networking software to achieve performance improvements.

4.4.3 QoS-Aware Resource Management on Server Clusters

Server clustering provides an additional dimension, the number of nodes, for QoS-aware resource management. It is to partition the nodes into a number of sets, one for one request class. The set size of a request class is determined by the nodal capacities as well as the class QoS requirement in terms of priority or rate-allocation weight. In the case that the cluster nodes are homogeneous in processing power, priority-based scheduling ensures a higher priority request class is assigned more nodes and rate-based scheduling guarantees the partition size of a class be proportional to its precalculated weight.

QoS-aware resource management is an extension of a layer-7 load balancer, as discussed in Chapter 2. A load balancer diverts each incoming request to a server in the cluster for the objective of balancing the server workloads and meanwhile preserving the requests' session persistency. A QoS-aware resource manager works on a new abstraction of request class, ensuring requests of the same class to receive the same level of QoS and the QoS of different classes to be differentiated in a controlled manner. The manager partitions the cluster nodes and installs a load balancer for each partition.

The node partitioning can be done statically when the servers are initially configured or reconfigured. For example, most Web hosting service providers allow a content provider to install all its services in dedicated servers at a premium charge rate. This type of static partitioning requires *a priori* knowledge of workload characteristics of request classes. Due to its work nonconserving nature in resultant request scheduling, static partitioning often leads to low resource utilization. More importantly, this type of organization deviates from the objectives of scalability and high availability of server clustering.

An alternative to static node partitioning is dynamic partitioning. The QoS-aware resource manager first generates initial partitions, based on projected workload of different classes and their QoS requirements, as in static node partitioning. The partitions are not fixed. The server nodes are transferred from one partition to the other in response to the change of workload and resource configuration so as to provide a predictable and controllable QoS differentiation between the request classes. Zhu *et al.* [361] proposed a periodic repartitioning approach to ensure the QoS of different classes are guaranteed to be proportional to their predefined weights. The QoS is measured by a stretch factor metric — the ratio of response time of a request to its service time. It is a variant of slowdown.

Since repartitioning is a cluser-wide global operation, it is too costly to perform frequently. This limits the controllability of the resource manager for service differentiation and its adaptability to the changing workload of different traffic classes in short timescales. To overcome this drawback, Cardellini *et al.* [48] suggested a dynamic partitioning strategy that keeps adjusting the processing rates of the partitions by transferring server nodes between the partitions dynamically. When a server is selected for migration to a new partition, it will continue to service the accepted requests and meanwhile starts to accept requests of the new traffic class corresponding to the target partition.

Shen *et al.* [285] proposed a two-level request distribution and scheduling approach. In this approach, each node can service requests from any class. On receiving a request, a load balancer in the front end of the cluster first switches it to a node that has the smallest number of active and queued requests among a small number of randomly polled servers. At each node, a QoS-aware resource manager allocates resources to assigned requests belonging to different classes in a similar way to what we discussed in Section 4.4.2. Unlike the static and dynamic partitioning approaches that multiplex the the cluster nodes in space-sharing, this two-level resource management approach multiplexes the cluster nodes in time-sharing.

The above approaches require no modifications of nodal operating systems. They are implemented in a middleware level. Aron *et al.* extended the concept of resource container to a new abstraction of cluster reserve for performance isolation in server clusters [14]. Cluster reserve represents a cluster-wide resource principal. Resources assigned to cluster reserves A and B are dynamically split into resource assignments for the corresponding resource containers on each node. Resources allocated to the resource containers on each node add up to the desired allocation for the associated cluster reserve. Distribution of the resources between the resource containers in a cluster reserve is determined by a cluster-wide QoS resource manager depending on the resource usage on each individual server. The allocation optimization is independent of the request distribution strategy deployed in the cluster.

4.5 Content Adaptation

The access control and resource management technologies we discussed so far all assume requests have constant resource demands. The key was on finding a spacewise resource allocation and timewise request schedule to meet the aggregated resource requirements in a 2D space–time domain.

Content adaptation technologies define a third dimension of QoS-aware resource management for the provisioning of service differentiation. It regulates the resource requirements of requests for media-rich contents by delivering the contents of different qualities. Multimedia objects such as images and audio and video clips are becoming a prevalent part of Web content. They can be delivered in different formats, resolutions, sizes, color depths, and other quality control options.

Content adaptation has a twofold objective. One is to offer degraded levels of QoS when the server becomes highly loaded so as to maximize the server throughput. This type of server-centric adaptation extends admission control from its essential binary acceptance/rejection model to one that a heavily loaded server admits a request with a negotiated and degraded QoS instead of simply rejection. The other objective of content adaptation is to meet the diversity of client preferences and the resource constraints of access devices and networks. It supports client-centric service differentiation.

Adaptation of a content (or Web object more generally) can be performed by its origin server or proxies, either statically before the context is accessed or dynamically when a request for the content is received [205, 206]. Static adaptation needs to store multiple copies of a content object that differ in quality and size. The multiple versions are often organized in a quality tree. At the top of the tree is an original content with the highest possible quality. A child node is a degraded version of its parent in certain aspects. In client-centric content adaptation, the multiple versions can also be represented in different languages, such as HTML and Wireless Markup Language (WML), to make the content accessible from any devices.

We note that static adaptation is limited to static contents. It cannot do anything for dynamic contents that are generated on demand. Another limitation is due to storage constraint. An original media-rich content often keeps only a few number of copies of different qualities. The predictively adapted contents may not necessarily meet the changing client capabilities and preferences. There were studies on transcoding techniques for dynamic content adaptation; see [45, 118, 145] for examples. They transform multimedia objects from one form to another, trading off object quality for resource consumption.

Fox *et al.* developed a client-centric adaptation approach, in which transcoding proxies placed between clients and servers perform aggressive computation and storage on behalf of clients [118]. Adaptation is achieved via application-level data type specific transcoding techniques, which preserve information of image and text objects that has the highest semantic value. The Power Browser project introduces text summarization methods for Web browsing on handheld devices such as PDAs and cell phones [45]. Each Web page is broken into text units that each can be hidden, partially displayed, fully visible, or summarized. The methods can provide significant improvements in access times. Han *et al.* developed dynamic image transcoding in a proxy for mobile Web browsing [145]. The proxy-based approaches enable themselves to transparent incremental deployment, since a proxy appears as a server to existing clients and as a client to existing servers. Furthermore, proxy-based adaptation provides a smooth path for rapid prototyping of new client devices, content formats, and network protocols. The case studies showed that on-the-fly adaptation by transformational proxies can be a widely applicable, cost-effective, and flexible technique for addressing client variations and network constraints.

Lum and Lau investigated the space–time trade-off problem in static and dynamic content adaptation and proposed a context-aware decision engine for a good balance between transcoding overhead and spatial consumption [205, 206].

Chandra *et al.* developed a server-centric image transcoding technique to provide differentiated multimedia Web services [51]. Transcoding is used by the origin server or a server proxy to customize the size of JPEG image objects and hence manage the available network I/O bandwidth at the server. The primary performance metric is the image quality factor. Transcoding provides graceful degradation of image quality factors so that the preferred clients are served at quality factors that closely follow the original images and non-preferred clients are served at a lower image quality factor. The transcoding-enabled network I/O bandwidth management schemes allow the server to provide acceptable access latency to different clients by trading off image quality for size.

A key QoS factor with continuous media is streaming bit rate. It is an integrated quality factor of video frame (audio sample) rate, video frame (audio sample) size, and color depth. The idea of performing bandwidth management through adaptation of video frame rate and color depth was demonstrated by early experimental video gateway systems [7]. To provide differentiated streaming services, the continuous stream must be encoded in a layered coding format, or amenable to adaptive transcoding.

Layered coding techniques represent continuous media in a quality enhancement layer hierarchy so that medias of different quality levels can be extracted dynamically. Like static adaptation, layered coding is limited to support coarse-grained service differentiation. Adaptive transcoding adjusts the bit rate of a video stream on the fly according to the allocated network I/O bandwidth and makes the bandwidth usages controllable by the server's request scheduler at application level. It supports differentiated streaming services in a fine-gained level. Transcoding presents parameterized workload to streaming servers. In [360], we proposed a harmonic proportional bandwidth allocation mechanism for service differentiation on streaming servers. The harmonic allocation not only maximizes the bandwidth utilization, but also guarantees proportional sharing between classes with various prespecified differentiated weights. Chapter 5 draws on this paper.

Chapter 5

Service Differentiation on Streaming Servers

In this chapter, we investigate the problem of service differentiation provisioning on media streaming servers, with a focus on outbound network I/O bandwidth allocation. We formulate the bandwidth allocation problem as an optimization of a harmonic utility function of the stream quality factors and derive the optimal streaming bit rates for requests of different classes under various server load conditions. We prove that the optimal allocation not only maximizes the system utility function, but also guarantees proportional fair sharing between classes with different prespecified differentiation weights.

5.1 Introduction

Scalable streaming servers must be able to provide different levels of quality of service (QoS) to clients. It is because clients are different in their visiting interests, access patterns, service charges, and receiving devices. They can connect to a streaming server using a wide variety of devices, ranging from set-top boxes to PDAs. Their capabilities to receive, process, store, and display continuous medias (i.e., video and audio) can vary greatly. Given the diversity of client devices and their needs, the server has to tailor the content and provide different QoS levels accordingly. From the perspective of the server, the request arrival rate changes with time and hence it is nonstationary. Allocating resources to accommodate the potential peak arrival rate may not be cost-effective, if not impossible. In other words, it is desirable to provide different QoS levels to different requests during various access periods. To that end, there is a growing demand for replacing the current same-service-to-all paradigm with a model that treats client requests differently based on the client access patterns and the server resource capacities. Service differentiation aims to provide predictable and controllable per-class QoS levels to requests of different classes. It can also enhance the service availability because a request may be admitted and processed with a negotiated and degraded service quality by a heavily loaded server rather than being simply rejected.

For conventional Web applications, CPU cycle and disk I/O bandwidth are major resource constraints and response time is a primary QoS metric. Current response time differentiation strategies mostly rely on admission control and priority-based resource scheduling on individual servers [5, 57, 194] and node partitioning in server clusters [14, 361]. The characteristics and requirements of streaming services differ significantly from those of conventional Web services. I/O-intensive streaming services like video-on-demand (VoD) are usually constrained by disk I/O and especially network I/O bandwidth at the server communication ports. The service quality is measured not only by *start-up latency*, but also by allocated stream bandwidth — *streaming bit rate*. The response time oriented strategies are not sufficient for service differentiation on streaming servers. For example, priority-based request scheduling approaches are not sufficient because they cannot differentiate streaming bit rate. Dynamic node-partitioning strategies are not applicable because streaming services are continuous and long-lived.

There were studies on QoS-aware bandwidth management for rich media Web services; see [51, 118] for examples. Their focuses were on transformation of the format, color depth, and sizes of images as well as rich-texts to make a good trade-off between user-perceived document quality and transmission time. This chapter shares the objective of content adaptation, but focuses on server-side bandwidth management for continuous streaming applications. The characteristics of streaming media vary tremendously according to the content, the compression scheme, and the encoding scheme. At an application level, a primary performance metric perceived by clients is the media quality (i.e., streaming bit rate).

Without loss of generality, we consider videos as an example of streaming media. As in Chapter 3, we consider videos encoded in the constant bit rate (CBR) scheme. The idea of performing bandwidth management through adaptation of video frame rate and color depth was demonstrated by early experimental video gateway systems; see [7] for an example. Recent advances in transcoding technology make it possible to dynamically transform a video stream from its original encoding bit rate to degraded ones at a fine-grained level [249, 267, 346]. In other words, current transcoding techniques can adjust the bit rate of a CBR-encoded video stream on the fly according to the allocated network I/O bandwidth and make the bandwidth usages controllable by the server's request scheduler at application level. For example, Roy and Shen exploited the multimedia instruction set of the new Itantium processor to save computational power such that one transcoding unit can support many concurrent sessions in real time [267].

In this chapter, we study the transcoding-enabled QoS-adaptive bandwidth allocation problem for service differentiation on streaming servers. It complements the existing efforts on QoS-fixed bandwidth allocation for system scalability. Figure 5.1 shows a queueing network with N client request classes in a streaming cluster with M servers. It assumes a dispatcher to make admission decisions and determine the server for each incoming request. For each admitted and classified request by the dispatcher, two key quality factors are the streaming bit rate and the delay in the listen queue. To provide differentiated services, the dispatcher-based request scheduler needs to determine: (1) what network I/O bandwidth (bit rate) should be allocated to the stream for the request; and (2) when the stream should be scheduled to deliver.

In [360], we formulated the problem of network I/O bandwidth allocation at the

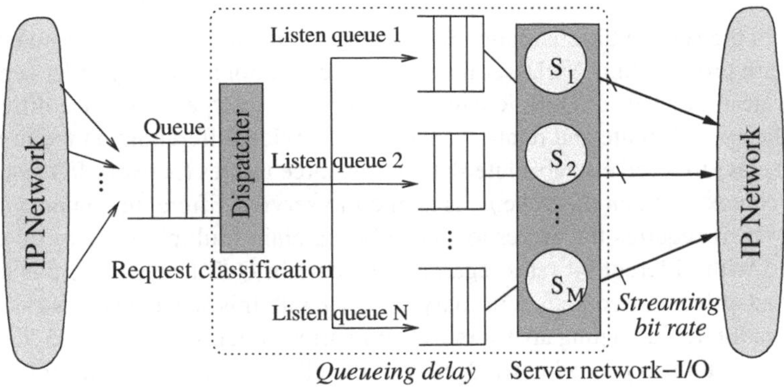

Figure 5.1: A queueing network model for a streaming cluster.

application level for service differentiation as an optimization of a harmonic utility function of the stream quality factors. We derived an optimal bandwidth allocation scheme, called harmonic proportional allocation, for request classes with different priorities under various server load conditions. The harmonic proportional allocation scheme not only maximizes the utility function, but also guarantees proportional fair bandwidth sharing between the classes with respect to their predefined differentiation weights. Simulation results showed the harmonic proportional allocation scheme achieves the objective of relative DiffServ in both short and long timescales and enhances the service availability to a great extent when the server load is high. In the following, we present the details of the allocation scheme and its impact on service availability.

5.2 Bandwidth Allocation for Differentiated Streaming Services

The objective of the network I/O bandwidth allocation problem is to determine stream bandwidth of each request class in such a way that the overall QoS is optimized and meanwhile the stream qualities are guaranteed to be proportional to their prespecified differentiation weights. The weights can be determined by clients' priorities, receiving devices, payments, etc. Divide the scheduling process into a sequence of short intervals of bandwidth allocation and request scheduling. The bandwidth allocation decision needs to be carried out in each interval, based on the measured bandwidth release rate and the predicted arrival rate of request classes.

5.2.1 Service Differentiation Models and Properties

Within the service differentiation infrastructure, incoming requests from different clients are grouped into N classes that are sorted in a nonincreasing order according to their desired levels of QoS. Recall that there are two types of service differentiation schemes: absolute and relative. Absolute service differentiation requires each request class to receive an absolute share of resource usages (network I/O bandwidth in this context). When this scheme is applied to service differentiation on a streaming server, it requires the server to statically maintain multiple replicas of a video encoded with different bit rates regarding the certain QoS levels if there is no support of adaptive transcoding. A primary concern with this scheme is its weak ability of adaptation to fluctuating arrival rates from various clients. In Section 5.5, we will show that without *a priori* knowledge about the clients' access patterns, this scheme could lead to a low resource utilization.

In relative service differentiation, service quality of class i is better or at least no worse than class $i+1$ for $1 \leq i \leq N-1$. The term "or no worse" is necessary, since in heavy-load conditions all request classes will tend to experience their minimum QoS levels. In this context, applications and clients do not get an absolute service quality assurance. Instead, this differentiation scheme assures that the class with a higher desired level of QoS (referred to as a higher class) will receive relatively better service quality than the class with a lower desired level of QoS (lower class). So it is up to the applications and clients to select appropriate QoS levels that best meet their requirements, cost, and constraints. For relative service differentiation, a streaming server has to assign different QoS levels to clients based on the dynamic load conditions. Therefore, it needs the support of adaptive content adaptation techniques such as transcoding since it is highly costly, if not impossible, to have a sufficient number of replicas *a priori* for the differentiation needs. There are layered video coding techniques that represent videos in a quality enhancement layer hierarchy so as to support different levels of video quality. Compared to such scalable video coding techniques, video transcoding techniques can support differentiated streaming services at a much finer grained level.

In order for a relative service differentiation scheme to be effective, it should satisfy two basic properties:

1. *Predictability:* The differentiation should be consistent. Higher classes should receive better or no worse services than lower classes independent of variations of the class load distributions.

2. *Controllability:* The scheduler must contain a number of controllable parameters, which are adjustable for quality spacings between classes.

In streaming services, a unique QoS metric of streams is the stream bandwidth (streaming bit rate). It has a lower bound and an upper bound. For example, 1 Mbps could be referred to as the lower bound of streaming bit rate of an MPEG-I movie for general clients. Different client classes may expect different lower bounds. The upper bound is the encoding bit rate of the video because today's transcoding

Table 5.1: Notations in the network I/O bandwidth allocation model.

Symbol	Description
N	Number of the request classes
λ_i	Arrival rate of request class i
δ_i	Quality differentiation weight of request class i
μ_i	Rate of channel allocation to request class i
q_i	Quality factor of request class i in terms of streaming bit rate
B	Upper bound of the aggregate channel allocation rate
U	Upper bound of streaming bit rate for all classes
L_i	Lower bound of streaming bit rate for class i

technologies can only dynamically degrade the streaming bit rate. We argue that an effective relative service differentiation scheme on streaming servers should have the following additional requirements:

1. *Upper and lower bounds:* Quality guarantees should be provided for all requests. Admission control is needed to prevent the system from being overloaded.

2. *Availability:* One goal of offering differentiated services on streaming servers is to serve as many requests as possible at sufficiently acceptable QoS levels to gain and retain the business. If the available network I/O bandwidth at server side is enough to provide the lower bound of QoS level to all requests, rejection rate could be minimized.

3. *Fairness:* Requests from lower classes should not be overcompromised for requests of higher classes.

5.2.2 Network I/O Bandwidth Allocation

The basic idea of network I/O bandwidth allocation for providing differentiated streaming services is to divide the allocation and scheduling process into a sequence of short intervals. In each interval, based on the measured bandwidth release rate and the predicted arrival rate of request classes, the stream bandwidth (bit rate) of each request class is determined based on differentiation requirements. The streams for the request classes are allocated and then scheduled according to various scheduling approaches. During a sufficiently short interval, it is expected that there is no large gap between the real results and derived ones [60].

In the following, we consider the bandwidth allocation problem for N request classes. Table 5.1 summarizes the notations used in the formulation.

We define a *channel* as the bandwidth unit allocated to a stream. It is the lower bound of streaming bit rate for the class which has the minimum QoS expectation. Let λ_i be the arrival rate of requests in class i in an allocation interval. Let μ_i be the rate of channel allocation to requests in class i. We define a *quality factor* q_i of requests in class i $(1 \leq i \leq N)$ as

$$q_i = \frac{\mu_i}{\lambda_i}. \tag{5.1}$$

It represents the quality of request class i in the current bandwidth allocation interval. For example, if the rate of channel allocation to class i is 8 per time unit and the request arrival rate of class i is 4 per time unit, the quality for requests in class i in the current interval is two channels.

Let B denote the bound of the aggregate channel allocation rate during the current bandwidth allocation interval. It is the ratio of the number of channels to be released in the current interval plus the unused ones from previous intervals to the length of the allocation interval. We have the resource constraint of

$$\sum_{i=1}^{N} \mu_i \leq B. \tag{5.2}$$

From the system's perspective, service availability is an essential objective. Suppose the request arrival rate of class i in the current scheduling period is 4 per time unit and the rate of channel allocation to class i is 3 per time unit because of heavy system load. Although the calculated quality factor for the class i calculated is 0.75, due to the existence of the lower bound of streaming bit rate, at least one request from this class may be rejected and other requests receive the lower bound of streaming bit rate.

Let L_i denote the lower bound of streaming bit rate for class i. Assume all video objects have the same initial encoding bit rate U. It represents the upper bound of streaming bit rate of all classes. We consider the network I/O bandwidth allocation for service differentiation when $\sum_{i=1}^{N} L_i\lambda_i \leq B < \sum_{i=1}^{N} U\lambda_i$. That is, the total number of available channels is enough to guarantee the minimum QoS level but not enough to support the maximum QoS level for all contending classes. Otherwise, the problem is either trivial or infeasible. For example, when $B \geq U\lambda_i$, we can simply give each class a quality factor U, the upper bound of streaming bit rate. When $B < L_i\lambda_i$, the admission control must do rejection and we would not consider service differentiation. Hence, for service availability, we have an additional constraint:

$$L_i \leq q_i \leq U \qquad 1 \leq i \leq N. \tag{5.3}$$

5.3 Harmonic Proportional-Share Allocation Scheme

In this section, we first present a proportional-share bandwidth allocation scheme tailored from proportional-share scheduling in packet networks and operating systems. Then, we propose a harmonic proportional share allocation scheme that not only optimizes an overall system utility function, but also ensures proportional bandwidth sharing between request classes. Consider N contending request classes with predefined quality differentiation weights $\delta_1, \delta_2, \ldots, \delta_N$. Since relative service differentiation requires class i would receive better or no worse services than class $i+1$, without loss of generality, weights δ_i are sorted in a nonincreasing order as $\delta_1 \geq \delta_2 \geq \ldots \geq \delta_N$. Note that the actual differentiation weights can be determined by the clients' priority, receiving device, payment, etc.

5.3.1 Proportional-Share Bandwidth Allocation

Proportional-share bandwidth allocation borrows its idea from proportional-share scheduling in networking and operating systems. At the operating systems level, there has also been a renewal of interest in fair-share schedulers [170], which we also refer to as proportional-share scheduling. For example, Waldspurger and Weihl proposed lottery scheduling for fair-share resource management for CPU utilization [317]. At the network level, it is usually called fair queueing. There are many fair queueing algorithms that ensure the per-hop queueing delay of the packets in different classes to be proportional to their predefined differentiation weights; see [88, 320] for examples. The proportional model is interesting because it is fair. It has been accepted as an important relative QoS differentiation model in the network core.

Our proportional-share bandwidth allocation scheme assigns quality factors to request classes in proportion to their quality differentiation weights δ_i. Recall that a request class i needs to maintain a lower bound of quality factor L_i. For service availability, in each bandwidth allocation interval, the proportional-share allocation scheme for streaming services states that for any two classes i and j, $1 \leq i, j \leq N$,

$$\frac{q_i - L_i}{q_j - L_j} = \frac{\delta_i}{\delta_j}, \tag{5.4}$$

subject to constraints of (5.2) and (5.3).

According to the constraint (5.2), the objective of (5.4) leads to a proportional bandwidth allocation rate as

$$\mu_i = L_i\lambda_i + (B - \sum_{k=1}^{N} L_k\lambda_k)\frac{\delta_i\lambda_i}{\sum_{k=1}^{N}\delta_k\lambda_k}. \tag{5.5}$$

It follows that the proportional quality factor of class i is calculated as

$$q_i = L_i + (B - \sum_{k=1}^{N} L_i\lambda_i)\frac{\delta_i}{\sum_{k=1}^{N}\delta_k\lambda_k}. \tag{5.6}$$

The quality factor of (5.6) reveals that the proportional-share allocation scheme generates consistent and predictable schedules for requests of different classes on streaming servers. The classes with higher differentiation weights δ_i receive better or no worse services than the classes with lower δ_i independent of variations of the class loads. The quality factor of each request class i is controlled by its own channel allocation rate μ_i.

5.3.2 Harmonic Proportional Allocation

Note that the proportional-share allocation scheme that aims to control the inter-class quality spacings does not necessarily yield best overall system QoS. To optimize the overall system QoS and meanwhile ensure quality spacings between the classes, we define a weighted harmonic function of the quality factors of all the classes, which ensures the lower bounds of streaming bit rate, as the optimization function. Specifically, we formulate the bandwidth allocation for service differentiation as the following optimization problem:

$$\text{Minimize} \quad \sum_{i=1}^{N} \delta_i \frac{1}{q_i - L_i} \tag{5.7}$$

Subject to constraints (5.2) and (5.3).

The minimization of the harmonic objective function (5.7) requires that requests of higher classes would be allocated more bandwidth, but this biased allocation should not overcompromise the share of requests from lower classes. We refer to the optimal allocation scheme as *harmonic proportional allocation.* Note that when $\sum_{i=1}^{N} L_i\lambda_i = B$, the quality factor q_i of each class i is equal to its lower bound L_i and there is no need for optimization and differentiation due to the service availability requirement.

The objective function (5.7) is continuous and it is convex and separable in its variables. Resource allocation constraint (5.2) describes the total amount of resource to be allocated. Constraint (5.3) ensures the positivity of variables. This optimization becomes a special case of the resource allocation problem. This kind of resource allocation optimization has been applied in many systems [4, 253, 328]. We define the optimization function as the weighted harmonic function of the quality factors of all the classes. It implies that the classes with higher differentiation weights get higher QoS factors and hence differentiation predictability is achieved. Interestingly, the derived allocation scheme also guarantees a proportional share allocation between the classes. The rationale behind the objective function is its feasibility, differentiation predictability, and proportional fairness properties, as we discussed in Section 5.2.1.

The optimization above is essentially a continuous convex separable resource allocation problem. According to theories with general resource allocation problems [155], its optimal solution occurs only if the first order derivatives of each component function of (5.7) over variables $\mu_1, \mu_2, \ldots, \mu_N$ are equal. Specifically, the optimal solution of (5.7) occurs when

$$-\frac{\delta_i\lambda_i}{(\mu_i - L_i\lambda_i)^2} = -\frac{\delta_j\lambda_j}{(\mu_j - L_j\lambda_j)^2}, \tag{5.8}$$

for any classes i and j, $1 \leq i, j \leq N$.

Combining with the constraint (5.2), the set of equations (5.8) leads to the optimal allocation scheme

$$\mu_i = L_i\lambda_i + (B - \sum_{k=1}^{N} L_k\lambda_k)\frac{\sqrt{\delta_i\lambda_i}}{\sum_{k=1}^{N}\sqrt{\delta_k\lambda_k}} \qquad 1 \leq i \leq N. \tag{5.9}$$

As a result, the optimal quality factor of class i, $1 \leq i \leq N$, is

$$q_i = L_i + (B - \sum_{k=1}^{N} L_k\lambda_k)\frac{\sqrt{\delta_i\lambda_i}}{\lambda_i\sum_{k=1}^{N}\sqrt{\delta_k\lambda_k}}. \tag{5.10}$$

To show the implications of the derived bandwidth allocation scheme on system behavior, we give the following basic properties regarding the controllability and dynamics due to the allocation scheme:

1. If the differentiation parameter of a class increases, the quality factor of the class increases at the cost of other classes.
2. The quality factor of a class i decreases with the increase of arrival rate of each class j.
3. Increasing the load of a higher class causes a larger decrease in the quality factor of a class than increasing the load of a lower class.
4. Suppose that a fraction of the class i load shifts to class j, while the aggregate load remains the same. If $i > j$, the quality factor of class i increases, while that of other classes decreases. If $i < j$, the quality factor of class j decreases, while that of other classes increases.

In addition, we can show that the optimal allocation scheme has the property of fairness. That is,

Theorem 5.1 *The harmonic proportional allocation scheme guarantees a square root proportional-share distribution of excess bandwidth over the minimum requirements between classes with different differentiation weights.*

Proof Define $\tilde{\mu}_i$ as the allocation of excess bandwidth to class i over the minimum requirement $L_i\lambda_i$. According to (5.9), we have

$$\tilde{\mu}_i = \mu_i - L_i\lambda_i = (B - \sum_{k=1}^{N} L_k\lambda_k)\frac{\sqrt{\delta_i\lambda_i}}{\sum_{k=1}^{N}\sqrt{\delta_k\lambda_k}}. \tag{5.11}$$

The allocation of $\tilde{\mu}_i$ yields an increment of quality factor, $\tilde{q}_i = \tilde{\mu}_i/\lambda_i$, to the minimum quality factor of L_i. That is, $\tilde{q}_i = q_i - L_i$. By comparing the quality factor increments of two classes i and j, we obtain the quality spacing between the classes that

$$\frac{\tilde{q}_i}{\tilde{q}_j} = \frac{\lambda_j \cdot \sqrt{\delta_i \lambda_i}}{\lambda_i \cdot \sqrt{\delta_j \lambda_j}} = \sqrt{\frac{\lambda_j}{\lambda_i}} \cdot \sqrt{\frac{\delta_i}{\delta_j}}. \tag{5.12}$$

Eq. (5.12) indicates that the ratio of excess bandwidth allocation between the two classes with given arrival rates is square root proportional to their predefined differentiation weights. This completes the proof. □

Note that when system load is heavy, that is, $\sum_{i=1}^{N} L_i \lambda_i$ is close to the bound B, all requests are going to be allocated their lower bounds of the bit rate, which would in turn minimize the rejection rate when the system is heavily loaded.

Recall that the quality factor q_i must be less than or equal to the upper bound of streaming bit rate U. According to (5.10), the computed q_i could be greater than U when the system is lightly loaded. As a result, certain request classes may not be able to use all their allocated channels. To improve the channel utilization, we can redistribute the excess channels to other request classes or simply leave them for calculating the channel release rate in the next allocation interval.

We also note that according to (5.11), given fixed λ_i, the classes with higher δ_i get more portion of available network I/O bandwidth. However, by (5.12), $\tilde{q}_i \geq \tilde{q}_j$ if and only if $\frac{\delta_i}{\lambda_i} \geq \frac{\delta_j}{\lambda_j}$ holds. Otherwise, the property of predictability of service differentiation becomes violated. For differentiation predictability, one solution is temporary weight promotion, as suggested in the context of server node partitioning [361]. That is, the request scheduler temporarily increases quality differentiation weight δ_i, based on the current request arrival rates, so as to ensure $\frac{\delta_i}{\lambda_i} \geq \frac{\delta_j}{\lambda_j}$. The derived proportional-share bandwidth allocation scheme (5.5) does not have this requirement.

5.4 Implementation Issues

We built a simulation model for the evaluation of the network I/O bandwidth allocation schemes with two popular request schedulers on streaming servers. The model divides the simulation process into a sequence of short intervals and performs bandwidth allocation and scheduling functions based on the predicted arrival rates of the request classes and measured bandwidth release rate of the servers in each interval. A fairly accurate estimation of the parameter is needed so that the proposed bandwidth allocation schemes can adapt to the dynamically changing values. Figure 5.2 shows the framework of bandwidth allocation.

Estimation of request arrival rate. As shown in Figure 5.2, the request arrival rate of each class (λ_i) is estimated by counting the number of requests from each class occurring in a moving window of certain immediate past periods. A

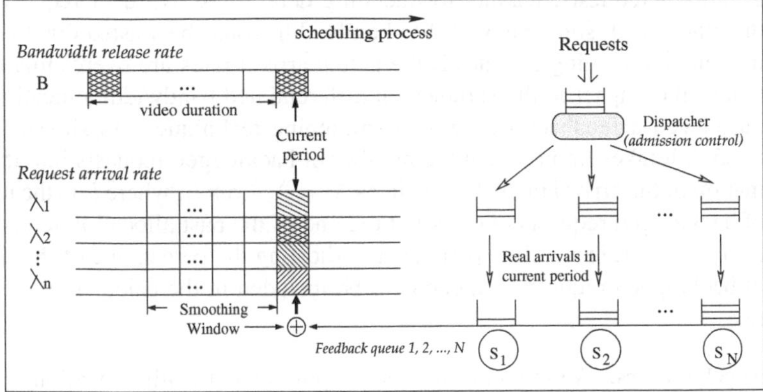

Figure 5.2: A moving window based network I/O bandwidth allocation framework.

smoothing technique based on a decaying function is applied to take weighted averages over past estimates.

Measure of bandwidth allocation rate bound (B). Since the focus of this chapter is on service differentiation provisioning, we assume that the streaming servers provide support for a video playback function only. Because of the continuous nature of streaming services, it is feasible to measure the bandwidth to be released in the current allocation interval. As shown in Figure 5.2, we employed a similar smoothing window to calculate the bound of bandwidth allocation rate in each allocation interval. It takes into account the bandwidth to be released in the current allocation interval and the excess bandwidth in the past allocation intervals.

Admission control. The derived bandwidth allocation schemes in (5.5) and (5.9) ensure that the requests in higher classes always get better services in terms of streaming bit rate. However, streaming services usually have a maximum acceptable waiting time [4]. If the aggregate channel requirement of all classes ($\sum_{i=1}^{N} L_i\lambda_i$) exceeds the bound of bandwidth allocation rate (B) in the current allocation interval due to some bursty traffic, the dispatcher in the streaming cluster imposes admission control and drops those requests which have waited in the queue for more than the maximum acceptable waiting time. The requests to be rejected are first from the lowest differentiated class, and then the second lowest class, and so on.

Feedback queue. We note that accuracy of the arrival rate estimations in an allocation interval is affected by the variance of interarrival time distributions. It fluctuates due to the existence of bursty traffic. When the actual arrival rates exceed estimated arrival rates, streaming bandwidth would be overallocated to current requests, leading to queueing delay of subsequent requests. On the other hand, some network I/O bandwidth would be wasted due to under-measured streaming bit rates if the actual arrival rates are overestimated. To reduce the impact of the variance on differentiated bandwidth allocations, we introduced a feedback queue as a smoothing technique. As shown in Figure 5.2, it takes into account the number of backlogged requests into the estimation of the arrival rates. It calculates $\lambda_i = \lambda_i + \alpha \times l_i$, where l_i is the number of backlogged requests of class i at the end of the past allocation interval and α, $0 \leq \alpha \leq 1$, is a scaling parameter, indicating the percentage of the number of backlogged requests in a queue to be included in the calculation of arrival rate.

Request schedulers. We implemented the proposed bandwidth allocation schemes with two popular request scheduling approaches. A strict priority scheduler picks requests from higher classes before processing those from lower classes. Requests of lower classes are only executed if no request exists in any higher classes in the current scheduling interval. A first-come-first-service with first-fit backfill (FCFS/FF) policy [190] picks requests of different classes in the order they arrive, as long as sufficient system resources are available to meet their requirements. In the case that the head-of-the-line request is blocked due to the lack of sufficient resources, first-fit backfilling searches further down the listen queue for the first request that is able to be scheduled immediately.

5.5 Service Availability Evaluation

In this section, we examine the impact of service differentiation on system performance and the impact of various bandwidth allocation schemes with the two request schedulers on service differentiation provisioning in terms of streaming bit rate and queueing delay.

The experiments assumed that the videos were encoded with a bit rate of 8 Mbps, a typical value for high-quality MPEG movies. The minimal acceptable bit rate was assumed to be the same for all request classes. It was set to 1 Mbps, a typical value for low-quality MPEG movies. Thus, 8 Mbps and 1 Mbps were the upper and lower bound of streaming bit rate for transcoding, respectively. Video transcoding was used to degrade the streaming bit rate on the fly according to the results of network I/O bandwidth allocation in (5.5) and (5.9). The streaming cluster consisted of eight servers and each server had a network I/O bandwidth of 1.8 Gbps. In total, the streaming capacity of the cluster ranged from 20 requests at 8 Mbps per minute to 160 requests at 1 Mbps per minute. The aggregate arrival rate (λ) ranged from 20 to 160 requests per minute. The access patterns were generated by a Poisson process and each video lasted 2 hours. The maximum acceptable queueing delay was set to 4 minutes. If the queueing delay of a request exceeded 4 minutes, either the admission

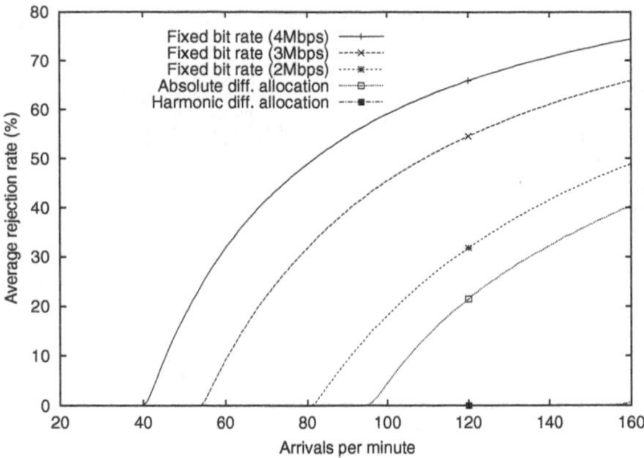

Figure 5.3: Impact on average rejection rate in systems of different loads.

control rejected the request or the client dropped the request.

Due to the lack of service differentiation workload on streaming servers, we adopted a service differentiation workload tailored from an eBay on-line auction that was used for analysis of queueing-delay differentiation in [361]. It consisted of three classes of client requests, 10% requests from registered clients for bidding and posting (class A), 40% requests from registered clients for browsing and searching (class B), and 50% requests from unregistered clients (class C). That is, their request arrival ratios $(\lambda_a, \lambda_b, \lambda_c)$ were $(1, 4, 5)$. Their quality differentiation weights $(\delta_a, \delta_b, \delta_c)$ were assumed to be (4, 2, 1). In the simulation, one allocation and scheduling interval was 5 minutes. The workload was estimated as the average in past four intervals. We ran the simulation for many times. Each representative result plotted in this section is an average of 500 runs.

5.5.1 Impact of Service Differentiation

We first examine the system performance due to various differentiated bandwidth allocation schemes. In addition to the proposed proportional-share and harmonic proportional allocation schemes, two static bandwidth allocation schemes are included. A static nonuniform bandwidth allocation scheme provides absolute service differentiation. It allocates the fixed streaming bit rate 4 Mbps, 2 Mbps and 1 Mbps to requests from class A, B, and C, respectively. A static uniform bandwidth allocation scheme supports no service differentiation; we considered three uniform encoding bit rates: 2 Mbps, 3 Mbps and 4 Mbps for all the requests. The streams are delivered according to FCFS/FF scheduling policy, unless otherwise specified.

Figure 5.3 shows the impact of service differentiation on the average rejection rate in systems of different load conditions. It shows that the harmonic allocation scheme guarantees service availability. This is achieved by degrading the streaming bit rates

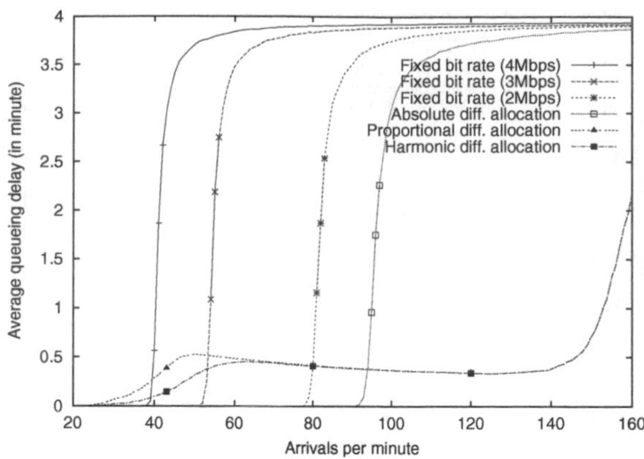

Figure 5.4: Impact on average queueing delay in systems of different loads.

adaptively with transcoding according to system load. The proportional-share allocation scheme achieves similar results to the harmonic allocation. The absolute differentiation allocation scheme cannot adapt to system load dynamically and so that it cannot guarantee service availability. Like the system with the absolute differentiation allocation, the average rejection rate in the system without service differentiation increases abruptly after arrival rate exceeds corresponding knee points. Figure 5.3 also reveals that both the harmonic proportional and proportional-share bandwidth allocation schemes can achieve high throughput and high service availability when servers are heavily loaded.

Figure 5.4 shows the impact of service differentiation on the average queueing delay. It can be seen that without service differentiation or with the absolute service differentiation, the average queueing delay increases abruptly after arrival rate exceeds certain knee points and rejection occurs at the corresponding levels, as shown in Figure 5.3. The average queueing delay is approaching and bounded by the maximum acceptable waiting time (4 minutes). In contrast, the harmonic proportional and proportional-share allocation schemes maintain the average queueing delay in acceptable degrees at various arrival rates. The queueing delay is due to the variance of interarrival time distributions. It also shows that the harmonic allocation scheme yields slightly lower queueing delay than the proportional-share allocation when the system load is light.

Figure 5.5 shows that the harmonic proportional allocation scheme enables the streaming system to efficiently and adaptively manage its available network I/O bandwidth. The proportional-share allocation scheme obtains similar results, which were omitted for brevity. On the other hand, the absolute differentiation allocation does not provide such an adaptivity. Like the system in a best effort service model, the absolute differentiation allocation wastes considerable streaming bandwidth when system load is light. This type of system can fully utilize its bandwidth

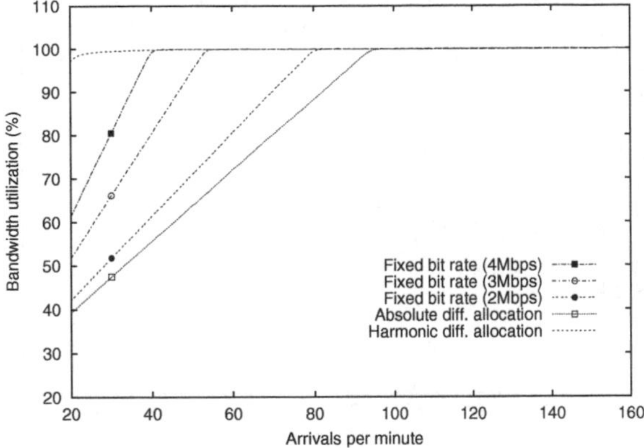

Figure 5.5: Impact on bandwidth utilization in systems of different loads.

when arrival rate exceeds some knee points. However, as shown in Figure 5.3, this utilization comes at the cost of driving rejection rate to unacceptable levels.

In summary, in comparison with the absolute service differentiation strategy, both the harmonic and proportional-share allocation schemes make it possible for a streaming server to achieve high throughput, high service availability, and low queueing delay when the server is heavily loaded.

5.5.2 Impact of Differentiated Bandwidth Allocation

The second experiment was on the impact of the various differentiated network I/O bandwidth allocation schemes on service differentiation provisioning in details.

Figure 5.6 shows a microscopic view of the streaming bit rate of individual requests in the three classes due to the harmonic bandwidth allocation scheme, when arrival rate is low (50 requests/minute), medium (80 requests/minute), and high (110 requests/minute), respectively. The simulation at each arrival rate was run for 60 minutes. Each point represents the streaming bit rate of individual requests in the classes in consecutive recording time units. It can be seen that the bandwidth allocation scheme consistently enforced prespecified quality spacings in terms of streaming bit rate between the request classes even in short time scales, although it was devised for maximizing the overall quality factor of streams.

Figure 5.7 shows the average streaming bit rate of requests from each class due to the harmonic proportional and proportional-share allocation schemes at various arrival rates. The transcoding-enabled allocators degrade the streaming bit rate of each request class adaptively with varying system load. When system load is light, requests from class A tend to receive the upper bound of streaming bit rate (i.e., 8 Mbps). When system load is moderate, all request classes get their fair shares. When system load is heavy, all request classes tend to receive the lower bound of

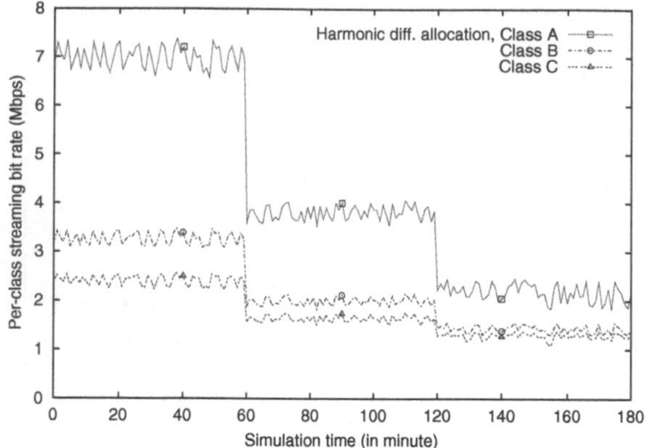

Figure 5.6: A microscopic view of streaming bit rate changes in harmonic allocation.

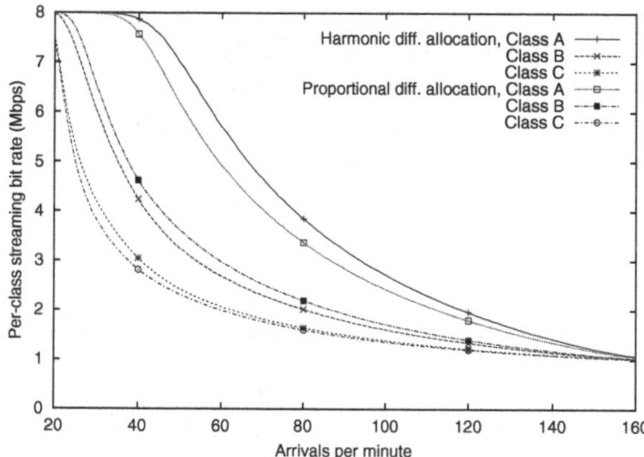

Figure 5.7: Differentiated streaming bit rates due to various allocation schemes.

streaming bit rate (i.e., 1 Mbps). Furthermore, requests from class A receive higher streaming bit rate by the use of the harmonic allocation scheme than by the use of the proportional-share allocation scheme. In general, the harmonic allocation scheme favors requests from higher classes more than the proportional-share allocation scheme. The proportional-share allocation scheme adjusts the quality levels of request classes in proportion to their differentiation weights. The harmonic allocation scheme is also proportional, as shown by (5.12) and proved in Theorem 5.1. In all cases, requests from higher classes consistently receive better or no worse service quality than requests from lower classes. Evidently, both the harmonic proportional and proportional-share allocation schemes can achieve long-term objectives of service differentiation in terms of streaming bit rate.

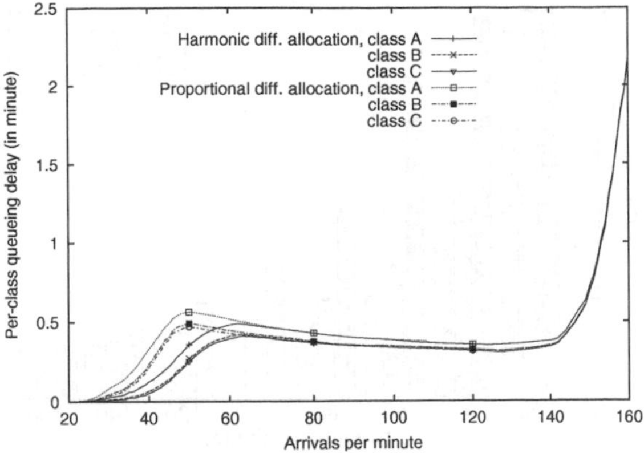

Figure 5.8: Queueing delay of request classes.

Figure 5.8 shows the queueing delay of requests from the three classes due to the harmonic proportional and proportional-share allocation schemes at various arrival rates. The queueing delay is due to the variance of interarrival time distributions. When system load is light ($\lambda < 30$), the queueing delay of all request classes is trivial. It is because some network I/O bandwidth was unused during the past allocation intervals due to the existence of upper bound of streaming bit rate. When arrival rate exceeds 40 requests/minute, unexpectedly, we find a queueing-delay dip scenario. That is, the queueing delay initially increases and then marginally decreases as arrival rate increases and then increases suddenly as arrival rate is close to system's streaming capacity. Note that the queueing delay is not only affected by the variance of interarrival time distributions, but also affected by differentiated bandwidth allocations. Because the backlogged requests in queues are not considered in the calculation of arrival rate of classes, the streaming bit rates are overallocated to current requests, leading to higher queueing delay of subsequent requests. As the arrival rate further increases, the requests are allocated with lower streaming bit rates so that the impact of the backlogged requests on the bandwidth overallocation decreases and thus the impact on queueing delay of subsequent requests decreases. The impact of the variance of interarrival time distributions on queueing delay dominates and it is significant when system load is close to the system's streaming capacity ($150 < \lambda < 160$). In Section 5.5.4, we will show that the feedback queue can mitigate this kind of queueing-delay dip scenarios.

Figure 5.8 also shows that by the use of the two bandwidth allocation schemes, requests from class A have higher queueing delay compared to requests from classes B and C. Recall that the simulation assumed FCFS/FF request scheduling in combination with the bandwidth allocation schemes. Although FCFS/FF scheduling does not provide any queueing-delay differentiation between different classes, the QoS-aware bandwidth allocation schemes affect the performance metric of queueing de-

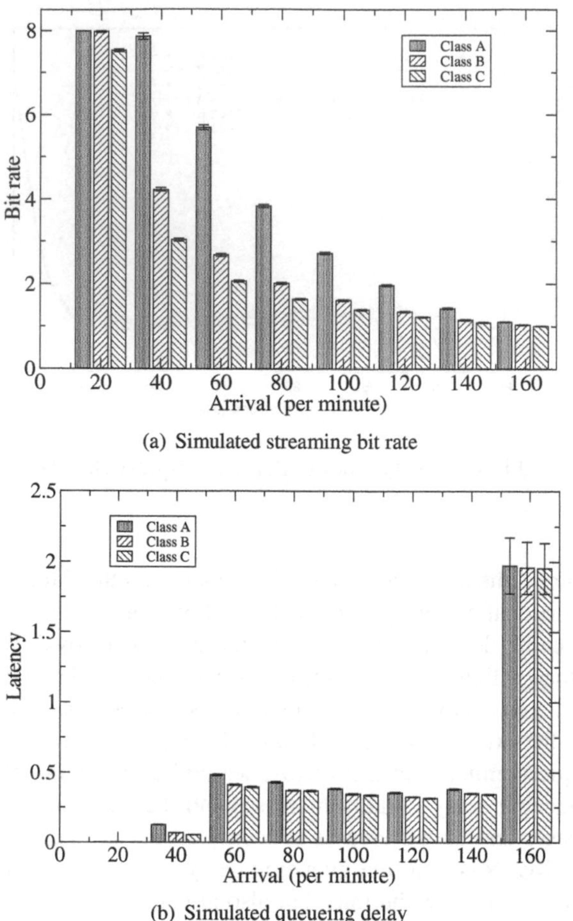

(a) Simulated streaming bit rate

(b) Simulated queueing delay

Figure 5.9: Confidence intervals of differentiated quality-of-service.

lay, as well. Requests from higher classes tend to be allocated with higher streaming bit rates. This results in higher queueing delay. Due to the first-fit feature of the FCFS/FF scheduler, on the other hand, requests from lower classes have higher probabilities to be processed with the differentiated streaming bit rates. Figure 5.8 also shows that compared with the proportional-share allocation scheme, the harmonic allocation postpones the emergence of queueing delay.

Figures 5.7 and 5.8 illustrate the average values of QoS metrics. In Figures 5.9(a) and (b), we present 95% confidence intervals of QoS metrics, i.e., streaming bit rate (Mbps) and queueing latency (minute), together with their mean values for the three classes when the arrival rate is 20, 40, 60, 80, 100, 120, and 159 requests/minute. When the arrival rate is 160 requests/minute, all requests receive the uniform lower bound of streaming bit rate (1 Mbps). The intervals were obtained by the method of independent replication [308]. It can be seen that these bounds are uniformly tight. This can be explained by the fact that the variances are only due to the exponential interarrival time distributions. Results of the proportional-share allocation are similar. Due to the space limit, we do not present confidence intervals for all experiments. The data in Figures 5.9 are entirely representative.

In summary, we observed that both the harmonic proportional and proportional-share allocation schemes can achieve the objective of service differentiation provisioning (in terms of streaming bit rate) in long and short timescales. They have positive side effects on the request queueing delay and the harmonic allocation scheme leads to lower queueing delay when the system is lightly loaded.

5.5.3 Impact of Request Scheduling

We have analyzed the proposed QoS-aware bandwidth allocation schemes in combination with the FCFS/FF request scheduling. The preceding experiments have shown that the FCFS/FF scheduler does not differentiate queueing delay of requests from different classes, although the bandwidth allocation schemes can affect the queueing delay of various request classes, as shown in Figure 5.8. In this experiment, we investigate the relationship between the harmonic proportional bandwidth allocation with request scheduling approaches.

Figure 5.10(a) shows the average queueing delay of requests from three classes due to the strict priority request scheduler. In such scheduling, requests in a queue cannot be serviced until the higher priority queues are all empty in the current allocation interval. It shows that the priority scheduler imposes certain degrees of control over queue-delay differentiation between the request classes. The queueing delay of requests from class A is rather limited since the priority scheduler favors the requests of higher classes. This is achieved at the cost of higher queueing delay of requests of lower classes. Note that the strict priority scheduling itself cannot guarantee quality spacings between various classes. Time-dependent priority scheduling schemes, widely addressed in the packet scheduling in the network side [88, 320], deserve further studies in providing queueing-delay differentiation in streaming services.

Figure 5.10(b) shows the streaming bit rate of request classes generated by an FCFS/FF scheduler and a priority scheduler in combination with the harmonic proportional allocation scheme. It shows that various scheduling schemes generate marginally different results by the use of the same bandwidth allocation scheme. Requests from higher classes have higher probabilities to be scheduled by the priority scheduler than by the FCFS/FF scheduler. Hence, their arrival rates calculated by the priority scheduler are less than those calculated by the FCFS/FF scheduler. This leads to higher streaming bit rate for requests from higher classes.

5.5.4 Impact of Queueing Principle with Feedback Control

Figures 5.8 and 5.10(a) illustrate the queueing-delay dip scenarios. That is, the queueing delay increases and then marginally decreases as arrival rate varies within a certain range. As we discussed above, it is due to the impact of the backlogged

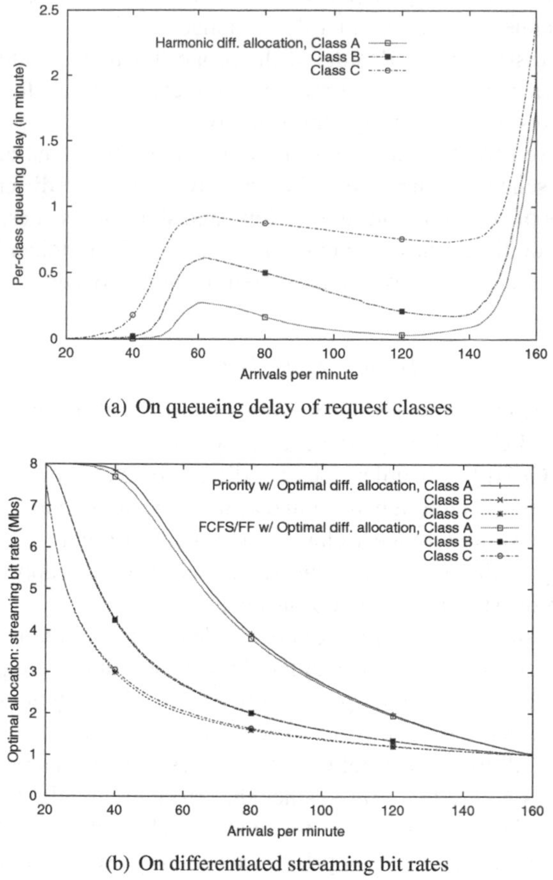

(a) On queueing delay of request classes

(b) On differentiated streaming bit rates

Figure 5.10: Impact of request schedulers with the harmonic allocation scheme.

requests on the bandwidth allocation, together with the variance of interarrival time distributions. In this experiment, we investigate the impact of the feedback queue technique on mitigating the queueing-delay dip scenarios. It includes various percentages of the number of backlogged requests in listen queues into the calculation of arrival rate of different classes.

Figure 5.11 shows the impact of the feedback queueing principle on queueing delay due to the harmonic proportional allocation with the FCFS/FF scheduler and the strict priority scheduler, respectively. Evidently, the feedback queue technique can reduce queueing delay significantly because it reduces the impact of the variance of interarrival time distributions on the calculation of arrival rate of request classes. As 100% of the number of backlogged requests in the listen queues are included in the calculation of the arrival rates, the queueing-delay dip scenarios almost disappear. Figure 5.11(a) shows that by the use of FCFS/FF scheduler, requests from all classes

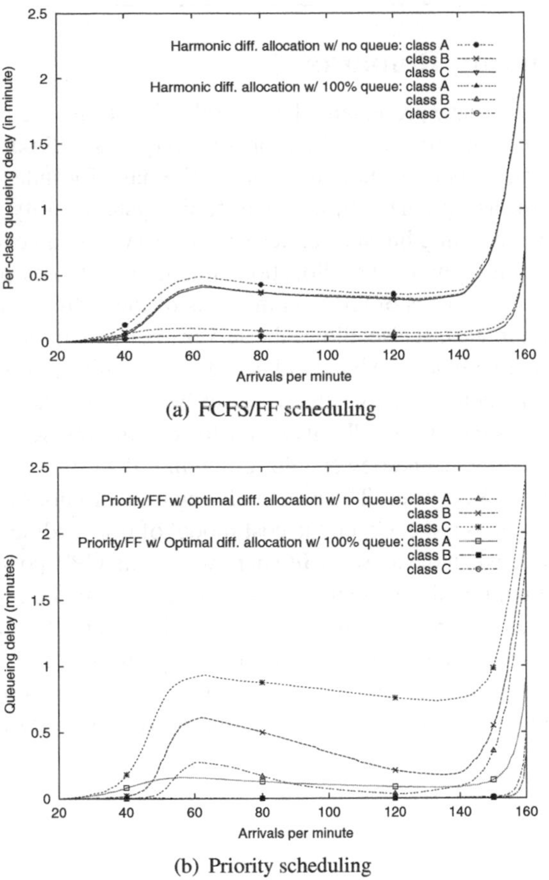

(a) FCFS/FF scheduling

(b) Priority scheduling

Figure 5.11: Impact of the feedback queue with different scheduling policies.

receive lower queueing delays. Figure 5.11(b) shows that the priority scheduler favors requests from higher classes, which almost have no queueing delay.

In [360], we conducted sensitivity analyses of the impact of the feedback queue on queueing delay. Recall that the number of backlogged requests to be counted in the estimation of arrival rate is determined by a parameter α. As α increases, the estimated arrival rate approaches to the real one. It leads to a reduction of the requests' average queueing delay and a mitigation of the queueing delay dip scenarios. In Chapter 7, we will extend the feedback queueing approach and present an integrated queueing method with feedback control for robust QoS provisioning.

5.6 Concluding Remarks

In this chapter, we have presented a bandwidth allocation scheme to facilitate the delivery of high bit rate streams to high-priority requests without overcompromising low-priority requests on streaming servers. We have formulated the bandwidth allocation problem as optimization of a harmonic system utility function and derived the optimal streaming bit rates under various server load conditions. We have proven that the optimal harmonic allocation scheme guarantees proportional sharing between different classes of requests in terms of their streaming bit rate in both short and long timescales. It also allows the server to efficiently and adaptively manage its constrainted network I/O bandwidth and achieve high-service availability by maintaining low queueing delay when the server becomes overloaded.

This chapter considered no CPU overhead for on-line transcoding. As a matter of fact, the transcoding technology enables efficient utilization of network I/O bandwidth at the cost of CPU cycles. The cost modeling of continuous media transcoding is still an open research issue. Given the cost model of transcoding, it is important to make a trade-off between the network I/O bandwidth and CPU power. We addressed the problem of providing differentiated streaming services in the context of CBR encoding scheme. Another popular encoding scheme is variable bit rate (VBR), which ensures constant video quality by varying streaming bit rate. But VBR streams exhibit a variability in their resource requirements. Therefore, transcoding-enabled service differentiation provisioning on streaming servers deserves further study.

Chapter 6

Service Differentiation on E-Commerce Servers

Service differentiation aims to provide differentiated quality of service (QoS) to requests from different clients. Service differentiation on an e-commerce should also treat the requests in the same session from the same client differently according to his/her access patterns. For example, a checkout request is obviously more important than catalog browsing and should be responded with little delay. In this chapter, we propose a two-dimensional (2D) service differentiation model for on-line transactions to provide different levels of QoS to sessions from different customer classes and to requests in different states of a session.

6.1 Introduction

E-commerce applications are characteristic of session traffic. A session is a sequence of individual requests of different types made by a single customer during a single visit to an e-commerce site. During a session, a customer can issue consecutive requests of various functions such as browse, search, select, add to shopping cart, register, and pay. It has been observed that different customers exhibit different navigation patterns. Actually, only a few customers are heavy buyers and most are occasional buyers or visitors. Recent studies on customer behaviors of some e-commerce sites showed that only 5% to 10% of customers were interested in buying something during the current visit and about 50% of these customers were capable of completing their purchases [219, 220]. Although it is important to accommodate the remaining 90% to 95% customers in order to turn them into loyal customers in future, the 5% to 10% premium customers should be preferential. This requires a scalable e-commerce server to provide different levels of QoS to sessions from different customers. We refer to this as *intersession service differentiation.*

An e-commerce session contains a sequence of requests for various functions in different states. The request sequence can be modeled as a customer behavior graph that describes state transitions between the different states of a session. For example, as shown in Figure 6.1, Menascé *et al.* proposed such a graph called customer behavior model graph (CBMG) [219, 220]. This kind of graph has one node for each possible state and transitions between these states. Each transition is assigned a

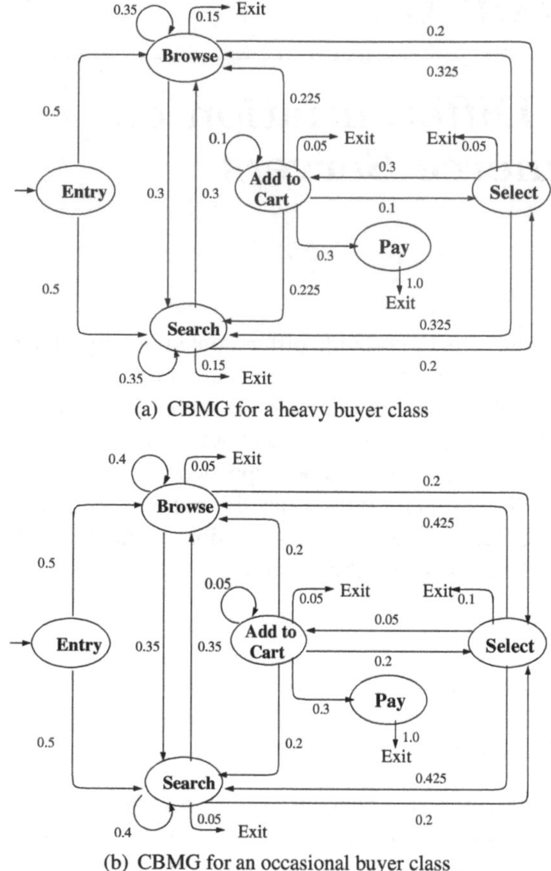

(a) CBMG for a heavy buyer class

(b) CBMG for an occasional buyer class

Figure 6.1: Customer behavior model graph for different customer classes.

probability based on customer navigation patterns. It tells that the average number of visits to various states is different in a single session and different states have various probabilities to end in buy. That is, requests in different states have different opportunities to turn themselves to be profitable. E-commerce servers should also provide different levels of QoS to requests in different states in each session so that profitable requests like order and checkout are guaranteed to be completed in a timely manner. We refer to this as *intrasession service differentiation.*

In this chapter, we investigate the problem of the 2D service differentiation provisioning on e-commerce servers, where the customers (and their requests) can be classified according to their profiles and shopping behaviors. User-perceived QoS of on-line transactions is often measured by a performance metric of response time. It refers to the duration of a request between its arrival and departure times, including waiting time in a backlogged queue and actual processing time. Network transmission delay is not considered because it is beyond the control of service providers. Response time reflects user-perceived absolute performance of a server. It is not suitable for comparing the quality of requests that have a large variety of resource demands. Customers are likely to anticipate short delays for "small" requests like browsing, and are willing to tolerate long delays for "large" requests like search. It is desirable that a request delay be kept proportional to its processing requirement. *Slowdown* is a normalized queueing delay, measuring the ratio of its delay in a backlogged queue relative to its service time. Since slowdown translates directly to user-perceived system load, it is more often used as a performance metric of responsiveness on Internet servers [30, 146, 261, 361]. Previous performance studies with respect to slowdown focused on resource management to minimize the overall system slowdown [146, 261, 361]. For example, Harchol-Balter evaluated on-line request scheduling algorithms in terms of mean slowdown of the requests [146].

In Chapter 4, we reviewed two main approaches for service differentiation provisioning: priority-based scheduling and rate allocation. In [356], we proposed a rate-based allocation strategy for 2D service differentiation provisioning with respect to slowdown between intersession and intrasession requests. It is complementary to load balancing technologies for preserving session integrity on server clusters in Chapter 2 and QoS-aware admission control in Chapter 4 for maximizing the server throughput in terms of completed sessions, instead of requests. This chapter presents the details of the 2D service differentiation model.

The knowledge about customers' navigation patterns is important to provisioning service differentiation on e-commerce servers. Menascé *et al.* designed a CBMG model to describe customers' navigation patterns through an e-commerce site [219, 220]. Based on CBMGs, the authors presented a family of priority-based resource management policies for e-commerce servers [220]. Three priority classes were used: high, medium, and low. Priorities of sessions changed dynamically as a function of state a customer was in and as a function of the amount of money the shopping cart had accumulated. Resource management strategies were geared toward optimizing business-oriented metrics, such as revenue per second. Their objective was to maximize a global system utility function and the resource allocation strategies cannot control quality spacings among different requests.

6.2 2D Service Differentiation Model

6.2.1 The Model

For service differentiation, incoming requests from different clients need to be classified into multiple classes according to their desired levels of QoS. Because different customers have different navigation patterns, the 2D service differentiation model classifies the customers into m classes according to statistics of their shopping behaviors, such as buy to visit ratio. Customers in the same class have similar

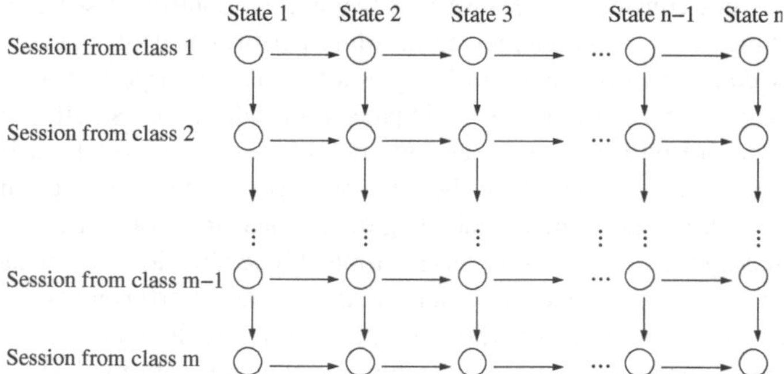

Figure 6.2: A 2D service differentiation model on e-commerce servers.

navigation patterns. Actually, some e-commerce sites request customers to log in before they start to navigate through the site. Their profiles help simplify the customer classification. The 2D service differentiation model assumes that each session of customer class i ($1 \le i \le m$) has n states, each corresponding to a request function. A customer's request is classified as being in different states according to the type of function requested. Figure 6.2 presents an illustration of the 2D service differentiation model. Each vertical arrow means a nonincreasing order of QoS levels between two requests in the same state from different sessions. Each horizontal arrow means a nonincreasing order of QoS levels between two states in a session.

Let α_i be the normalized quality differentiation weight of sessions from class i. That is, $\alpha_i > 0$ and $\sum_{i=1}^{m} \alpha_i = 1$. Because sessions from class i should receive better or no worse service quality than sessions from class $i+1$ according to intersession service differentiation, without loss of generality, we assume $\alpha_1 \ge \alpha_2 \ge \cdots \ge \alpha_m$. The values of α_i can be determined according to the shopping behaviors of class i, such as their buy to visit ratio [219, 220]. Let β_j be the normalized quality differentiation weight of state j in a session. That is, $\beta_j > 0$ and $\sum_{j=1}^{n} \beta_j = 1$. The values of β_j can be determined according to the characterization of transition probability from state j to state pay in sessions from all classes. Without loss of generality, we assume $\beta_1 \ge \beta_2 \ge \cdots \ge \beta_n$.

In general, a proportional resource allocation scheme assigns quality factors to request classes in proportion to their quality differentiation weights. In the 2D service differentiation model for on-line transactions on e-commerce servers, the quality factors of request classes are represented by their slowdown. Let $s_{i,j}$ denote the slowdown of a request from a session of class i in state j. Consequently, the proportional allocation model imposes the following constraints for intersession service differentiation and intrasession service differentiation, respectively:

$$\frac{s_{i_2,j}}{s_{i_1,j}} = \frac{\alpha_{i_1}}{\alpha_{i_2}}, \qquad \text{for all } j = 1, 2, \ldots, n \tag{6.1}$$

$$\frac{s_{i,j_2}}{s_{i,j_1}} = \frac{\beta_{j_1}}{\beta_{j_2}}, \qquad \text{for all } i = 1, 2, \ldots, m. \tag{6.2}$$

In Chapter 5, we discussed two basic requirements of relative differentiated service, *predictability* and *controllability*. Predictability requires that higher classes receive better or no worse service quality than lower classes, independent of the class load distributions. Controllability requires that the scheduler contain a number of controllable parameters that are adjustable for the control of quality spacings between classes. In the 2D service differentiation, each dimension should meet the basic requirements. An additional requirement on e-commerce servers is *fairness*. That is, requests from lower classes should not be overcompromised for requests from higher classes. For example, it is important to provide preferential treatments to sessions from premium customers and to requests that are likely to end with a purchase. E-commerce servers should also handle other nonbuying sessions that account for about 90% to 95% of visits if one wants to turn them into loyal customers [219, 220].

We assume that session arrivals from each customer class meet a Poisson process. We note that requests in each state from sessions of different customers are independent because the session head requests are independent. However, a customer may visit a state many times in a session. For example, a customer can submit a search request at time t_1, select a commodity at time t_2 (after some think time), and submit another search request at time t_3. Evidently, these requests at the search state are dependent and their dependency degree is determined by the navigation pattern of that customer. We notice that an e-commerce server can accommodate many concurrent sessions from independent customers and that the number of revisits in a session is limited (on average, the maximum number of visits at a state in a session is 2.71 and 6.76 for a heavy buyer class and for an occasional buyer class, respectively, according to [219, 220]). That is, all the requests at the same state are weakly dependent. Therefore, we assume that request arrivals in each state from sessions of a customer class still meet a Poisson process. We allocate the e-commerce server's processing rate into $m \times n$ task servers. A task server is an abstract concept in the sense that it can be a child process in a multiprocess server, or a thread in a multithread server. The requests are scheduled in a processor-sharing manner by storing them into $m \times n$ queues, each associated with a state and handled by a task server.

Assume requests in Poisson process arrive at a rate λ. Denote μ the request processing rate. It follows that the traffic intensity $\rho = \lambda/\mu$. Let s be a request's slowdown. According to queueing theories [173], when $\rho < 1$ ($\lambda < \mu$), we have the expected slowdown as

$$s = \frac{\rho}{1-\rho}. \tag{6.3}$$

6.2.2 Objectives of Processing Rate Allocation

The basic idea of the processing rate allocation for provisioning 2D service differentiation on an e-commerce server is to divide the scheduling process into a sequence of short intervals. In each interval, based on the measured resource utilization and the predicted workload, the available processing resource usages are allocated to requests in different states from different sessions.

Table 6.1: Notations in the 2D service differentiation model.

Symbol	Description
m	Number of customer classes
n	Number of states in a session
λ_i	Session arrival rate of customer class i
α_i	Quality weight of session i
β_j	Quality weight of requests in state j
C	Available processing resource
$c_{i,j}$	Allocated resource to requests in state j of session i
r_j	Average resource demand of a session in state j
$v_{i,j}$	Average number of visits to state j in session i
$d_{i,j}$	Demanded resource by requests in state j of session i
$s_{i,j}$	Slowdown of a request in state j of session i

Let C be the total amount of the processing resource available during the current resource allocation interval. The server's request scheduler has to determine the amount of the resource usages allocated to requests in each queue so that 2D service differentiation can be achieved and resource utilization can be maximized. Table 6.1 summarizes the notations used in the problem formulation.

A session in different states usually demands different processing resource usages [12, 219, 220, 293]. Let r_j be the average resource demand of a session in state j. Note that requests in the same state from different sessions generally demand the same amount of the processing resource. Let $c_{i,j}$ be the amount of the resource allocated to requests from sessions of class i $(1 \leq i \leq m)$ in state j $(1 \leq j \leq n)$ in the current allocation interval. Thus, $c_{i,j}/r_j$ is the processing rate of requests in state j from sessions of class i. Let $v_{i,j}$ denote the average number of visits to state j in a session from class i. Let $d_{i,j}$ denote the resource requirement of a session from class i in state j. That is, $d_{i,j} = v_{i,j} r_j$. According to (6.3), the slowdown of a request from a session of class i in state j, $s_{i,j}$, is calculated as:

$$s_{i,j} = \frac{\lambda_i v_{i,j}}{c_{i,j}/r_j - \lambda_i v_{i,j}} = \frac{\lambda_i d_{i,j}}{c_{i,j} - \lambda_i d_{i,j}}, \tag{6.4}$$

where λ_i is the session arrival rate of class i. Note that the term $\lambda_i d_{i,j}$ represents the resource demand of all sessions of class i in state j. For service availability, it must be smaller than the allocated resource $c_{i,j}$.

We consider the processing rate allocation for 2D service differentiation when the following constraint holds in each resource allocation interval:

$$\sum_{i=1}^{m} \sum_{j=1}^{n} \lambda_i d_{i,j} < C. \tag{6.5}$$

That is, the request processing rate of the server is higher than the request arrival rate.

Otherwise, a request's slowdown can be infinite. Service differentiation provisioning would be infeasible.

Based on the concept of slowdown for individual requests, we define a metric of *session slowdown* as $\sum_{j=1}^{n} \beta_j s_{i,j}$ to reflect the weighted slowdown of requests in a session from class i. We further define a metric of *service slowdown* as $\sum_{i=1}^{m} \sum_{j=1}^{n} \alpha_i \beta_j s_{i,j}$ to reflect weighted session slowdown of sessions from all classes.

We then formulate the processing rate allocation problem for 2D service differentiation as the following optimization problem:

$$\text{Minimize} \quad \sum_{i=1}^{m} \sum_{j=1}^{n} \alpha_i \beta_j s_{i,j} \tag{6.6}$$

$$\text{Subject to} \quad c_{i,j} > \lambda_i d_{i,j}, \quad \text{for any } i \text{ and } j \tag{6.7}$$

$$\sum_{i=1}^{m} \sum_{j=1}^{n} c_{i,j} \leq C. \tag{6.8}$$

The objective function (6.6) is to minimize the service slowdown of the server. It implies that sessions from higher classes get lower slowdown (higher QoS) and hence intersession differentiation is achieved. It also implies that sessions in high states get lower slowdown (higher QoS) and hence intrasession differentiation is achieved. The rationale behind the objective function is its feasibility, differentiation predictability, controllability, and fairness. Constraint (6.7) implies the allocated resource at each state must be more than requested so as to ensure service availability; constraint (6.8) means the total amount of allocated resource in different sessions and states must be capped by the available resource.

The processing rate allocation problem (6.6) bears much resemblance to the bandwidth allocation problem in Section 5.3. In fact, the network I/O bandwidth can be viewed as a measure of processing rate in I/O intensive applications. Similarly, we derive the optimal rate allocation in the next section.

6.3 An Optimal Processing Rate Allocation Scheme

The processing rate allocation problem is a continuous convex separable resource allocation problem. Like the bandwidth allocation problem in Chapter 5, its optimal solution occurs when the first order derivatives of the objective function (5.7) over variables $c_{i,j}, 1 \leq i \leq m$ and $1 \leq j \leq n$ are equivalent. Specifically, the optimal solution to (5.7) occurs when

$$-\frac{\alpha_{i_1} \beta_{j_1} \lambda_{i_1} d_{i_1,j_1}}{(c_{i_1,j_1} - \lambda_{i_1} d_{i_1,j_1})^2} = -\frac{\alpha_{i_2} \beta_{j_2} \lambda_{i_2} d_{i_2,j_2}}{(c_{i_2,j_2} - \lambda_{i_2} d_{i_2,j_2})^2} \tag{6.9}$$

for $1 \leq i_1, i_2 \leq m$ and $1 \leq j_1, j_2 \leq n$.

It follows that

$$\frac{c_{i,j} - \lambda_i d_{i,j}}{c_{1,1} - \lambda_1 d_{1,1}} = \sqrt{\frac{\alpha_i \beta_j \lambda_i d_{i,j}}{\alpha_1 \beta_1 \lambda_1 d_{1,1}}} \tag{6.10}$$

for $1 \leq i \leq m$ and $1 \leq j \leq n$.

Let $\tilde{\lambda}_{i,j} = \alpha_i \beta_j \lambda_i d_{i,j}$. Under the constraint (6.8), the set of equations (6.10) leads to the optimal processing rate allocation as

$$c_{i,j} = \lambda_i d_{i,j} + \frac{\tilde{\lambda}_{i,j}^{1/2}}{\sum_{i=1}^{m} \sum_{j=1}^{n} \tilde{\lambda}_{i,j}^{1/2}} (C - \sum_{i=1}^{m} \sum_{j=1}^{n} \lambda_i d_{i,j}). \tag{6.11}$$

Accordingly, the slowdown of a request is

$$s_{i,j} = \frac{\lambda_i d_{i,j} \sum_{i=1}^{m} \sum_{j=1}^{n} \tilde{\lambda}_{i,j}^{1/2}}{\tilde{\lambda}_{i,j}^{1/2} (C - \sum_{i=1}^{m} \sum_{j=1}^{n} \lambda_i d_{i,j})}. \tag{6.12}$$

From (6.12), we have the following three basic properties regarding the predictability and controllability of service differentiation given by the optimal processing rate allocation scheme:

1. If the session weight or the state weight of a request class increases, slowdown of all other request classes increases, while slowdown of that request class decreases.

2. Slowdown of a request class increases with the increase of session arrival rate and the number of visits to that state of each request class.

3. Increasing the workload (session arrival rate and number of visits to a state in a session) of a higher request class causes a larger increase in slowdown of a request class than increasing the workload of a lower request class.

Recall $d_{i,j} = v_{i,j} r_j$. From (6.12), we further have the following service differentiation ratios:

$$\frac{s_{i_2,j}}{s_{i_1,j}} = \sqrt{\frac{\lambda_{i_2} v_{i_2,j}}{\lambda_{i_1} v_{i_1,j}}} \sqrt{\frac{\alpha_{i_1}}{\alpha_{i_2}}} \quad \text{for } j = 1, 2, \ldots, n \tag{6.13}$$

$$\frac{s_{i,j_2}}{s_{i,j_1}} = \sqrt{\frac{d_{i,j_2}}{d_{i,j_1}}} \sqrt{\frac{\beta_{j_1}}{\beta_{j_2}}} \quad \text{for } i = 1, 2, \ldots, m \tag{6.14}$$

$$\frac{s_{i_2,j_2}}{s_{i_1,j_1}} = \sqrt{\frac{\lambda_{i_2} d_{i_2,j_2}}{\lambda^{i_1} d_{i_1,j_1}}} \sqrt{\frac{\alpha_{i_1} \beta_{j_1}}{\alpha_{i_2} \beta_{j_2}}}. \tag{6.15}$$

From (6.13), (6.14), and (6.15), we can see that the optimal processing rate allocation has the property of fairness, as well. That is,

Theorem 6.1 *The optimal allocation (6.11) guarantees relative service differentiation between the requests in both intersession and intrasession dimensions and their quality spacings with respect to slowdown are square root proportional to their predefined differentiation weights.*

Notice that if session arrival rate λ_i and the resource requirement of a session from a customer class in a state ($d_{i,j}$) are fixed, a request class (i, j) with a higher session weight α_i or with a higher state weight β_j gets more portion of available processing rate of the e-commerce server. However, we note that the predictability of intersession service differentiation holds if and only if $\sqrt{\frac{\lambda_{i_1} v_{i_1,j}}{\lambda_{i_2} v_{i_2,j}}} \leq \sqrt{\frac{\alpha_{i_1}}{\alpha_{i_2}}}$ for all $j = 1, 2, \ldots, n$. Also, the predictability of intrasession service differentiation holds if and only if $\sqrt{\frac{d_{i,j_1}}{d_{i,j_2}}} \leq \sqrt{\frac{\beta_{j_1}}{\beta_{j_2}}}$ for all $i = 1, 2, \ldots, m$. Otherwise, the essential requirement of 2D service differentiation, predictability, will be violated. The differentiated schedules will be inconsistent. As in the harmonic bandwidth allocation scheme in Chapter 5, we can temporarily promote weights for providing predictable differentiated services. Based on the current session arrival rates and the number of visits to a state in sessions, the scheduler increases session weights α_i and state weights β_j in the current resource allocation interval so that the predictability of 2D service differentiation holds. In this case, the allocation scheme is heuristic.

We also note that without the constraint (6.5), a request's slowdown could be infinite and provisioning slowdown differentiation would be infeasible. Session-based admission control mechanisms can be applied to drop sessions from low classes so that constraint (6.5) holds.

The processing rate allocation (6.11) minimizes the service slowdown, while ensuring square root proportional quality spacings between the requests of different sessions in different states with respect to their differentiation weights. A proportional share allocation scheme on e-commerce servers can be derived, in a similar way to the proportional bandwidth allocation scheme on streaming servers in Chapter 5, as

$$c_{i,j} = \lambda_i d_{i,j} + \frac{\tilde{\lambda}_{i,j}}{\sum_{i=1}^{m} \sum_{j=1}^{n} \tilde{\lambda}_{i,j}} (C - \sum_{i=1}^{m} \sum_{j=1}^{n} \lambda_i d_{i,j}). \tag{6.16}$$

In [356], we proved that the scheme satisfies the predictability, controllability, and fairness requirements of differentiated services, as well.

6.4 Effectiveness of 2D Service Differentiation

6.4.1 A Simulation Model

To evaluate the proposed processing rate allocation schemes on provisioning 2D service differentiation, we built a simulation model for e-commerce servers. We used

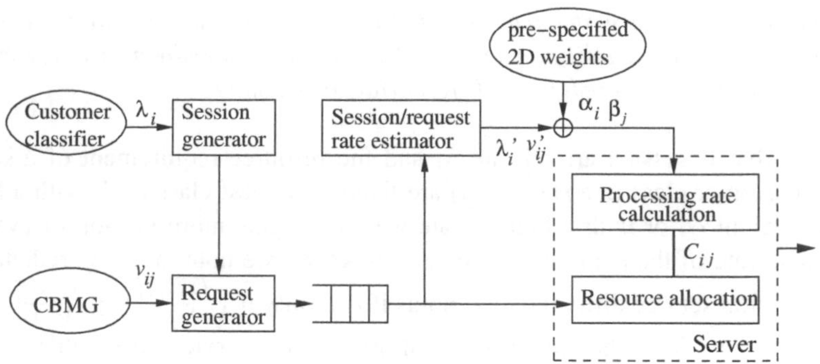

Figure 6.3: The architecture of the e-commerce server simulation model.

a synthetic workload generator derived from the real traces [54, 219, 220]. It allowed us to perform sensitivity analysis in a flexible way. Figure 6.3 outlines the basic architecture of the simulation model. It consists of a customer generator, a session generator, a request generator, a session/request rate estimator, a listen queue, and an e-commerce server.

Based on the customer classification (e.g., heavy buyer or occasional buyer), the customer generator assigns session arrival rate for each customer class (λ_i). The session generator then produces head requests that initiate sessions for the class. The session generation follows a Poisson process. The subsequent requests of a session are generated by the request generator according to its CBMG. That is, based on the current state, transition probability, and think time associated with each state, requests are generated for the session. Figure 6.1 shows two CBMGs for a heavy buyer class and an occasional buyer class, respectively. The CBMGs were derived from an on-line shopping store trace [219, 220]. We implemented both profiles in our simulations. Think time between state transitions of the same session is a part of customer behavior. It is generated by an exponential distribution with a given mean.

Each session request is sent to the e-commerce server and stored in a listen queue. The listen queue is limited to 1024 entries, which is a typical default value [60]. If the listen queue is full, a new request to the server is rejected and both the request and the whole session is aborted. According to recent TPC-W e-commerce workload characterization studies [95], most request service times are small and very few requests take time two or three orders of magnitude larger. In the simulations, we simulated a request mix of tiny, small, medium, large, and super in size with a ratio of 1:2:10:20:50. The distribution for different request sizes is 45%, 35%, 15%, 4%, and 1%. This request mix is also consistent with the embedded objects profile in the TPC-W benchmark [293]. The average resource demand of sessions in each state is assumed to be the same.

We simulated the proposed processing rate allocation schemes on the e-commerce server by dividing the request scheduling process into a sequence of short intervals of processing rate calculation and resource allocation. The calculation of processing rate for each class was based on the measured session arrival rate of each class, the number of visits to a state in a session from each class, the average resource demand of a request in a state in a single session, as well as the prespecified 2D differentiation weights α_i and β_j. A fairly accurate estimation of these parameters is required so that the proposed processing rate allocation schemes can adapt to the dynamically changing workloads. We utilized history information to estimate these values in the session/request rate estimator. The estimate of session arrival rate of each customer class (λ'_i) was obtained by counting the number of new sessions from each customer class occurring in a moving window of the past allocation intervals. As a simple way of calculating request arrival rate based on history information, the moving window is widely used in many similar experiments [60]. Smoothing techniques were applied to take weighted averages over past estimates. Similarly, the number of visits to a state in a session from each customer class ($v'_{i,j}$) was estimated.

Like others in [5, 60, 146, 352], we assumed CPU processing rate to be the single resource bottleneck in the e-commerce server. The interval of CPU processing rate allocation was set to 5 seconds, as CPU utilization is updated on a 5-second interval in some operating systems [60]. The e-Commerce server maintained $m \times n$ listen queues. Given the processing rate for each class $c_{i,j}$ according to the results of (6.11) and (6.16), a generalized proportional-share scheduling (GPS) algorithm [170] was simulated to allocate CPU resource between $m \times n$ threads. Each thread processed requests from a request class stored in the corresponding queue.

In the following, we present representative simulation results. We considered two customer classes: heavy buyer (class A) and occasional buyer (class B), Their CBMG are shown in Figure 6.1 [219, 220]. Because the buy to visit ratio of two customer classes is 0.11 and 0.04, respectively, we assigned session differentiation weight as 11:4 to classes A and B. We defined six states for each session: pay, add to shopping cart, select, search, browse, and entry. They were sorted in a nonincreasing order according to their number of state transitions needed to end in the pay state. We assigned state differentiation weight as 5:4:3:2:2:1 to the states correspondingly. The average number of visits to each state in a session is derived from the CBMGs. It is 0.11, 0.37, 1.12, 2.71, 2.71, and 1 for customer class A, and 0.04, 0.14, 2.73, 6.76, 6.76, and 1 for customer class B, respectively. The ratio of session arrival rate of customer classes A and B was set to 1:9, according to [219, 220]. Each simulation result is an average of 200 runs.

6.4.2 Effect on Service Slowdown

Figure 6.4 shows the results of service slowdown with the increase of server load. Service slowdown is defined to be the optimization objective function in (5.7). The results were obtained by the use of the optimal and the proportional allocation schemes. For comparison, Figure 6.4 also includes the results without service differentiation. When the server load is below 10%, slowdown of all request classes is very small. When the server load is above 90%, slowdown of some request classes is very large. Actually, due to the limitation of listen queue size, some requests were rejected

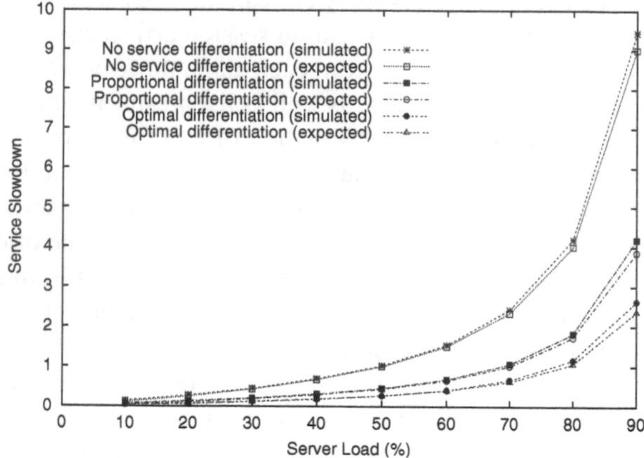

Figure 6.4: Service slowdown with the increase of server load due to different differentiation approaches.

and their sessions were aborted. Session-based admission control mechanisms are required when the server is heavily loaded. The focus of this work is on provisioning 2D service differentiation by the use of proposed processing rate allocation schemes. Thus, we varied the server load from 10% to 90%.

From Figure 6.4, we can see that simulation results agree with the expected results before the server is heavily loaded ($\leq$70%). The agreement verifies the assumption made in the modeling (Section 6.2.1) that request arrivals in each state from sessions of a class can be seen to be a Poisson process if session arrivals of that class meet a Poisson process. The gap between the simulated results and the expected ones increases as the server load increases. This is due to the variance of arrival distributions.

Figure 6.4 also shows that the QoS degrades as the server load increases. System support for service differentiation reduces the performance degradation rate by a factor of 2 or more. The optimal allocation scheme further slows down the performance degradation due to the proportional allocation by a factor of up to 80%. Detailed sensitivity analysis will be followed.

Figure 6.5 shows slowdown of individual requests in different states due to intersession and intrasession service differentiation when the server has 50% (medium) load. The results were due to the optimal allocation scheme and the proportional allocation scheme, respectively. In this figure, each bar shows slowdown of a request class. It can be seen that the requests in the browse state and search state almost have the same slowdown. This is because the customers have similar behaviors at these two states in terms of their state transition probabilities and resource demands. They were assigned the same state weight. In the following discussions, we will use the search state to represent both and use the results of the search state to address inter-session service differentiation.

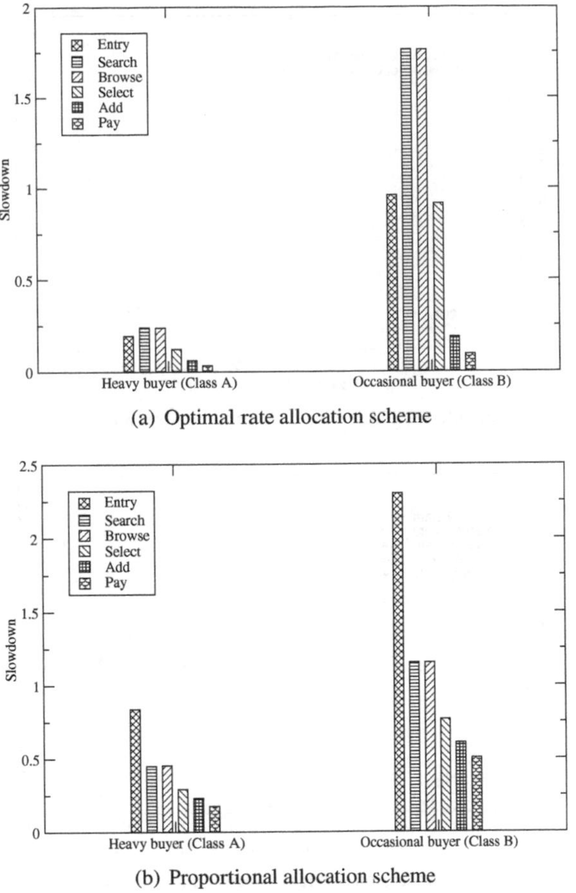

(a) Optimal rate allocation scheme

(b) Proportional allocation scheme

Figure 6.5: Intersession and intrasession slowdown differentiation when the server load is 50%.

Both Figure 6.5(a) and (b) show that the objective of intersession service differentiation is achieved. In each state, sessions from class A always have lower slowdown than those of class B. From Figure 6.5(a), it can be seen that the requirement of intrasession differentiation predictability between the entry state and the search state is violated in both class A and B categories. Sessions in the entry state should have higher slowdown than sessions in the search state. Although the optimal allocation scheme minimizes service slowdown, the requirement of $\sqrt{\frac{d_{i,j_1}}{d_{i,j_2}}} \leq \sqrt{\frac{\beta_{j_1}}{\beta_{j_2}}}$ can be violated between the corresponding states, as we discussed in Section 6.3. This violation scenario provides an intuition into the fact that the predictability of the optimal processing rate allocations depends on class load distributions. It demonstrates that to provide predictable service differentiation, a scheduler must be able to control the

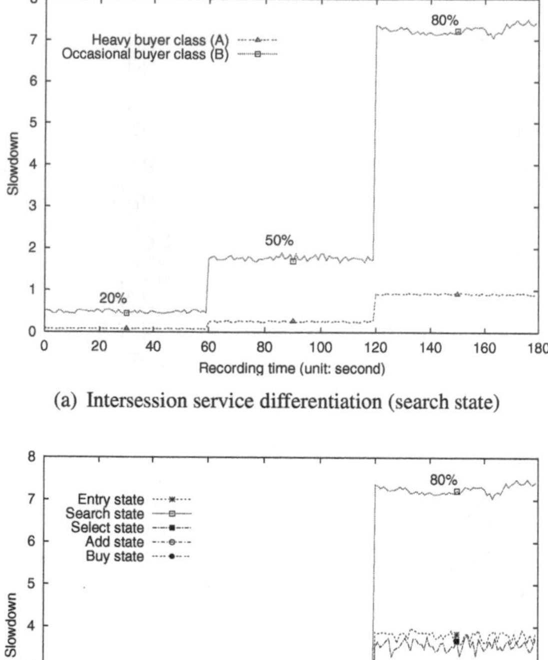

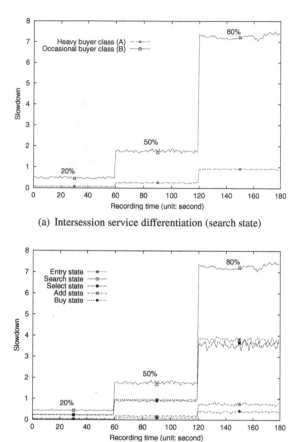

(a) Intersession service differentiation (search state)

(b) Intrasession service differentiation (class B)

Figure 6.6: A microscopic view of slowdown of individual requests due to the optimal allocation scheme.

settings of some parameters (e.g., promoting differentiation weights). In contrast, there are no such violations in Figure 6.5(b). This is because the proportional allocation scheme guarantees differentiation predictability. The differentiation ratios between request classes are proportional to the ratios of their weights and independent of their load variations. However, this is achieved at the cost of much higher service slowdown, as shown in Figure 6.4.

Figure 6.6 shows a microscopic view of slowdown of individual requests due to the optimal allocation scheme when the server has 20% (light), 50% (medium), and 80% (heavy) load. Each case recorded the results for 60 seconds. Each point represents slowdown of a request class in consecutive recording time units (seconds). Figure 6.6(a) illustrates intersession service differentiation. All sessions are at search state. The plots from other states have the similar shapes. They show that the objec-

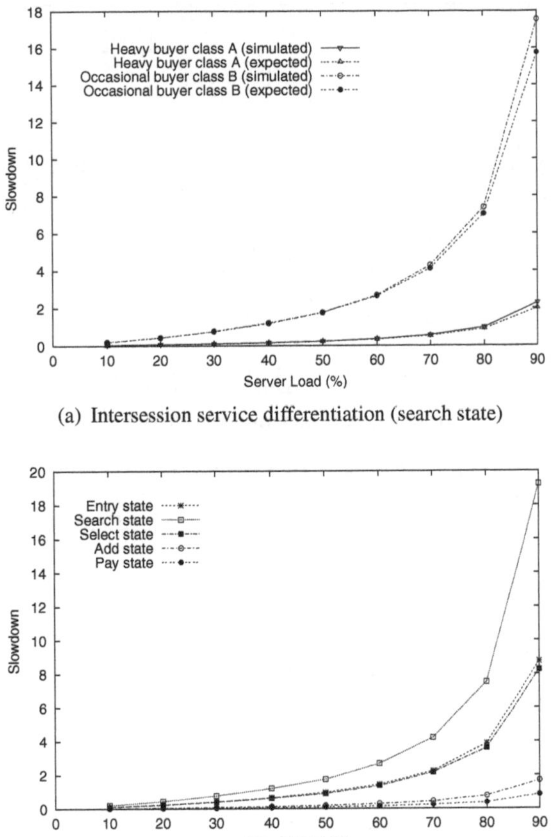

(a) Intersession service differentiation (search state)

(b) Intrasession service differentiation (class B)

Figure 6.7: A long-term view of slowdown of request classes due to the optimal allocation scheme.

tive of intersession differentiation is achieved consistently in the short run. Figure 6.6(b) illustrates intrasession service differentiation over sessions of class B. The results of class A have similar patterns. In the following, we use results of class B to address intrasession service differentiation.

Figure 6.7 shows the average slowdown of request classes with the increase of server load from 10% to 90%. Figure 6.7(a) and (b) are about inter-session differentiation over sessions in the search state and intrasession service differentiation over sessions of class B, respectively. From this figure, we can see that the optimal allocation scheme can consistently achieve 2D service differentiation at various workloads in the long run. Figure 6.7(a) shows that the simulated results meet the expectations according to (6.12) before the server is heavily loaded ($\leq$70%) in intersession service differentiation. The gap between the simulated results and the expected ones

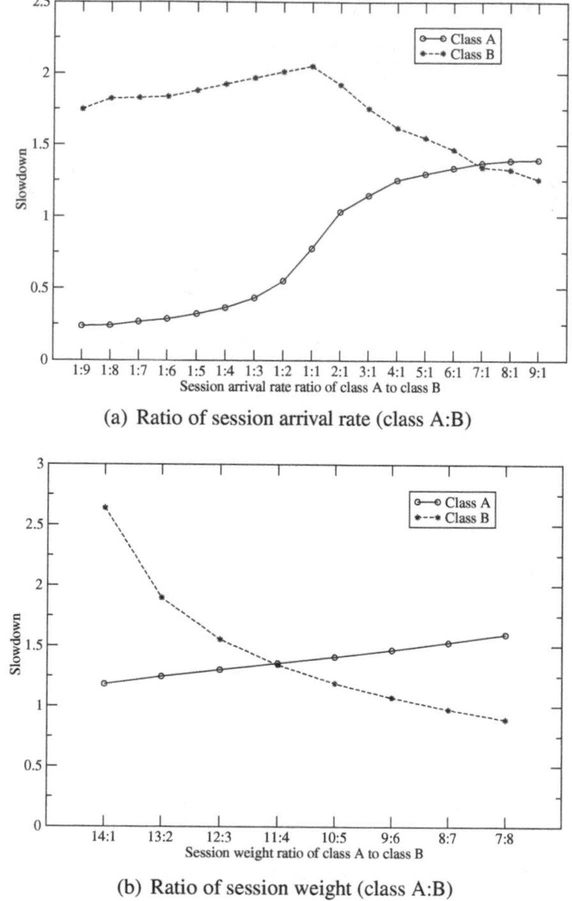

(a) Ratio of session arrival rate (class A:B)

(b) Ratio of session weight (class A:B)

Figure 6.8: Differentiation violations at various ratios of session arrival rates and session weights.

increases as the load increases because the slowdown variance increases. The gap in intrasession service differentiation scenarios has similar shapes.

6.4.3 Controllability of Service Differentiation

In Section 6.3, we remarked the optimal rate allocation scheme might lead to a possible violation of differentiation predictability. Figure 6.8 illustrates violations of intersession service differentiation at various ratios of session arrival rate and session weight between customer classes A and B, when server has 50% load.

In Figure 6.8(a), the x-axis shows the ratios between session arrival rates of the two classes (A:B). The session weight ratio is fixed to be 11:4. As the ratio of session arrival rates increases, the gap between the slowdown of class A and that of class B decrease. This is under our expectation according to (6.13). When the ratio goes up to 7:1, the predictability of intersession service differentiation becomes violated. To study the effect of session weight promotion, we assumed a fixed ratio of session arrival rate between classes A and B as 7:1. Figure 6.8(a) shows that there is no differentiation violation until the session weight ratio (A:B) goes down to 11:4. The degree of violation increases as the ratio continues to drop.

6.5 Concluding Remarks

There is a growing demand for replacing the current best-effort service paradigm with a model that differentiates requests' QoS based on clients' needs and servers' resource limitations. In this chapter, we have presented a 2D service differentiation model with respect to slowdown for session-based e-commerce applications, namely, intersession and intrasession service differentiation. We have defined a performance metric of service slowdown as weighted sum of slowdown of requests in different sessions and different session states. We have formulated the 2D service differentiation model as an optimization problem of the processing rate allocation with the objective of minimizing service slowdown. We have derived optimal rate allocations and showed that the optimal allocations guaranteed square root proportional slowdown differentiation between the requests in both intersession and intrasession dimensions. Simulation results have shown that the scheme can consistently achieve 2D service differentiation in the short run and long run. It guarantees 2D service differentiation at a minimum cost of service slowdown.

The rate allocation was based on M/M/1 GPS model. In servers with a single queue, a more practical scheduling principle is FCFS. In such scenarios, the expected slowdown formula of (6.3) no longer holds. Moreover, the rate allocation derived in (6.11) and (6.16) are derived from mean-value analysis, based on expected workload arrival rate of requests and their average resource demands. In Chapter 7, we will address these two issues in a general context of rate allocation.

In this study, we assumed that the e-commerce server had a single processing resource bottleneck. To provide 2D service differentiation, the server's scheduler determines how much the critical resource should be allocated to handling sessions in different states coming from different customer classes. Processing a request often needs to consume resources of different types and the most critical resource type may vary in different scenarios. As we reviewed in Chapter 4, accurate realization of a rate allocation is a challenge. Integrated multiple resource management for rate allocation also deserves further investigations.

Chapter 7

Feedback Control for Quality-of-Service Guarantees

A desirable behavior of a heavy-loaded Internet server is that the queueing delay of a request be proportional to its service time. Measuring quality of service in terms of slowdown is in line with the objective. Chapter 6 presented a queueing model-based rate allocation scheme for proportional slowdown differentiation (PSD) on an M/M/1 server. This chapter revisits the problem in a general M/G/1 server model. The queueing theoretical approach provides predictable PSD services on average in the long run. However, it comes along with large variance and weak predictability in short timescales. This chapter presents an integrated feedback control strategy that adjusts the processing rate allocated to each class, according to measured deviations from the target slowdown ratios. It results in highly consistent PSD services in both short and long timescales.

7.1 Introduction

A proportional differentiation model states that the service qualities different request classes should be kept proportional to their prespecified differentiation parameters, independent of the class loads. The model has been applied to the control packet or request queueing delays in both network core and network edges in the form of proportional delay differentiation (PDD).

Queueing delay is an important metric of server responsiveness. It is not suitable for comparing the service qualities of different requests that have a large variety of resource demands. Previous studies on Internet traffic characterization agreed that the request size of Internet traffic obeys a heavy-tailed distribution. That is, Internet objects are of drastically different sizes and a majority of accesses are to small objects. It is observed that clients often anticipate short delays for "small" requests although they can tolerate long delays for "large" requests. Slowdown and its variants are becoming important alternate metrics of server responsiveness [72, 105, 146, 147, 148, 261, 361]. For example, in [148], the authors proposed a size-based scheduling algorithm that assigned high priorities to requests with small service time so as to improve the performance of Web servers in terms of mean slowdown and mean response time. In [72], the authors presented a size-aware scheduling algorithm to reduce the mean slowdown in a distributed systems.

A PSD model is to maintain prespecified slowdown ratios between different classes. Assume incoming requests are classified into N classes. Let $S_i(k)$ denote the average slowdown of class i computed at sampling period k, and δ_i its prespecified differentiation parameter. For class i and class j, the PSD model requires that

$$\frac{S_i(k)}{S_j(k)} = \frac{\delta_i}{\delta_j}, \quad 1 \leq i, j \leq N. \tag{7.1}$$

The PDD model has been studied extensively in both network core and network edges, as reviewed in Chapter 4. The existing PDD algorithms cannot be tailored for slowdown differentiation because PSD provisioning requires information of both queueing delay and service time. The service time of a request is costly, if not impossible, to predict *a priori*. In Chapter 6, we presented a two-dimensional (2D) model to provide differentiated levels of quality of service (QoS), in terms of slowdown, between intersession and intrasessions requests. The model is limited to servers in a restrictive $M/M/1$ queueing system.

In this chapter, we investigate the issues of slowdown differentiation in terms of slowdown for servers in a more practical $M/G/1$ queueing system with bounded Pareto service time distribution. We refer to this model as $M/G_P/1$ in the rest of this chapter. In [357], we derived a closed analytical form of the expected per-class slowdown. Based on the expression, we developed a queueing theoretical strategy for processing rate allocation. The queueing theoretical strategy can guarantee the controllability of PSD services on average. It, however, comes along with large variance and weak predictability due to the nature of bursty pattern of Internet traffic.

A key challenge to providing fine-grained QoS is to amortize the effect of disturbances from unpredictable and dynamic server workloads. A general approach to address this issue is feedback control. It adjusts the allocated resource of a class according to the difference between the target QoS and the achieved one in previous scheduling epochs. Because of its self-correcting and self-stabilizing behavior, the feedback control approach has recently been applied as an analytic method for providing differentiated services; see [322] for a concise review. The effectiveness of a feedback control approach is largely dependent on its agility to the workload changes. Since the response time is affected by the moving average of system workload, the feedback control approach needs to be augmented by a workload predictor. Queueing model provides such a good estimator based on input traffic. In [322], we developed an integrated queueing model with feedback control to provide PSD services at a fine-grained level. In the following, we present the details of the approach.

7.2 Slowdown in an $M/G_P/1$ Queueing System

We assume that an Internet server is modeled as an $M/G_P/1$ queueing system as suggested in [146]. Many empirically measured computer workloads are heavy-tailed, such as sizes of documents on a Web server [11, 13], and sizes of FTP transfers on the Internet [243]. In general, a heavy-tailed distribution is one for which

$$Pr\{X \leq x\} \sim x^{-\alpha}, \quad 0 < \alpha < 2,$$

where X denotes the service time density distribution.

The most typical heavy-tailed distribution is Pareto distribution. Its probability density function is

$$f(x) = \alpha k^{\alpha} x^{-\alpha-1}, \quad \alpha, k > 0, x \geq k, \tag{7.2}$$

where α is the shape parameter. With respect to the fact that there is some upper bound on a client request's required processing resource in practice, a bounded Pareto distribution has been adopted to characterize the workloads on Internet servers [13, 146]. Let x be a request's service time. The probability density function of the bounded Pareto distribution is defined as

$$f(x) = \frac{\alpha k^{\alpha} x^{-\alpha-1}}{1-(k/p)^{\alpha}} \quad \alpha, k, p > 0,$$

where $k \leq x \leq p$. We refer to an $M/G/1$ queueing system where the service time follows a bounded Pareto distribution as an $M/G_P/1$ system.

7.2.1 Slowdown Preliminaries

Since α, k, and p are parameters of the bounded Pareto distribution, for simplicity in notation, we define a function

$$\mathcal{K}(\alpha, k, p) = \frac{\alpha k^{\alpha}}{1-(k/p)^{\alpha}}.$$

The probability density function $f(x)$ is rewritten as

$$f(x) = \mathcal{K}(\alpha, k, p) x^{-\alpha-1}. \tag{7.3}$$

Let m_1, m_{-1}, m_2 be the first moment (mean), the moment of its reverse, and the second moment, respectively. Let w denote a request's queueing delay in an $M/G_P/1$ queueing system with first-come-first-served (FCFS) scheduling principle and λ the arrival rate of the requests. From (7.3), we have:

$$\begin{aligned} m_1 &= E[X] \\ &= \begin{cases} \frac{\mathcal{K}(\alpha,k,p)}{\mathcal{K}(\alpha-1,k,p)} & \text{if } \alpha \neq 1, \\ (\ln p - \ln k)\mathcal{K}(\alpha, k, p) & \text{if } \alpha = 1; \end{cases} \end{aligned} \tag{7.4}$$

$$m_{-1} = E[X^{-1}] = \frac{\mathcal{K}(\alpha, k, p)}{\mathcal{K}(\alpha+1, k, p)}; \tag{7.5}$$

$$m_2 = E[X^2] = \frac{\mathcal{K}(\alpha, k, p)}{\mathcal{K}(\alpha-2, k, p)}. \tag{7.6}$$

According to the Pollaczek–Khinchin formula, we obtain the expected slowdown shown in the following lemma.

Lemma 7.1 *Given an $M/G_P/1$ FCFS queue where the arrival process has rate λ on a server. Let w be a request's queueing delay. Then its expected queueing delay and slowdown are*

$$E[w] = \frac{\lambda m_2}{2(1-\lambda m)}, \tag{7.7}$$

$$S = E[\frac{w}{x}] = E[w]E[x^{-1}] = \frac{\lambda m_2 m_{-1}}{2(1-\lambda m_1)}. \tag{7.8}$$

Note that the slowdown formula follows from the fact that w and x are independent random variables in an FCFS queue.

7.2.2 Slowdown on Internet Servers

We assume that the processing rate of an Internet server can be proportionally allocated to a number of *virtual servers* using proportional-share resource scheduling mechanisms, such as generalized proportional-share scheduling (GPS) [241] and lottery scheduling [317]. The processing rate of a server is measured in terms of *virtual servers*. Each virtual server is an abstraction of processing units that process requests from the class in an FCFS manner. It is a process (or thread) in a process (or thread) per request servers like Apache or a processing node in a cluster of servers.

Lemma 7.2 *Given an $M/G_P/1$ FCFS queue on a virtual server i with processing capacity c_i, let m_1^i, m_{-1}^i, and m_2^i be the first moment (mean), the moment of its reverse, and the second moment, respectively. Then,*

$$m_1^i = \frac{m_1}{c_i}, \tag{7.9}$$

$$m_{-1}^i = c_i m_{-1}, \tag{7.10}$$

$$m_2^i = \frac{m_2}{c_i^2}. \tag{7.11}$$

Proof On the virtual server i, the lower bound and upper bound of the bounded Pareto distribution is k/c_i and p/c_i, respectively. According to (7.2), we have

$$\int_{k/c_i}^{p/c_i} f(x)dx = c_i^{\alpha}(1-(k/p)^{\alpha}).$$

Thus, on the virtual server i, we define the probability density function of the bounded Pareto distribution as

$$f(x) = \frac{\alpha k^{\alpha}}{c_i^{\alpha}(1-(k/p)^{\alpha})} x^{-\alpha-1},$$

where $\alpha > 0$ and $k/c_i \leq x \leq p/c_i$. With the notation $\mathcal{K}(\alpha, k, p)$, the probability density function is rewritten as

$$f(x) = \mathcal{K}(\alpha, k, p) c_i^{-\alpha} x^{-\alpha-1}.$$

It follows that

$$m_1^i = \begin{cases} \frac{1}{c_i} \frac{\mathcal{K}(\alpha,k,p)}{\mathcal{K}(\alpha-1,k,p)} & \text{if } \alpha \neq 1, \\ \frac{1}{c_i}(\ln p - \ln k)\mathcal{K}(\alpha, k, p) & \text{if } \alpha = 1. \end{cases} \tag{7.12}$$

$$m_{-1}^i = c_i \frac{\mathcal{K}(\alpha, k, p)}{\mathcal{K}(\alpha+1, k, p)}. \tag{7.13}$$

$$m_2^i = \frac{1}{c_i^2} \frac{\mathcal{K}(\alpha, k, p)}{\mathcal{K}(\alpha-2, k, p)}. \tag{7.14}$$

By substitution with m_1, m_{-1}, and m_2 in (7.4), (7.5), and (7.6), we obtain the result of this lemma. This concludes the proof. □

According to Lemma 7.1 and Lemma 7.2 , we have the following theorem.

Theorem 7.1 *Given an $M/G_P/1$ FCFS queue on a virtual server i with processing capacity c_i, and arrival rate λ_i, its expected slowdown is*

$$S_i = E[s_i] = \frac{\lambda_i m_2 m_{-1}}{2(c_i - \lambda_i m_1)}. \tag{7.15}$$

Note that the $M/G_P/1$ queue reduces to an $M/D/1$ queue when the service time of all requests are the same. In an $M/D/1$ FCFS system, (7.15) becomes

$$S_i = \frac{\lambda_i b}{2(c_i - \lambda_i b)}, \tag{7.16}$$

where b is the constant service time. This fixed-time queueing model is valid in session-based e-commerce applications. A session is a sequence of requests of different types made by a single customer during a single visit to a site. Requests at some states such as home entry or register take approximately the same service time.

7.3 Processing Rate Allocation with Feedback Control

In Chapter 4, we reviewed two approaches for QoS differentiation: dynamic priority-based scheduling and processing rate allocation. In this section, we present a processing rate allocation scheme with feedback control for the PSD objective (7.1).

7.3.1 Queueing Theoretical Approach for Service Differentiation

7.3.1.1 Processing Rate Allocation

Due to the constraint of the conservation law, the sum of processing rates that are allocated to all classes must be equal to the total capacity of the underlying system. Let c_i denote the processing rate allocated to class i. The rate allocation is subject to constraint

$$\sum_{i=1}^{N} c_i = 1. \tag{7.17}$$

In addition, to ensure the feasibility of PSD services, the system should not be overloaded.

According to Theorem 7.1, we solve the problem and obtain the processing rate of class i as

$$c_i = \lambda_i m_1 + \frac{\lambda_i/\delta_i}{\sum_{i=1}^{N} \lambda_i/\delta_i}(1 - \sum_{i=1}^{N} \lambda_i m_1). \tag{7.18}$$

The first term of (7.18) is a baseline that prevents the class from being overloaded. By (7.15), if the class is overloaded, then its average slowdown becomes infinite according to queueing theory. The second term is a portion of excess rate according to its differentiation parameter and the load condition (i.e., its *normalized arrival rate*), after every class receives its baseline. This portion leads to a service differentiation.

From (7.18), we obtain the expected slowdown of class i as

$$S_i = \frac{\delta_i m_2 m_{-1} \sum_{i=1}^{N} \lambda_i/\delta_i}{2(1 - m_1 \sum_{i=1}^{N} \lambda_i)}. \tag{7.19}$$

Like (6.12), (7.19) implies three PSD basic properties: predictability, controllability and fairness.

7.3.1.2 Simulation Model

We built a simulation model of an $M/G_P/1$ queueing system. Like the e-commerce server simulation model in 6.4.1, this model consisted of a number of request generators, waiting queues, a rate predictor, a processing rate allocator, and a number of virtual servers.

The request generators produced requests with appropriate interarrival distribution and size distribution. We assume that a request's service time is proportional to its size. In the simulation model, the bounded Pareto service time distribution was generated using modified GNU scientific library [120]. Note that the number of classes in the PSD service model is usually rather limited. It varied from two to three in many similar experiments for proportional delay differentiation service provisioning. Each request was sent to the server and stored in its corresponding waiting queue. Requests from the same class were processed by a virtual server in an FCFS manner.

The rate predictor calculates the processing rate of a class using predicted load of the class. In the simulation, the load was predicted for every sampling period, which was the processing time of a thousand average-size requests. We investigated different methods, such as moving window and exponential averaging, for the load prediction. Because no qualitative difference was observed, we only present the results due to the moving window. In this method, the predicted load is the average of past five sampling periods. The rate allocator performs the proposed rate-allocation strategy according to (7.18) for every class. Meanwhile, the processing rate is reallocated for every sampling period.

Simulation parameters were set as follows. The shape parameter (α) of the bounded Pareto distribution was set to 1.5, as suggested in [88]. As indicated in [146], its lower bound and upper bound were set to 0.1 and 100, respectively. We also carried out experiments with larger upper bound settings to evaluate its impact. All classes were assumed to have the same load. In every experiment, the simulator was first warmed up for 10 sampling periods, the slowdown of a class was then measured for every following sampling period. After 60 sampling periods, the measured slowdown was averaged. Each reported result is an average of 100 runs.

7.3.1.3 Simulation Results

We conducted experiments with different number of classes (i.e., two classes and three classes). Figure 7.1 shows the simulation results, including the 95th, the 50th, and the 5th percentiles. Furthermore, if the percentile is too large, its value is indicated in the figure directly. In this figure, it can be observed that the strategy guarantees the achieved slowdown ratios to be around the target ones, which means a higher priority class has proportionally smaller average slowdown than a lower priority class in long timescales.

From this figure, it is obviously that the achieved slowdown ratios are not distributed equally around the 50th percentile. For example, Figure 7.1(a) shows that, when the target slowdown ratio δ_2/δ_1 is 4 and the system load is 10%, the 95th percentile of the achieved slowdown ratio is 12 while the 5th percentile is 1.2. We believe this kind of behavior is caused by the heavy-tailed property of the bounded Pareto distribution.

More important, it is shown that the variance of the achieved slowdown ratios is relatively large. For example, when the system load is 50% and the target slowdown ratio is 4, the difference between the 95th and the 5th percentile is 5. There are two reasons for this. First, although the strategy is able to control the average slowdown of a class by adjusting the processing rate based on queueing theory, it provides no way to control the variance of the slowdown simultaneously. Second, the workload of a class is stochastic and may change abruptly [26, 243, 325]. Because the load condition of a class is hard to be predicted from history, the rate allocation may not be always accurate. The inaccuracy results in a large variance along with the achieved slowdown ratio. In Section 7.3.2 we shall show that the integrated strategy is able to

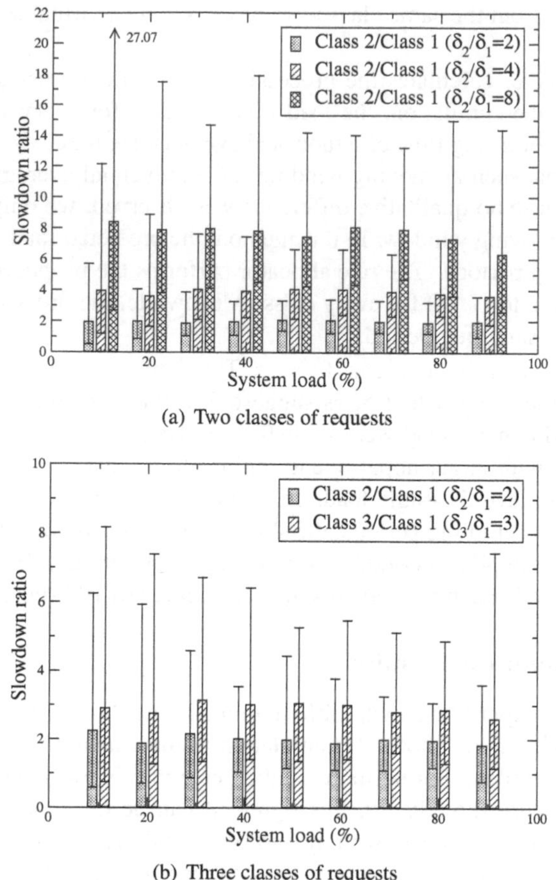

(a) Two classes of requests

(b) Three classes of requests

Figure 7.1: Achieved slowdown ratios due to the queueing model approach.

provide fine-grained PSD services.

Figure 7.1(a) also shows the provided PSD services violate the predictability requirement when the target slowdown ratio is small. For example, when δ_2/δ_1 is 2 and the system load is 10%, the 5th percentile of the achieved slowdown ratios is smaller than 1. It means that the higher priority classes received worse services than the lower priority class sometimes. In Section 7.3.2 we shall show that the integrated strategy is able to meet the requirement of service predictability.

7.3.2 Integrated Feedback Control Approach

As shown in Figure 7.1, the queueing theoretical strategy provides PSD services with large variance. For example, when the system load is 50% and the target slowdown ratio is 4.0, the achieved average slowdown is 4.04. Its 5th percentile, however,

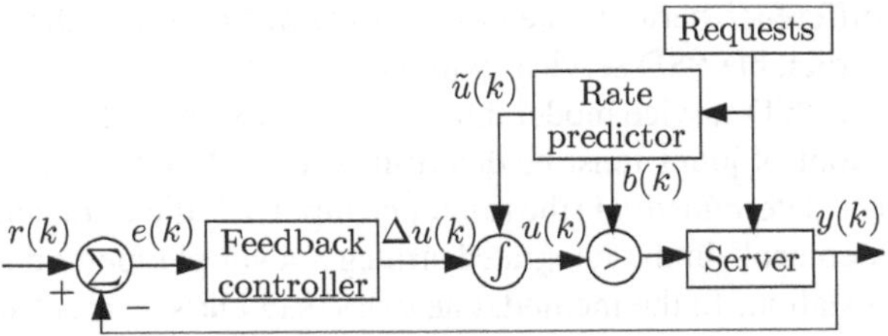

Figure 7.2: The structure of integrated processing rate allocation strategy.

is 2.15 while the 95th percentile becomes as large as 6.63. In addition, the provided services are unpredictable when the target ratio is small. For example, when $\delta_2/\delta_1 = 2$ and the system load is 10%, the 5th percentile of the achieved slowdown ratios is smaller than 1. To address this issue, we develop and integrate a feedback controller into the rate allocation strategy.

Figure 7.2 illustrates the basic structure of the integrated strategy. There are two major components in the system: a rate predictor and a feedback controller. The rate predictor is used to estimate the processing rate of a class according to its predicted load condition and differentiation parameter based on queueing theory. The difference (error) between the target slowdown ratio and the achieved one is fed back to the feedback controller, which then adjusts the processing rate allocation using integral control.

The feedback controller is based on integral control. Proportional integral derivative (PID) control is one of the most classical control design techniques [119]. The advantage of proportional controller lies on its simplicity. The processing rate is adjusted in proportion to the difference between the target slowdown ratio and the achieved one. It, however, cannot eliminate the steady-state error introduced in the process. In other words, the target slowdown ratio of two classes cannot be accurately achieved. Essentially, we are controlling a probability when implementing PSD services in an $M/G_P/1$ queueing system. The probability is not an instantly measurable quantity, such as the temperature or the pressure, and there exist measurement noises. Derivative control takes into account the change of errors in adjusting the processing rate of a class and hence responds fast to errors. Because it is very sensitive to measurement noises, tuning of the feedback controller becomes very difficult [119]. In contrast, an integral controller is able to eliminate the steady-state error and avoid overreactions to measurement noises.

Recall that the PSD service model aims to maintain prespecified slowdown ratios between different classes. By (7.19), we have

$$\frac{S_{i+1}}{S_i} = \frac{\lambda_{i+1}}{\lambda_i} \frac{c_i - \lambda_i m_1}{c_{i+1} - \lambda_{i+1} m_1}. \tag{7.20}$$

Accordingly, the slowdown ratio between two classes is actually controlled by their processing rate ratio. Therefore, the feedback controller adjusts their rate ratios according to the difference between the target slowdown ratio and the achieved one so as to provide predictable PSD services with small variance.

To translate the PSD service model into the feedback control framework shown in Figure 7.2, the control loops must be determined first. In every loop, the following aspects need to be determined: (1) the reference input, (2) the error, and (3) the output of the feedback control. In the integrated strategy, a simple method is introduced to perform the translation. In the method, one class, say class 1, is selected as the base class, and a control loop is associated with every other class. This leads to a total of $N-1$ control loops in the system.

In the control loop associated with class i, the reference input in the kth sampling period is

$$r_i(k) = \frac{\delta_i(k)}{\delta_1(k)}. \tag{7.21}$$

The output of the server is the achieved slowdown ratio of class i to the base class

$$y_i(k) = \frac{S_i(k)}{S_1(k)}. \tag{7.22}$$

If the rate allocation suggested by the rate predictor is accurate enough, the slowdown ratio error can be reduced to zero. In reality, because of the variance of traffic and the error in load prediction, no rate allocation could completely eliminate the difference between the target slowdown ratio and the achieved one. We define the error associated with class i as

$$e_i(k) = r_i(k) - y_i(k) = \frac{\delta_i(k)}{\delta_1(k)} - \frac{S_i(k)}{S_1(k)}. \tag{7.23}$$

The integral feedback controller sums up the errors caused by previous rate allocation and adjusts the processing rate of every class accordingly. The output of the feedback controller is

$$\Delta u_i(k) = \Delta\frac{c_i(k)}{c_1(k)} = \Delta\frac{c_i(k-1)}{c_1(k-1)} + g \cdot e_i(k), \tag{7.24}$$

where g is the control gain of the feedback controller. The control gain determines the adjustment of the processing rate ratio corresponding to the error.

The adjustment is incorporated with the estimation fed forwarded from the rate predictor. The processing rate allocation of the next sampling period $k+1$ is defined as

$$\frac{c_i(k+1)}{c_1(k+1)} = \frac{c_i(k)}{c_1(k)} + \Delta u_i(k). \tag{7.25}$$

By (7.17) and (7.25), the processing rate of class i of sampling period $k+1$ thus is

$$c_i(k+1) = \frac{c_i(k+1)/c_1(k+1)}{1 + \sum_{i=1}^{N} c_i(k+1)/c_1(k+1)}. \tag{7.26}$$

This implies that the processing rate of a class is proportional to its processing rate ratio to the base class.

7.4 Robustness of the Integrated Approach

According to (7.20), the slowdown ratio of two classes in an $M/G_P/1$ FCFS system has a nonlinear relationship with their allocated processing rates . It is also dependent upon their class load conditions. Implementation of the feedback controller involves two more control parameters. One is an upper bound (or limit) of the processing rate c_i allocated to each class and the other is the control gain g, as defined in (7.24).

Recall that PSD services should be fair and predictable; it implies that the processing rate limit allocated to any class must be bounded so that no other classes will be overloaded. According to (7.18), we know the basic resource requirement of a class i, $i \neq 1$, is $\lambda_i m_1$. Let b_i denote the processing rate limit (or bound) of class i. It follows that

$$b_i = 1 - \sum_{j=1...N, j \neq i} \lambda_j m_1. \tag{7.27}$$

A large rate limit implies a long settling of the control; a small rate limit would compromise the ability of the feedback controller to correct errors.

By its definition in (7.24), the control gain parameter determines the adjustment process of the processing rate allocated to each class to correct errors. A large gain may cause excessive slowdown oscillation due to error amplification; a small gain can tolerate reducing overshoot and make tuning the controller easier.

Both the processing rate limit and the control gain parameters allow us to turn the settling time of the controller for error correction. In [322], we studied their impacts on the agility of the adaptive controller. The controller measures the difference between the target slowdown ratio and the achieved one for each class and then calculates rate allocations based on (7.26). The error information is then combined with predicted rates for adjustment. In the case that the new rate exceeds its rate limit, the actual rate is set to the rate limit. Simulation results suggested a rate limit of 0.6 and a control gain of 1 for two request classes with equal loads.

Figure 7.3 shows the achieved slowdown ratios due to the integrated approach, when the target slowdown ratios were set to 2, 4, and 8. In comparison with Figure 7.1, we can see that the integrated strategy provides PSD services at a much finer grained level than the queueing theoretical strategy. For example, when the system load is 50% and the target slowdown ratio is 4, the difference between the 95th and the 5th percentile is 0.5 and 5 due to the integrated strategy and the queueing theoretical strategy, respectively. It is because the difference between the target slowdown ratio and the achieved one is taken into account in the integrated strategy. Whenever an error happens, the rate allocation is adjusted accordingly. Therefore, it is able to reduce the variance of the achieved slowdown ratios.

Moreover, the PSD services due to the integrated strategy are predictable under various system conditions (i.e., a higher priority class always receives better service than a lower priority class). This comes with the capability of the strategy in reducing

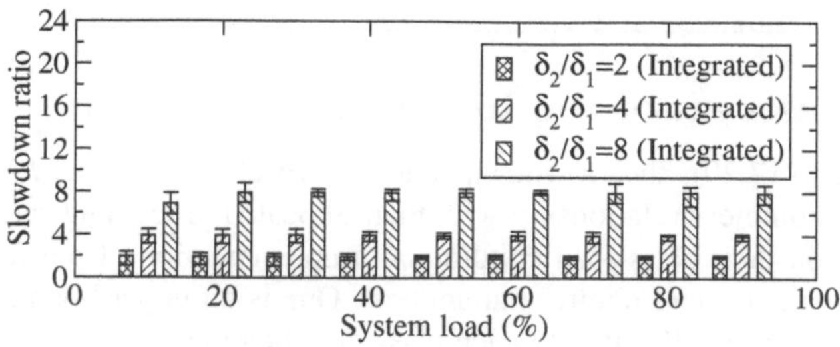

Figure 7.3: Achieved slowdown ratios due to the integrated strategy.

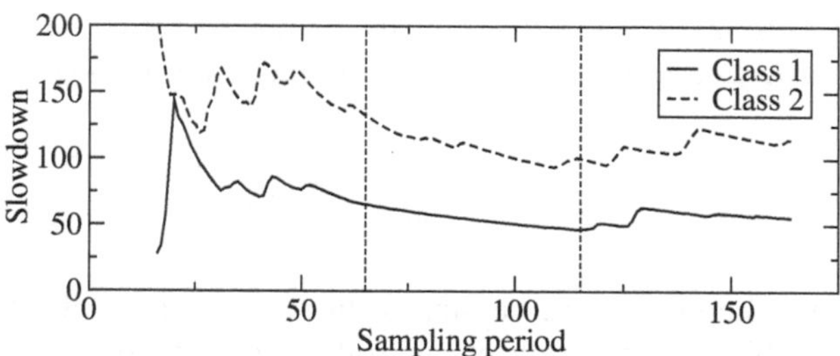

Figure 7.4: Average slowdown over all requests due to the integrated strategy under changing load conditions ($\delta_2/\delta_1 = 2$).

the variance. In summary, the integrated strategy is able to provide predictable and finer grained PSD services than the queueing theoretical strategy.

We conducted simulation to further demonstrate the robustness of the integrated strategy under changing load conditions. The initial load of both classes was assumed to be 0.4. At the 65th sampling period, the load of class 1 was decreased to 0.2 and then changed back after the 115th sampling period. Figure 7.4 shows the average slowdown of all processed requests. For example, at the 65th sampling period, the result is the achieved slowdown average over all requests processed between the 15th and the 65th sampling periods. It can be seen that the average slowdown of both classes changes according to their load conditions. The changes of class load conditions have little effect on the slowdown ratio. This is because the rate predictor can quickly capture the change of load conditions and calculate new processing rates of every class. Based on the results from the rate predictor, the feedback controller is able to reduce the errors caused by changing load conditions effectively. These results illustrate that the integrated strategy is robust in providing fine-grained PSD services.

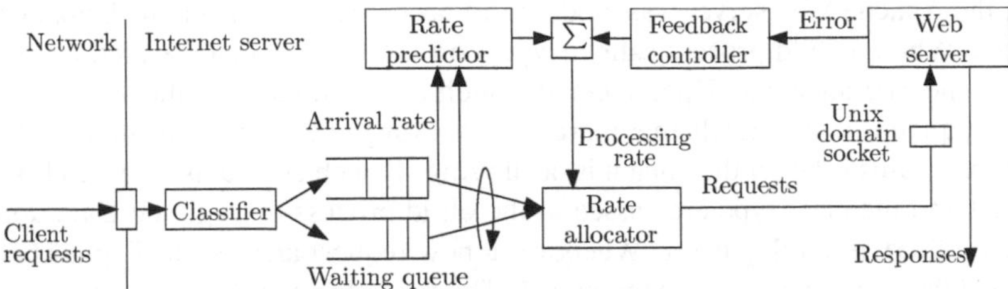

Figure 7.5: The implementation structure of the integrated strategy for PSD service provisioning.

7.5 QoS-Aware Apache Server with Feedback Control

We implemented the integrated strategy on Apache Web server 1.3.31 running on Linux 2.6. Apache Web server is normally executed with multiple processes (or threads). At the start-up, a parent server process sets up listening sockets and creates a pool of child processes. The child processes then listen on and accept client requests from these sockets. Recall that, by (7.25), we can control the slowdown ratio between two classes by manipulating their processing-rate ratio. Since every child process in Apache Web server is identical, in the implementation, we realize the processing-rate allocation by controlling the number of child processes that a class is allocated.

7.5.1 Implementation of a QoS-Aware Apache Server

The implementation structure of the integrated strategy is presented in Figure 7.5. It consists of a classifier, a rate predictor, a feedback controller, and a rate allocator.

The classifier determines a request's class according to rules defined by service providers. In our implementation, the rules are based on the request's header information (e.g., IP address and port number). To make simulation simple, a request's class type can also be determined randomly according to the arrival ratio between classes. For example, for two classes, the probability that a request to be classified as class 1 is 25% if the arrival ratio of class 1 to class 2 is 1:3. After being classified, a request is stored in its corresponding waiting queue. Associated with each waiting queue is a record of traffic information, such as arrival rate and service time.

The rate predictor obtains the arrival rate and service time from waiting queues and estimates the processing rate that a class should be allocated according to (7.18).

The feedback controller influences the achieved slowdown ratios between different classes and calculates the processing rate of a class according (7.26) by incorporating the rate estimation from the rate predictor. It then sets the processing rate (number of allocated processes) on the rate allocator.

The rate allocator enforces the processing-rate allocation. In our implementation, the Apache Web server was modified to accept requests from the Unix domain socket. When a child process calls *accept()* on the Unix domain socket, a signal is sent to the rate allocator. Upon receiving such a signal, the rate allocator scans the waiting queues to check whether there is a backlogged class whose number of allocated processes is larger than that it is occupying. If such class exists, its head-of-line request and the class type are passed to the child process through the Unix domain socket; otherwise, a flag is set. Whenever a new request arrives, the flag is checked first and the waiting queues are rescanned. The rate allocator also increases a counter that records the number of occupied processes by the class. After handling a client request, the child process sends the request's class type back to the rate allocator, which then decreases the corresponding counter. In addition, the Apache Web server was modified so that the processing time of an object is proportional to its size.

In practice, client-perceived performance of a Web server is based on a whole Web page, which can be a single file or an HTML file with multiple embedded objects. According to HTTP 1.1, a client first establishes a persistent connection with the Web server and sends requests for the Web page and possible embedded objects. The queueing delay of a Web page thus is defined as the time interval between the arrival of a connection request and the time the established connection is passed to a child process of Apache Web server. The service time is defined as the time interval between the passing of established connection and the time the response of the whole Web page is sent to the client. We determine the end of response based on the "refer" header field of an HTTP request. An embedded object uses this to indicate which Web page it is contained. A similar method is also proposed in [233]. Both the queueing delay and the service time are measured on the Web server.

7.5.2 QoS Guarantees in Real Environments

We conducted experiments to measure the performance of the integrated strategy. The experimental environment consisted of four PCs running on a 100 Mbps Ethernet. Each PC was a Dell PowerEdge 2450 configured with dual-processor (1 GHz Pentium III) and 512 MB main memory. We installed one Apache Web server on one PC while a commonly used Web traffic generator SURGE [27] ran on other PCs. In the generated Web objects, the maximum number of embedded objects in a given page was 150 and the percentage of base, embedded, and loner objects were 30%, 38%, and 32%, respectively. The workload of emulated requests was controlled by adjusting the total number of concurrent user equivalents (UEs) in SURGE. Note that the fixed number of UEs does not affect the representativeness (i.e., self-similarity and high dynamics) of the generated Web traffic [27].

The Apache Web server was set up with support of HTTP 1.1. The maximal concurrent child server processes were set to be 128. The system was first warmed up for 180 seconds. After that, the controller was turned on. Since the sampling period should be reasonable large so that the scheduler can determine the effect of resource allocation on the achieved slowdown ratio, it was set to 3 seconds and the rate predictor estimates processing rate every 10 sampling periods. The control gain in these experiments was set to 1, as suggested in [322].

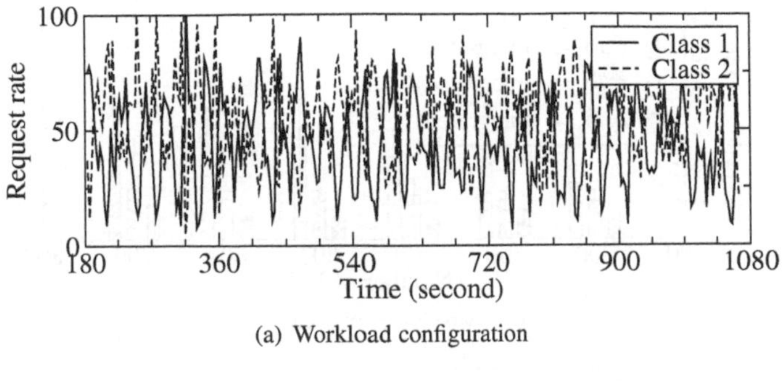

(a) Workload configuration

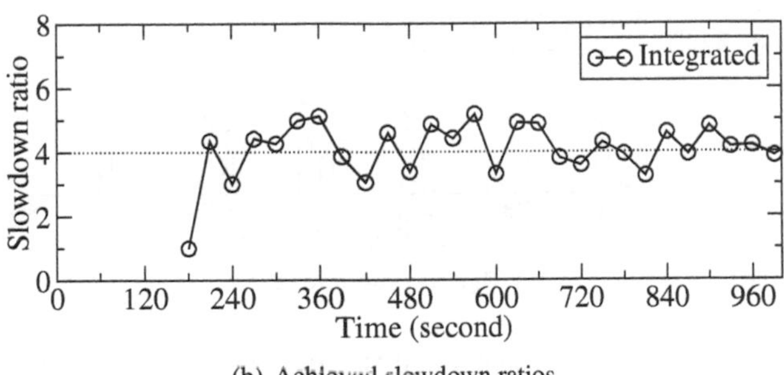

(b) Achieved slowdown ratios

Figure 7.6: Performance of the integrated strategy where the target slowdown ratio is set to 4.

We first conducted an experiment to evaluate the effectiveness of the integrated strategy. Figure 7.6 shows the workload configurations and corresponding results. In this experiment, we assumed there were two classes and the target slowdown ratio was set to 4. The number of UEs was set to 300. The arrival ratio of class 1 to class 2 is randomly changed within 1:9 and 9:1 for every 3 seconds. Figure 7.6(a) depicts the arrival rate of both classes for every 3 seconds. It indicates that the workload of a class changes frequently and abruptly. From Figure 7.6(b) we can observe that, although the workload generated by SURGE does not strictly meet the $M/G_P/1$ model, the achieved slowdown ratios can still be kept around the target. This demonstrates that the integrated strategy is effective in providing PSD services under realistic workload configurations. It also shows that the suggested small control gain is a good choice in a real system.

We carried out more experiments to evaluate the performance of the integrated strategy under different target slowdown ratios, different workloads, and multiple classes. Figure 7.7 shows the 5th, 50th, and 95th percentiles of achieved slowdown ratios. In these experiments, the workload has a similar configuration as shown in Figure 7.6(a). In Figure 7.7(a) we can observe that the target ratios (2, 4, and 8) can be achieved under different workload conditions. Figure 7.7(b) suggests that the tar-

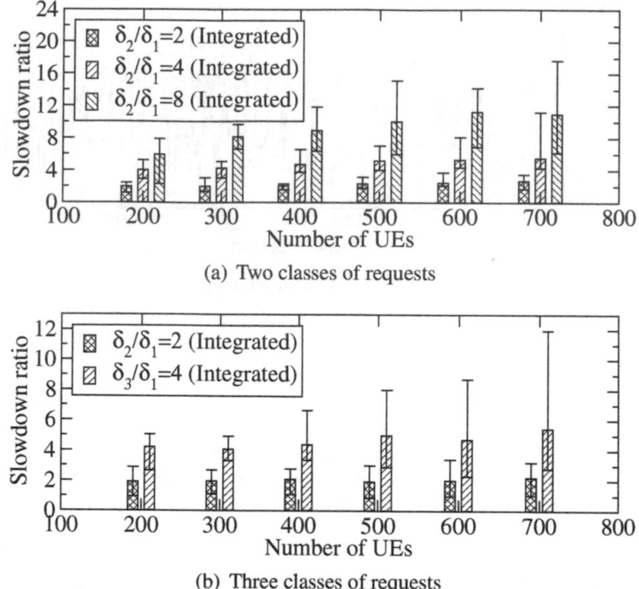

Figure 7.7: Performance of the integrated strategy under different target ratios, different workloads, and multiple classes.

get ratios can also be achieved under multiple classes. Furthermore, in Figure 7.7(a) and (b), we can observe that the variance of achieved ratios becomes relatively large with the increase of workloads. For example, in Figure 7.7(b), when the numbers of UEs are 600 and 700, the differences between the 95th and the 5th percentiles are 6.5 and 9.3, respectively. In the implementation, to control the average slowdown of a class, we control its average queueing delay by adjusting the number of its allocated processes. Because all processes have the same priority, this method has little impact on the service time of an individual request. With the increase of workload, a class's average queueing delay increases and so do its slowdown and variance. One possible solution is to control the service time of each individual request by adjusting the priority of child processes according to workload conditions.

7.6 Concluding Remarks

Slowdown is an important performance metric of Internet services because it takes into account the delay and service time of a request simultaneously. Although delay and loss rate have been studied in QoS-aware design for a long time, there has been little work in providing differentiated services in terms of slowdown. In this chapter, we have investigated the problem of processing rate allocation for PSD provisioning on Internet servers. We have proposed a queueing theoretical strategy for processing rate allocation in an $M/G/1$ FCFS queueing system with bounded Pareto service time distribution. In order to improve the predictability of the services and reduce the variance, we have developed and integrated a feedback controller into the strategy. Simulation results have showed that the queueing theoretical strategy can guarantee PSD services on average. The integrated strategy achieves the objective at a fine-grained level. We have modified the Apache Web server with an implementation of the processing-rate allocation strategy. Experimental results have further demonstrated its feasibility in practice.

We note that the reponse time measured in this work is a server-perceived latency from the arrival time of a request to the delivery time of its response. User-perceived QoS control should also take into account transmission time of the request and response in network side. Although an Internet server has no control over the network condition, the server can measure client-perceived response time by profiling the details of each HTTP transaction. Recent advance in real-time nonintrusive measurement of network performance makes the control of client-perceived QoS possible. eQoS [321] represents such a first resource management system with the capability of providing client-perceived end-to-end QoS guarantees.

Chapter 8

Decay Function Model for Server Capacity Planning

Server-side quality of service (QoS)-aware resource configuration and allocation is a challenge in performance critical Internet services. This chapter presents a decay function model of resource allocation algorithms to overcome the difficulties caused by highly variable and bursty patterns of Internet traffic. In the model, resource allocation is represented as a transfer-function-based filter system that has an input process of requests and an output process of server load. Unlike conventional queueing network models that rely on mean-value analysis for input renewal or Markov processes, this decay function model works for general time-series-based or measurement based processes and hence facilitates the study of statistical correlations between the request traffic, server load, and quality of services.

8.1 Introduction

Performance critical Internet applications, such as on-line trading and streaming services, are heavily dependent on scalable servers for guaranteed QoS provisioning. The server scalability is mostly stress tested by the use of application-specific benchmarks or synthetic workload [218]. These empirical studies have developed plentiful practical knowledge about scalable Internet servers. However, they are insufficient for the study of the impact of general Internet traffic, particularly its inherent bursty traffic patterns and autocorrelations on the server performance. The following practical yet challenging problems, are largely remaining unsolved, or cannot be precisely resolved in theory: (1) What resource capacity should a server be configured for a requirement of certain levels of QoS? (2) What level of QoS can a server with certain resource capacity support? (3) How can a server provide QoS guarantees by doing effective scheduling? This chapter aims to develop a formalism for the study of these key resource configuration and allocation issues.

Queueing network (QN) models are widely used to address similar issues in packet scheduling in networks and job scheduling in closed computer systems [262]. Queueing theories are based on an assumption of input renewal or Markov processes. There were early studies that treated Internet requests as packets in routers or jobs in computer systems and simply applied the mean-value-analysis model for performance evaluation of Internet servers. They produced estimates for references in practical capacity planning and resource configuration; see [218, 291] for examples. But the model applicability and accuracy were found very limited because recent Internet workload characterization studies [13, 71, 160] all pointed to self-similarity and heavy tail as inherent properties of Internet traffic. Another inherent characteristic of Internet traffic is high variability. Impact of bursty traffic patterns is also beyond the capability of means value analysis of the traditional QN models.

Research on extending the QN models to overcome the above limitations in the design, control, and analysis of network routers has never stopped. Three recent representative enhanced models are effective bandwidth-based queueing analysis [31, 143, 171, 172], hierarchical renewal processes like Markov Modulated Poisson Process (MMPP) [18, 195], and fractional Gaussian noise (FGN) [78, 99, 231, 259]. The effective bandwidth approach is based on an assumption that the Internet traffic process has the normalized logarithmic moment generating function. It implies the traffic can be bounded by an "envelope" process in a much simpler mathematical representation. The envelope process is then studied under resource reservation-based scheduling policies. We note that the effective bandwidth model is good at providing bounds on the QoS domain. But the precision of the bounds heavily depends on the bursty pattern of underlying traffic. The MMPP-based queueing models can include correlation statistics and other second order moments into Markov chain transitive matrices. However, the solvability of these advanced models remains open. FGN was introduced to study the self-similarity impact on the queueing system. Most queueing analysis of the FGN traffic model was based on the assumption of constant service rate and the model was limited to Gaussian-based martingale distributions. There were recent researches on the impact of long-range-dependent arrival process on waiting time in FGN/GI/1 and MG∞/GI/1 typed hierarchical queues [333]. Their models assumed fixed queueing disciplines and provided no control parameters for QoS provisioning.

There are also established frameworks for modeling of highly variable traffic on network routers; see [187, 209, 241] for examples. To handle the high variability of Internet traffic, the frameworks assume the traffic smoother model (or shaper), which essentially reduces the burst of Internet traffic and enables subsequent rate-based scheduling/traffic analysis. To further compensate the high variance of Internet traffic, traditional single-rate leaky-bucket scheduling is extended to multiple leaky-bucket scheduling. These techniques for handling the traffic variance are not applicable to end-server modeling for two main reasons. First, routers take packets from the same connection as input, which can be buffered for smoothing the downstreaming traffic. In contrast, Internet servers take client requests as a scheduling unit and the processing cost for each individual request cannot be smoothed by any preprocessing. Second, from a QoS point of view, the drop or loss of a packet in routers may cause an intolerable loss (multimedia traffic is an exception because it can tolerate a loss rate up to $10^{-5} \sim 10^{-7}$ [209]). However, scheduling on Internet servers can have wide choices of adaptations for different quality levels. For example, a multimedia server has choices of different compression ratio, different encoding to balance the resource need, and QoS.

On the server side, in recent years we have seen the increasing popularity of feedback control approaches to deal with the impact of high-order moments of Internet traffic in QoS-aware resource management. As we reviewed in Chapter 4, Abdelzaher *et al.* treated requests as aperiodic real-time tasks with arbitrary arrival times, computation times, and relative deadlines [2]. They applied linear feedback control theory to admit an appropriate number of requests so as to keep system utilization bounded. This approach assumes a fixed-priority scheduling policy and considers no information about input request processes. Feedback control adjusts resource allocation in response to measured deviations from desired performance. Since it ignores the change of workload input, the linear controller may not work for a nonlinear system when the residual errors are too big. There were recent studies on integrated queueing models with feedback control in request admission [201, 204, 282], and resource allocation [322, 360]. Chapters 5 and 7 presented such integrated control approaches for QoS provisioning on Web and streaming servers.

We note that the control approaches aim to reduce completion time by controlling the server load mean through admission control. The server load mean can be estimated according to the first moment statistics of input request processes. The feedback control approaches deal with the impact of high-order moments of Internet traffic, in particular, high variability and autocorrelations in an algorithmic manner. Although the approaches ensure the robustness of the resource allocation mechanism, the impact of request scheduling on server performance under various input request processes is left undefined.

In [342], we took a different approach, by constructing a decay function model for time varying request scheduling (or resource allocation). The server system under consideration assumes a general request arrival process combined with a general request size distribution. Each request is scheduled by a decay function which determines how much resource is allocated dynamically during the request processing time. The server load changes in the process of resource allocation. The adaptation of the decay function model reflects in the relationship between request completion time, request size and autocorrelation, and server load. Its goal is to minimize the server load variance due to high-order moments of Internet traffic by controlling completion time of individual requests (through resource allocation). The decay function model was named for the fact that resource allocations of each request will eventually vanish to zero when the request exists the server. This "decay" concept differs fundamentally from the "decay" usage scheduling in literature [98]. We use the terms of "request scheduling" and "resource allocation" interchangeably.

In the decay function model, input request process and output server workload processes could be any general time-series-based or measurement-based processes. There is no need to reconstruct renewal queueing processes from the measurement-based models, as required by QN models and control theoretical approaches. We modeled the time varied scheduling function as a filter and applied filter design theories in signal processing in the optimality analysis of various scheduling algorithms. The decay function can be adapted to meet changing resource availability and the adaptation does not affect the solvability of the model. This chapter presents the details of the decay function model.

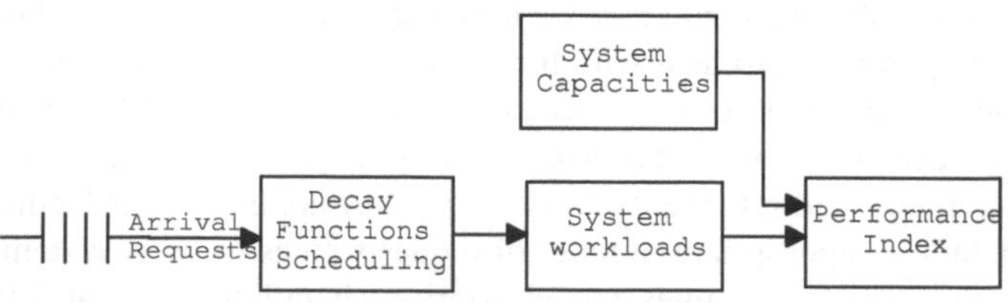

Figure 8.1: Illustration of the decay function model.

8.2 The Decay Function Model

Consider an Internet server that takes requests as an input process, passes them through scheduling — the decay function kernel, and generates a system load output process. Each incoming request needs to consume a certain amount of resource. We define the required resource amount as the request "size." An Internet request often needs to consume more than one type of resource (e.g., CPU time, memory space, network IO bandwidth, disk IO bandwidth, etc.). Since the server scalability is often limited by a bottleneck resource class, this work focuses on the management of this critical resource. The request size is a generic index and can be interpreted as different measures with respect to different types of resources. It is different from the size of its target file. In some cases, such as FTP and static content Web servers, the request size is proportional to the file size. For the system under consideration, we assume that the size distribution of requests is known. In a QoS-aware system, the size of a request can be derived from quality indices in different QoS dimensions [254].

We define the server capacity and server load accordingly. The output server load determines the level of QoS, in terms of the request completion time and rejection rate, for a given server capacity $\mathbf{c}$. Associated with the server is a scalable region $[c_l, c_h]$ for each request, $0 \leq c_l < c_h < \mathbf{c}$, in which the service quality of the request increases with the amount of its allocated resource. If the maximum amount of resource c_h is allocated and the request can still not meet its quality requirements, the server performance will not be improved by simply increasing the resource quantity. c_l is the lower bound of allocated resource which is determined by the request minimum quality requirement. In a scalable system, it is always true that more allocated resource will lead to nondecreasing QoS.

The load filtering model is a discrete time model, with t as server scheduling epoch index, t_s as arrival time of a request, and t_d for the deadline of the request. As shown in Figure 8.1, the model consists of three major components: request incoming process, decay function scheduling, and evaluation of system workload process. We model the request arrivals as a general fluid typed process:

$$\Big\{n(1), n(2), n(3), \ldots, n(t), \ldots\Big\}, \tag{8.1}$$

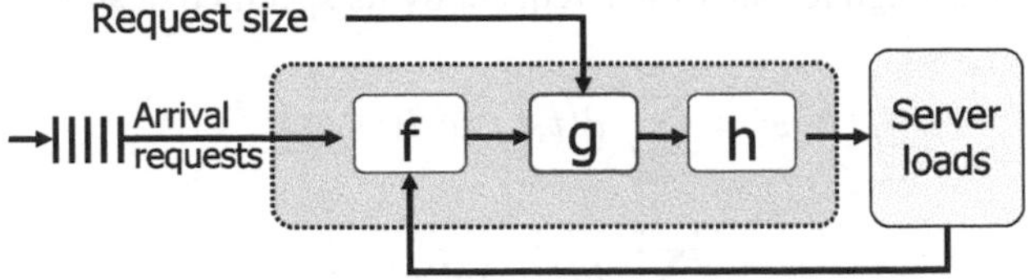

Figure 8.2: Decomposition of resource allocation.

where $n(t)$ is the number of requests arrived at time $t, t = 1, 2, \ldots$. In the simplest case, $n(t)$ at different scheduling epochs can be i.i.d. random integers from a general probabilistic distribution. Taking the request size into the model, process (8.1) can be rewritten as:

$$\left\{ \{w_i^t\}_{i=1,2,\ldots,n(t)} \right\}_{t=0,1,\ldots}, \tag{8.2}$$

where w_i^t is the size of ith request that arrived at time t.

Resource allocation activities of a computer system are represented in an algorithmic manner. In this work, we define a decay function to abstract the allocation on the Internet server. It is defined as a relationship between time and resource allocation for each request. Formally, the decay function is a function of system time t, request arrival time t_s, request size w, and current server workload $l(t)$. We denote it as $d\big(t, t_s, w, l(t)\big)$. Strictly speaking, the decay function should also depend on request deadline t_d. For tractability, we treat t_d as a model parameter, rather than a free variable.

In real Internet servers, the adaptation of resource allocation to request size and server workload does not change with time. That is, $\partial d/\partial w$ and $\partial d/\partial l$ are functions independent of time t. Under this assumption, by the use of variable separation techniques the decay function can be rewritten as a three-stepped process:

$$d\big(t, t_s, w, l(t)\big) = h(t, t_s)g(w)f\big(l(t)\big), \tag{8.3}$$

where $f(\cdot)$ measures the effects of workload on scheduling, $g(\cdot)$ represents the impact of request size w on scheduling, and $h(\cdot)$ is the function evolving with time to determine the resource allocation. This allocation decomposition is illustrated in Figure 8.2. We assume that resource allocation is a causal activity. That is, for all $t < t_s$, $h(t, t_s) = 0$. It means that no resources will be reserved in advance for a request before it arrives.

From the definition of decay function, the amount of resource actually consumed by a request must be equal to the request size w. In other words, the scheduler will always allocate enough resources to a request by its specified deadline t_d. That is,

$$\begin{aligned} w &= \sum_{t=t_s}^{t_s+t_d} d(t, t_s, w, l(t)) \\ &= \sum_{t=t_s}^{t_s+t_d} h(t, t_s) g(w) f(l(t)). \end{aligned} \tag{8.4}$$

We refer to (8.4) as a *resource allocation function* (or *scheduling function*).

The server workload at time t is equal to the sum of all resources allocated to the requests that arrive before t. Thus,

$$\begin{aligned} l(t) &= \sum_{t_s=0}^{t} \sum_{k=1}^{n(t_s)} d(t, t_s, w_k, l(t)) \\ &= \sum_{t_s=0}^{t} \sum_{k=1}^{n(t_s)} h(t, t_s) g(w_k) f(l(t)). \end{aligned} \tag{8.5}$$

We refer to (8.5) as a *workload function*. The workload and scheduling functions together determine the dynamics of resource scheduling on the Internet server.

The decay function model defined above is applicable to any scheduling algorithms. To make the model tractable, we classify the allocation algorithms by two dimensions: request size awareness and server load adaptation. Specifically,

1. Size-oblivious allocation, if $g(w) = 1$. It means allocation algorithms are oblivious to request sizes in the process of resource allocation, as long as the total allocated resource meets the resource demand of each request before its delay constraint.

2. Size-aware allocation, if $g(w)$ is an explicit function of w. This is the class of scheduling that takes into account the request size information to allocate resources at each scheduling epoch.

3. Nonadaptive allocation, if $f(l(t)) = 1$. The resource allocation algorithm is independent of server load $l(t)$.

4. Adaptive allocation, if $f(l(t))$ is a nondegenerating function. That is, the scheduler will adapt its resource allocation policy to the change of server load.

Best effort is an example of size oblivious allocation algorithms. It treats the requests of different sizes uniformly. Size aware allocation like fair-sharing scheduling is discussed extensively in QoS researches, where the feature of "controllable" is emphasized. Adaptive scheduling is to allocate resources in response to the change of system utilization so as to improve the system throughput. Nonadaptive scheduling is often oriented to applications where the QoS cannot be compromised. In this work, we focus on size-aware and nonadaptive scheduling algorithms. In particular, we assume $g(w) = w$ for size awareness. Recall that the request deadline t_d is considered as a predefined model parameter. It means the server can always finish a request in a fixed period of time. We hence refer to it as *fixed-time scheduling*, scheduling with fixed-time delay constraint, or time-invariant scheduling. Fixed-time scheduling decouples the deadline constraint from scheduling and simplifies the analysis of the decay function model.

We note that scheduling algorithms are normally time invariant. That is, $h(t, t_s) = h(t - t_s)$. Consequently, the workload function (8.5) in the case of fixed-time scheduling can be simplified as

$$\begin{aligned} l(t) &= \sum_{t_s=t-t_d}^{t} \sum_{k=1}^{n(t_s)} h(t-t_s) \cdot g(w_k) \\ &= \sum_{t_s=t-t_d}^{t} h(t-t_s) \cdot \sum_{k=1}^{n(t_s)} g(w_k) \\ &= \sum_{t_s=t-t_d}^{t} h(t-t_s) \cdot \tilde{w}(t_s), \end{aligned} \tag{8.6}$$

where

$$\tilde{w}(t_s) = \sum_{k=1}^{n(t_s)} g(w_k). \tag{8.7}$$

Given the arrival process in (8.2) and predefined impact of request size $g(w)$, the properties of the compounded process $\tilde{w}(t_s)$ are derivable. In the case that $n(t)$ is i.i.d. random number from a distribution, the size w is i.i.d. random number from another distribution; and $g(w) = w$, the compounded random process $\tilde{w}(t_s)$, is a simple random process following a distribution of random sum [107].

By introducing the convolution operator "$*$" on two vectors $a(n)$ and $b(n)$, we have

$$a(n) * b(n) = \sum_{m=-\infty}^{\infty} a(m)b(n-m).$$

It is known that $h(t-t_s) = 0$ when $t < t_s$ for causality consideration. Also, we note that no scheduling will be made before a system starts running. That is, $h(t-t_s) = 0$ when $t_s < 0$. As a result, the simplified workload function (8.6) can then be rewritten as a convolution:

$$l(t) = h(t) * \tilde{w}(t). \tag{8.8}$$

Equation (8.8) presents fixed-time scheduling on Internet servers as a perfect format of a linear system model with a transfer function $h(\cdot)$. Meanwhile, for fixed-time scheduling, the scheduling function (8.4) can be simplified as:

$$\sum_{t=t_s}^{t_s+t_d} h(t-t_s)g(w) = w.$$

That is,

$$\sum_{t=0}^{t_d} h(t) = \frac{w}{g(w)}.$$

In the case of $g(w) = w$, the scheduling function becomes

$$\sum_{t=0}^{t_d} h(t) = 1. \tag{8.9}$$

According to theorems of random signal processing [212], when the input "signal" $\tilde{w}(t)$ is a wide sense stationary (WSS) — the mean of $\tilde{w}$ is constant and its autocorrelation function depends only on the time difference; the statistical relationships with respect to their mean, variance, autocorrelation, and cross-correlation between input and output of (8.8) can be established. The assumption of WSS is quite general in the modeling of random signal. This assumption is also valid for Internet traffic. In [298], the authors proved that the Internet arrival process could be at least modeled as a phase-pieced WSS processes.

8.3 Resource Configuration and Allocation

8.3.1 Resource Configuration

In this section, we apply the analytical results derived from the preceding section to solve the problems we posed in the Introduction. The first question is that given an arrival process and fixed-time scheduling, what server capacity should be configured so as to satisfy a predefined quality requirement?

With *a priori* known mean and variance of the server load $l(t)$, we can estimate the probability distribution tail by Chebyshev's inequality [107].

Lemma 8.1 *The upper bound of the probability of workload l exceeding capacity $\mathbf{c}$, $prob(l > \mathbf{c})$, is*

$$\frac{\sigma_l^2}{\sigma_l^2 + (\mathbf{c} - \mu_l)^2}. \tag{8.10}$$

Proof From Chebyshev's inequality, it is known that

$$F\{I\} \leq a^{-1} E(u(y)),$$

where y is a random variable, I is an interval, $F\{I\}$ is the distribution function, $u(y) \geq 0$, $u(y) > a > 0$ for all y in I.

Substitute y with workload l, and define $u(l) = (l + x)^2$ with $x > 0$. It can be verified that $u(l) \geq 0$ and $u(l) > (\mathbf{c} + x)^2$ for $l > \mathbf{c} > 0$. Therefore,

$$prob(l > \mathbf{c}) \leq \frac{1}{(\mathbf{c} + x)^2} E((l(t) + x)^2).$$

Since

$$E((l(t) + x)^2) = \sigma_l^2 + \mu_l^2 + 2x\mu_l + x^2,$$

we have

$$prob(l > \mathbf{c}) \leq \frac{1}{(\mathbf{c} + x)^2}(x^2 + 2x\mu_l + \sigma_l^2 + \mu_l^2).$$

It can be proved that the right side of the inequality takes the minimum value at $x = -\mu_l + \sigma_l^2/\mathbf{c}$. This proves the lemma. □

From this lemma, we have an estimate of server capacity as follows.

Theorem 8.1 *Let v be a predefined bound of the probability that workload exceeds the server capacity. The estimate of capacity is*

$$\mathbf{c} = \sqrt{\frac{1 - v}{v}} \cdot \sigma_l + \mu_l. \tag{8.11}$$

This theorem tells that the capacity is determined by the mean and variance of server load. Because of the high variability of the Internet traffic, the workload variance is often a dominating factor in the calculation formula of the resource capacity. The relationship between server capacity, the chance of overload (workload exceeding capacity), and workload mean and variance is illustrated in Figure 8.3. From this figure, it can be seen that the capacity requirement increases sharply with setting of high criteria v, and this increase is amplified in the situations of high variance of workload. Note that Chebyshev's inequality provides a loose upper bound of $prob(l > \mathbf{c})$. In reality, depending on different underlying distributions, there could exist more accurate estimates of the bound. A tighter bound would be able to provide more precise estimate capacity or deadline. In this work, we focus on general situations, where Chebyshev's inequality provides a uniform solution that applies to all distributions with finite first and second order moments.

8.3.2 Optimal Fixed-Time Scheduling

It is known that the requirement for server capacity can be relaxed by reducing server load variance so as to maintain a uniform server load timewise. The goal of scheduling is to minimize the workload variance.

Define $\mathbf{h}(t) = [h(t), h(t+1), \ldots, h(t + t_d - 1)]'$ as a vector form of the decay function, and $\mathbf{w}(t) = [\tilde{w}(t), \tilde{w}(t-1), \ldots, \tilde{w}(t - t_d + 1)]'$ as a vector form of request sizes. Let $\Omega(t) = E[\mathbf{w}(t)\mathbf{w}'(t)]$, being the correlation matrix of input process $\tilde{w}$ in

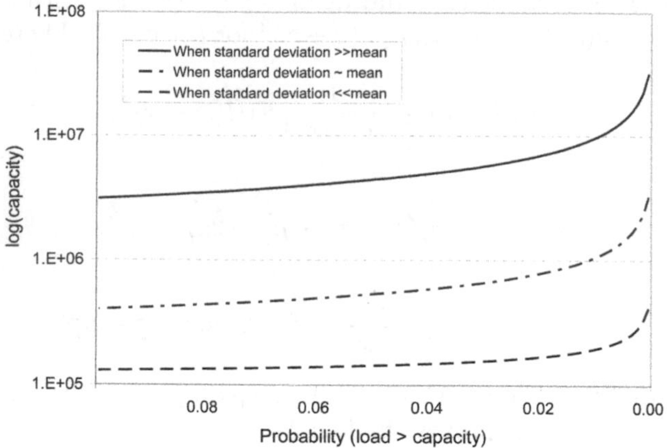

Figure 8.3: The capacity and chance of overload.

the order of t_d. We formulate the optimization problem in the following theorem. We also prove there exists one and only one solution to this problem.

Theorem 8.2 *If the compounded input process $\tilde{w}(t)$ is wide sense stationary (WSS), the optimal fixed-time scheduling is to find a* $\mathbf{h}$ *that*

$$\textit{Minimize } \mathbf{h}'\Omega\mathbf{h} \tag{8.12}$$

$$\textit{Subject to } \sum_{i=0}^{t_d} \mathbf{h}_i = 1, \textit{ and } \mathbf{h}_i \geq 0. \tag{8.13}$$

Moreover, the optimization problem (8.12) has a unique solution.

Proof Write (8.6) in vector format as

$$l(t) = \mathbf{h}'(t)\mathbf{w}(t). \tag{8.14}$$

Recall that $\tilde{w}$ is a WSS process and $\sum_{t=0,\ldots,t_d-1} h(t) = 1$. It follows that the mean of system load at time t

$$E[l(t)] = E[\mathbf{h}'(t)\mathbf{w}(t)] = E[w].$$

The explanation is that the server must finish all the work requirements for the requests. As we assume no degradation in the services (sum of h must be one), it has nothing to do with scheduling policy.

The variance of system load at time is

$$\begin{aligned} Var(l(t)) &= E[l^2(t)] - (E[l(t)])^2 \\ &= E[\mathbf{h}'(t)\mathbf{w}(t)\mathbf{h}'(t)\mathbf{w}(t)] - (E[w])^2 \\ &= E[\mathbf{h}'(t)\mathbf{w}(t)\mathbf{w}'(t)\mathbf{h}(t)] - (E[w])^2 \qquad (8.15) \\ &= \mathbf{h}'(t)\Omega\mathbf{h}(t) - (E[w])^2, \qquad (8.16) \end{aligned}$$

where $\Omega = E[\mathbf{w}(t)\mathbf{w}'(t)]$.

We point out that Ω is semipositive definite matrix. That is, for any nonzero vector $x \in \mathbf{R}^{t_d}$, $x'\Omega x \geq 0$. In real Internet environment, the covariance of input traffic random process should be nondegenerating — it is impossible to determine one component of $\mathbf{w}(t)$ from other components of $\mathbf{w}(t)$ with probability one [107]. This means the correlation matrix is strictly positive definite.

Since Ω is a symmetric positive definite matrix, there exists an orthogonal matrix $\mathbf{U}$, such that $\mathbf{U}^{-1}\Omega\mathbf{U} = \Lambda$ and

$$\Lambda = diag(\lambda_1, \lambda_2, \ldots, \lambda_{t_d}),$$

where λ_i are the eigenvalues of the matrix Ω and $\lambda_i > 0$.

Let $y = \mathbf{h}'\Omega\mathbf{h}$. It follows that

$$y = \mathbf{h}'\mathbf{U}\Lambda\mathbf{U}^{-1}\mathbf{h}.$$

Define $\mathbf{g} = \mathbf{U}^{-1}\mathbf{h}$. It follows $y = \mathbf{g}'\Lambda\mathbf{g}$. This is a standard form. It is minimized when $\lambda_i g_i^2 = \lambda_j g_j^2$ for any $1 \leq i, j \leq t_d$. Since $\mathbf{h}'\mathbf{1} = 1$ and $\mathbf{U}\mathbf{g} = \mathbf{h}$, there exits one and only one solution of $\mathbf{h}$ for the optimization problem (8.12). □

This theorem reveals the impact of correlation structure of Internet traffic on the scheduling policy. In the following, we give two optimal results in the case that the compound input process is homogeneous. The first result can be derived from Theorem 8.2 and its proof is omitted. We demonstrate the calculus of the results by examining input traffic with different autocorrelation functions.

Corollary 8.1 *If the compounded input process $\tilde{w}(t)$ is independent for different time t, the optimal fixed-time scheduling is to allocate an equal amount of resource to a request at each time unit before its deadline.*

Corollary 8.2 *If the compounded input process $\tilde{w}(t)$ is independent for different time t, then the relationship of length of deadline, capacity, chance of failure, and input statistics can be characterized by equation*

$$t_d = \frac{(1-v)\sigma_{\tilde{w}}^2}{v(\mathbf{c} - \mu_{\tilde{w}})^2}. \qquad (8.17)$$

Proof According to Corollary 8.1, the function $h(t)$ should be set to a constant value $1/t_d$ for period $0 < t \le t_d$ for minimizing the server load variance. Thus,

$$\sigma_l^2 = \sigma_{\tilde{w}}^2 \sum_{t=-\infty}^{t=\infty} h^2(t) = \frac{\sigma_{\tilde{w}}^2}{t_d}.$$

From Theorem 8.1, we know that

$$\mathbf{c} = \sqrt{\frac{1-v}{v}} \cdot \sigma_l + \mu_l.$$

Combining the two equations completes the proof. □

This corollary shows that QoS (in terms of completion time) can be adapted in response to the change of traffic intensity. To keep the same level of QoS for traffic with different means and variances, say (μ_1, σ_1^2) and (μ_2, σ_2^2), the ratio of their completion time is

$$\frac{t_{d_2}}{t_{d_1}} = \frac{\sigma_2^2 \cdot (\mathbf{c} - \mu_1)^2}{\sigma_1^2 \cdot (\mathbf{c} - \mu_2)^2}. \tag{8.18}$$

When $\mathbf{c} \gg \mu_1$ and $\mathbf{c} \gg \mu_2$, their ratio approximates σ_2^2/σ_1^2.

8.3.3 Examples

In the first example, we illustrate the above theorem by considering several traffic patterns with distinct autocorrelation functions (ACFs) $p(\tau)$, where τ is the lag of time unit. The server is assumed to finish each unit of task by 10 time units; that is, $t_d = 10$ for each request. The first two sample traffic models are multimedia traffic patterns introduced in [180]. One is *shifted exponential scene-length distribution* whose ACF is $p(\tau) = e^{-\beta|\tau|}$ and $\beta = 1/49$ and the other is *subgeometric scene-length distribution* with ACF in recursive form $p(\tau) = p(\tau - 1) - \alpha^{\sqrt{\tau}}/d$, where $\alpha = 0.8$ and $d = 40.67$. The third traffic model under consideration is an FGN process with a Hurst parameter $H = 0.89$, as described in [242]. We applied an exponential transformation to this traffic to eliminate negative values.

The last traffic model was generated from a real scene-based MPEG I video trace from the popular "Simpsons" cartoon clip [265]. It consists of around 20,000 frames which last for about 10 minutes at 30 frames per second. The video clip has a Hurst parameter 0.89 and thus possesses high degrees of long-range dependence and burstiness. The autocorrelation structure was produced at a coarse-grained group-of-pictures level under the assumption that the expected scene length was 50.

Figure 8.4 shows the ACFs of the traffic models, where ACFEXPON, ACFSUBGEO, ACFFGN, and SCFMPEG represent the four distinct traffic models, respectively. Their radical impacts on the optimality of scheduling functions (Theorem 8.2) are shown on Figure 8.5. A uniform decay function for independent traffic (Corollary 8.1) is included in Figure 8.5 for reference.

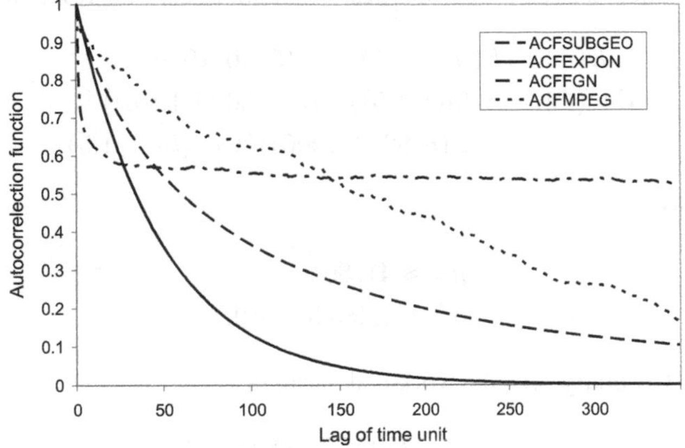

Figure 8.4: Four distinct patterns of autocorrelation function.

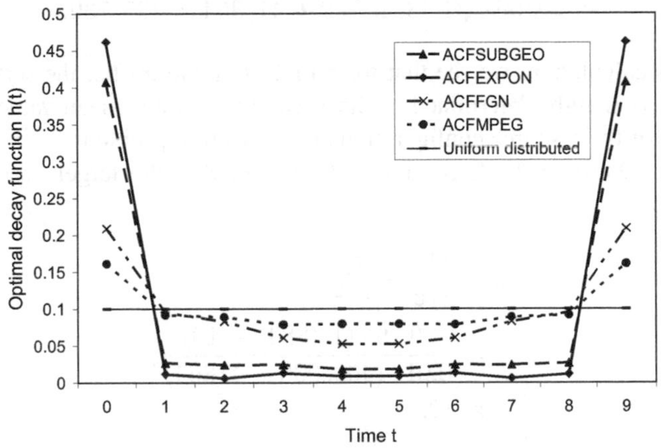

Figure 8.5: Optimal decay functions for traffic with different ACFs.

The second example assumes the number of request arrivals at each time conforms to a lognormal distribution $ln(1, 1.5)$, and the size of request fits a normal distribution $n(5000, 1000)$. Also, the server uses fixed-time scheduling.

1. What is the resource capacity necessary to be configured so as to keep the deadline of each request within 10 time units and the chance of failure below 1%?

2. Given a server capacity of 400,000, what kind of deadline can the server guarantee if it wants to keep the chance of failure below 1%?

Answer to Question 1: From properties of a random sum distribution, we know

$$\begin{aligned} E(\tilde{w}) &= E(ln(1,1.5))E(n(5000,1000)), \\ Var(\tilde{w}) &= E(ln(1,1.5))Var(n(5000,1000)) \\ &\quad +(E(n(5000,1000)))^2 Var(ln(1,1.5)), \end{aligned}$$

which leads to

$$\begin{aligned} \mu_{\tilde{w}} &\approx 41,864, \\ \sigma_{\tilde{w}}^2 &\approx 1.4884e+10. \end{aligned}$$

According to Corollary 8.1 , the workload variance

$$\sigma_l^2 = \sigma_{\tilde{w}}/10 \approx 1.4884e+9.$$

Thus the server capacity can be calculated according to Theorem 8.1 as

$$\mathbf{c} \approx \sqrt{99} \cdot \sqrt{1.4884e+9} + 41,864 \approx 425,730.$$

From this calculation, we see that for a high variable traffic, the server capacity requirement is mainly determined by its variance. Purely mean-value analysis is insufficient for the resource configuration and allocation problem.

Answer to Question 2: According to Corollary 8.2, the target deadline can be calculated as

$$\begin{aligned} t_d &= \frac{(1-v)\sigma_{\tilde{w}}^2}{v(\mathbf{c}-\mu_l)^2} \\ &\approx \frac{0.99 \cdot (1.4884e+10)}{0.01 \cdot (400,000-41864)^2} \\ &\approx 12. \end{aligned} \tag{8.19}$$

The above calculations show the applicability of our model in practical situations. In the next session, we will also show through more examples that when the workload distribution tail is a lognormal tail, the estimate of $t_{d_2}/t_{d_1} \approx \sigma_2^2/\sigma_1^2$ in (8.18) is very precise.

8.4 Performance Evaluation

To verify the above analytical results, we conducted simulations of the Internet server based on the decay function model. The simulation was conducted in three aspects: (1) verify the relationship between server capacity, scheduling and QoS for an i.i.d. input request process; (2) compare the decay function scheduling with

Table 8.1: Notations of performance metrics.

Symbols	Meanings
μ_t	The mean of completion time
σ_t^2	The variance of completion time
$\%_t$	The percentage of requests meeting the deadline t_d
μ_l	The mean of server workload
σ_l^2	The variance of workload
$\%_{\mathbf{c}}$	The percentage of server workload less than capacity **c**

another popular generalized proportional-share scheduling (GPS) [170]; (3) analyze the sensitivity of the decay function to request traffic.

We generated normal random numbers for request size and lognormal random numbers for the number of arrival requests at each scheduling epoch (unit time). All these random numbers are generated interindependently. The server was simulated with an infinite queue so that we can find exact distribution for the completion time of each request. This distribution serves as a common base for comparison of different scheduling algorithms. We did not consider admission control in the simulated server because the relationship between completion time and rejection rate is beyond the scope of this work. The simulated server treats all requests with the same priority. When a request arrives at the server, it first should be buffered in the queue. If the server has enough resource, the request will be scheduled for processing immediately without any queueing delay. Otherwise, the request will be blocked in the queue. We implemented two different schedulers on the server module: one for fixed-time scheduling and the other for GPS scheduling. GPS scheduling ensures fair sharing of resource between requests in processing at any scheduling epoch. Two performance metrics were used: completion time and server load. Completion time of a request includes waiting time in the backlogged queue, plus its processing time.

In summary, the simulation proceeds in this way: (1) generate the number of arrival requests $n(t)$ at time t from a lognormal distribution; (2) for each request, assign a size $w(t)$ from a normal distribution; (3) each request then enters the server for scheduling; (4) measure the completion time of each request and the server workload at each scheduling epoch as output. Because the number of arrival requests at each time and their sizes are i.i.d. random numbers, the compounded workload $\tilde{w}(t)$ is not only a WSS process, but also an ergodic process [212]. In this section, we use the notations listed in Table 8.1.

8.4.1 Capacity Configurations and Variances

In this experiment, we are intended to uncover the inherent properties of fixed-time scheduling. The simulated server deployed the optimal fixed-time scheduling algorithm with a request completion deadline of 10 time units ($t_d = 10$). We used the lognormal $ln(1, 1.5)$ and normal $n(5000, 1000)$ distributions to generate input

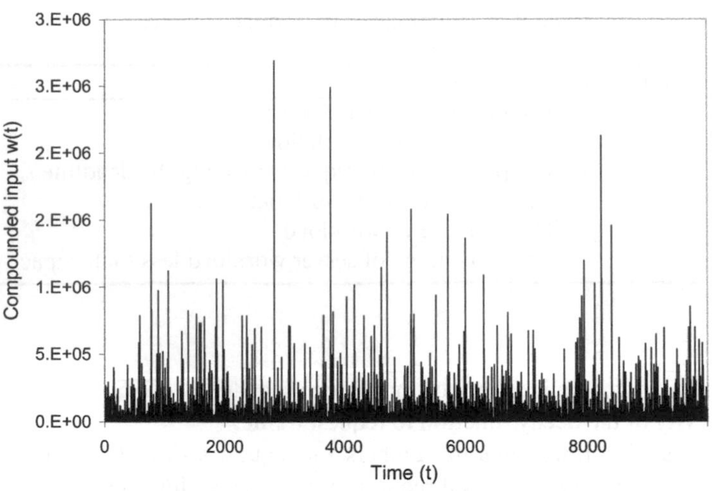

Figure 8.6: The compounded input $\tilde{w}(t)$.

Table 8.2: Statistics of server with optimal fixed-time scheduling ($\mathbf{c} = 425730$, $n(t) \sim ln(1, 1.5)$, $w \sim n(5000, 1000)$, $t_d = 10$).

Performance	Mean	Variance	Percentage
Server load	37534	1.0402×10^9	100%
Completion time	10	0	100%

traffic. We wish to keep the percentage of server load being less than its capacity $\%_{\mathbf{c}} \leq 99\%$, as the setting in the second example of Section 8.3.3. The compounded input pattern is shown in Figure 8.6.

Table 8.2 gives the statistics of the simulated server according to the Chebyshev estimation in Lemma 8.1. The 100% satisfactory rates in this table with respect to both of the server load and completion time show the conservativeness of this estimation because of the target 99% server capacity satisfactory rate.

As we mentioned in the preceding section, the Chebyshev estimation can be tightened by taking into consideration of the empirical distribution tail. Figure 8.7 shows the normal plots of compounded input $\tilde{w}(t)$ and server workload $l(t)$. Note that the normal plot for samples from a normal distribution is a straight line. This figure shows both $log(\tilde{w})$ and $log(l)$ are close to normal distributions in the tail. However, the slope of the workload plot is steeper than that of the compounded request. This indicates that a scheduling algorithm is essentially a "smoother" of fluctuating traffic that reduces server load variance and leads to better schedulability and QoS assurance.

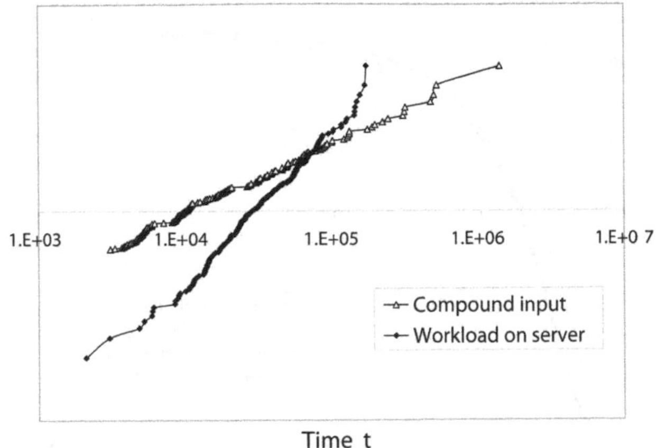

Figure 8.7: Normal plots of server load l and compounded input $\tilde{w}$.

Table 8.3: Statistics of server with various scheduling algorithms ($\mathbf{c} = 171000$, $n(t) \sim ln(1, 1.5)$, $w \sim n(5000, 1000)$, $t_d = 10$).

Scheduling	GPS	Incremental	Random	Optimal
Load mean μ_l	37534	37534	37534	37534
Load variance σ_l^2 (10^9)	1.68	1.22	1.00	0.894
Capacity satisfaction $\%_c$	99.8%	96.7%	99.2%	99.3%
Time mean μ_t	4.25	10.08	10.13	10.08
Time variance σ_t^2	5.38	0.70	0.51	0.33
Deadline satisfaction $\%_t$	96.7%	90.6%	97.0%	97.3%

The server workload plot can be approximated by a lognormal tail with distribution $ln(10.3, 0.74)$. From this distribution, the estimated capacity is 171,000. The simulation results are shown in the last column of Table 8.3. From the table, it can be seen that the 99.3% capacity satisfactory rate is in agreement with the initial 99% rate in simulation setting. Because of occasional overloads in the simulation, the queueing delay leads to only 2.7% of requests that missed their deadlines.

8.4.2 Decay Function versus GPS Scheduling

Using the same set of parameters as of the last experiment, we compared the optimal fixed-time scheduling with the GPS scheduling. In addition, we implemented two more heuristic fixed-time scheduling algorithms. One is randomized allocation that takes t_d to random numbers from a uniform distribution and then scales them to standard decay functions $h(t)$ so that $\sum_{i=1}^{t_d} h(t) = 1$. The other is incremental

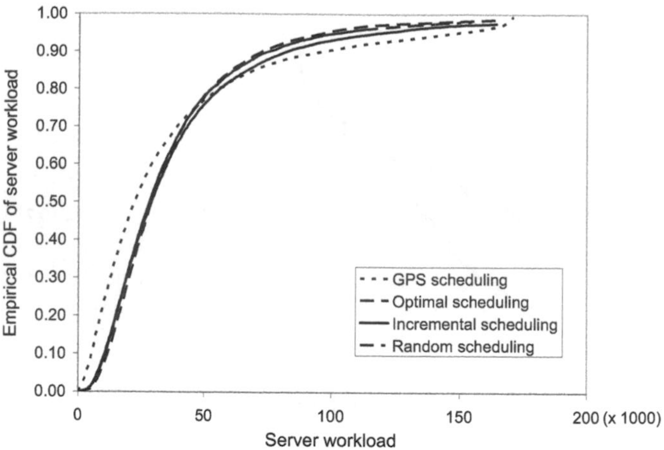

Figure 8.8: The empirical CDF of server load with various scheduling algorithms.

scheduling, in which the server increases resource allocation at a constant rate b to a request as its deadline approaches. That is, $h(t+1) = h(t) + b$. We assume the initial allocation in the first scheduling epoch $h(1) = b$, as well.

Table 8.3 presents their comparative results. From this table, it can be observed that the optimal fixed-time scheduling outperformed the others in two aspects. First, it dramatically lowers the variance of server workload by up to 50% in comparison with the GPS scheduling and more than 10% in comparison with the other fixed-time scheduling. Second, it provides better guaranteed completion time. GPS scheduling leads to a lower mean of completion time and a high variance. As a result, the percentage of requests that miss their deadlines is much higher than fixed-time scheduling. All the fixed-time scheduling algorithms guarantee completion time; the optimal scheduling algorithm yields a much smaller time variance, in comparison with incremental and randomized allocations.

Figure 8.8 shows the empirical cumulative distribution function (CDF) of the server load due to different scheduling algorithms. From this figure, we can see the tail of the optimal fixed-time scheduling is lighter than GPS and other heuristic fixed-time scheduling algorithms. This again indicates the superiority of the optimal fixed-time scheduling.

8.4.3 Sensitivity to the Change of Traffic Intensity

The Internet features high-variability traffic patterns. It is not uncommon for an Internet server to experience a heavier-than-expected incoming request traffic without any warning. In this simulation, we show that the optimality of fixed-time scheduling stands when the number of arrival requests $n(t)$ increases for each unit time. We

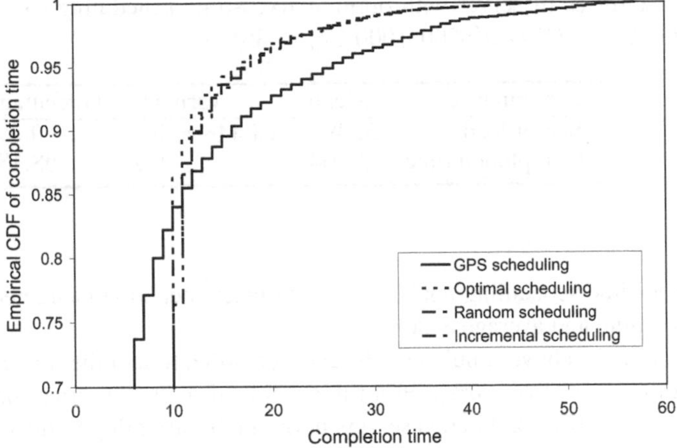

Figure 8.9: The CDF plot of completion time when $n(t) \sim ln(1, 1.7)$.

Table 8.4: Statistics server with various scheduling algorithms ($\mathbf{c} = 171000$, $n(t) \sim ln(1, 1.7)$, $w \sim n(5000, 1000)$, $t_d = 10$).

Scheduling	GPS	Incremental	Random	Optimal
Load mean μ_l	52497	52497	52497	52497
Load variance σ_l^2 (10^9)	2.92	2.31	2.04	0.189
Capacity satisfaction $\%_c$	99%	96.2%	96.8%	97%
Time mean μ_t	9.2	11.1	11.1	10.9
Time variance σ_t^2	65.7	14.8	14.5	13.2
Deadline satisfaction $\%_t$	84%	77%	81%	86%

increased the lognormal distribution parameter σ from 1.5 to 1.7, while keeping all the other parameters unchanged. This small increase in the parameter σ leads to a jump of server workload by almost 140%.

Table 8.4 shows that the variances of server load and request completion time due to the optimal fixed-time scheduling are significantly lower than those in GPS and other heuristic fixed-time scheduling. The 97% of capacity satisfactory rate $\%_\mathbf{c}$ reveals that the optimal fixed-time scheduling likely leads the server to fully loaded conditions. In other words, the optimal fixed-time scheduling is able to fully utilize existing resources on the server. This can be seen from the completion time index. It shows that more requests can be finished within a deadline in the optimal fixed-time scheduling than the other scheduling algorithms. Figure 8.9 shows that with a lower variance, the tail of completion time due to the optimal fixed-time scheduling is significantly lighter than GPS scheduling. With a light tail and small variance, the

Table 8.5: Statistics of server with optimal fixed-time scheduling ($\mathbf{c} = 171000$, $n(t) \sim ln(1, 1.7)$, $w(t) \sim n(5000, 1000)$, $t_d = 38$).

Performance	Mean	Variance	Percentage
Server load	52497	7.0448×10^8	99.3%
Completion time	38.04	0.2	98.1%

optimal fixed-time scheduling tends to provide better deadline guarantees in terms of both maximum and average delays.

Notice that in the above simulation, the average utilization of the server was maintained at a range of $[40\%, 60\%]$. From this point of view, the price for providing guaranteed service time on Internet servers under high-variability traffic is very high.

8.4.4 Quality-of-Service Prediction

In the last experiment, we show that the server performance will degrade with more arrival requests. Interesting questions are, How much degradation will there be and how much more resource does the server need to keep the same level of QoS? Answers to the first question will provide a higher customer satisfaction because people are likely to have more tolerance in performance degradation if they are informed of extra delay. Answers to the second question can enable Internet service providers to predict the needs for increasing capacity on their servers.

Although Chebyshev estimations of absolute server capacity and request completion time are conservative, their relative predictions are precise. Following the previous experiment with the number of requests coming in lognormal distribution $ln(1, 1.7)$, we wish to predict the server capability to assume service completion time.

From the remarks of Corollary 8.2, we can calculate the new request completion time t_{old} for $ln(1, 1.7)$ distribution based on the old completion time t_{old} for $ln(1, 1.5)$ as:

$$t_{new} = \frac{\sigma_{new}^2}{\sigma_{old}^2} \cdot t_1 \approx \frac{5.65e10}{1.4884e10} \cdot 10 \approx 38.$$

We ran simulations with a new deadline of 38 time units under the optimal fixed-time scheduling algorithm. The results are shown in Table 8.5. The simulation data show that in response to the change of incoming traffic, more than 98% of the requests can meet the adjusted deadline of 38 and that there were less than 1% of chances for the server getting overload.

If the server wants to keep the original deadline of 10 time units, it needs to be configured with more resource. According to 8.2, we have

$$1 = \frac{\sigma_{new}^2 \cdot (\mathbf{c}_{old} - \mu_{old})^2}{\sigma_{old}^2 \cdot (\mathbf{c}_{new} - \mu_{new})^2}.$$

Table 8.6: Statistics of server with optimal fixed-time scheduling ($\mathbf{c} = 313000$, $n(t) \sim ln(1, 1.7)$, $w(t) \sim n(5000, 1000)$, $t_d = 10$).

Performance	Mean	Variance	Percentage
Server load	52497	2.733×10^9	99.4%
Completion time	10.17	1.25	96.0%

As a result, the new server capacity $\mathbf{c}_{new} = 313000$.

We ran simulations based on this new predicted capacity. The experimental results in Table 8.6 show that with an adjusted server capacity 313000, 96% of the requests would meet their deadlines and the server would rarely become overloaded. Again, this demonstrates the prediction accuracy of Corollary 8.2.

8.5 Concluding Remarks

In this chapter, we have introduced a decay function model to study the QoS provisioning problems on Internet servers. The model abstracts QoS-aware scheduling into a transfer-function-based filter system that has a general time-series-based or measurement-based request process as input and a server load process as output. By using filter design theories in signal processing, it is able to reveal the relationships between Internet workloads, resources configuration and scheduling, and server capacity. The parameters of the model have strong physical meanings, especially the decay function that describes the detailed scheduling activities on the server, We have analyzed and solved the model for an important class of QoS-aware scheduling: fixed-time scheduling. We have also derived the optimality conditions with respect to the statistical properties of the input traffic. We verified the analytical results via simulations in both light and heavy loaded cases.

Note that the optimal fixed-time scheduling $h(t)$ is time invariant because it is not adaptive to input traffic. Derivation of $h(t)$ assumed *a priori* knowledge of the request interarrival distribution and correlation structure. The decay function model also assumed a uniform delay constraint t_d for all requests. In [354], we extended the decay function model to support requests with different delay constraints. We derived the optimal time-variant schedule to further reduce the server capacity requirement. The optimal time-variant schedule is a water-filling process based only on the current and past request arrivals.

The decay function model paves a new way for the derivation of optimal scheduling policies by investigating the impact of the second moments of input traffic on server load. We used Chebyshev's inequality for an estimation of required server capacity for general traffic and derived a relationship between request process and server load in a formalism for nonadaptive scheduling. We will further our study to extend the decay function model to include the feedback of server load in the analysis of adaptive scheduling. The current model assumes a proportional relationship between the request size and resource requirement. This may not be the case in real Internet servers due to the presence of caching and locality. The scheduling models need to be deliberated to take into account their effects as well.

Chapter 9

Scalable Constant-Degree Peer-to-Peer Overlay Networks

In the past few years, the immense popularity of peer-to-peer (P2P) resource sharing services has resulted in a significant stimulus to content-delivery overlay network research. An important class of the overlay networks is distributed hash tables (DHTs) that map keys to the nodes of a network based on a consistent hashing function. Most of the DHTs require $O(\log n)$ hops per lookup request with $O(\log n)$ neighbors per node, where n is the network size. In this chapter, we present a constant-degree DHT, namely, Cycloid. It achieves a lookup path length of $O(d)$ with $O(1)$ neighbors, where d is the network dimension and $n = d \cdot 2^d$. The DHT degree determines the number of neighbors with which a node must maintain continuous contact. A constant degree puts a cap on the cost for maintenance,

9.1 Introduction

The ubiquity of the Internet has made possible a universal storage space that is distributed among the participating end computers (peers) across the Internet. All peers assume equal role and there is no centralized server in the space. Such architecture is collectively referred to as the P2P system that is leading to new content distribution models for applications such as software distribution, distributed file systems, searching network, and static Web content delivery. The most prominent initial designs of P2P systems include Napster [66, 140], Gnutella [133], and Freenet [65]. Unfortunately, all of them have significant scaling problems. For example, in Napster a central server stores the index of all the files available within the Napster user community. To retrieve a file, a user queries this central server using the desired files well-known name and obtains the IP address of a user machine storing the requested file. The file is then downloaded directly from this user machine. Thus, although Napster uses a P2P communication model for the actual file transfer, the process of locating a file is still very much centralized. This makes it both expensive (to scale the central directory) and vulnerable (since there is a single point of failure). Gnutella decentralizes the file location process as well. Users in a Gnutella network self-organize into an application-level network on which requests for a file are flooded with a certain scope. Flooding on every request is clearly not scalable.

Freenet goes a step further by addressing the security issues and privacy issues, but it still employs a flooding-based depth-first search mechanism that incurs a heavy traffic for large systems. Both Gnutella and Freenet are not guaranteed to find an existing object because the flooding has to be curtailed at some point.

People started the investigation with the question: Could one make a scalable P2P file distribution system? It has been recognized that central to any P2P system is the indexing scheme used to map file names (whether well known or discovered through some external mechanism) to their location in the system. That is, the P2P file transfer process is inherently scalable, but the hard part is finding the peer from whom to retrieve the file. Thus, a scalable P2P system requires, at the very least, a scalable indexing mechanism. In response to these scaling problems, several research groups have independently proposed a second generation of P2P systems that support scalable routing and location schemes. Unlike the earlier work, they guarantee a definite answer to a query in a bounded number of network hops while maintaining a limited amount of routing information; among them are Chord [301], Tapestry [351], Pastry [266], CAN [256], and Kademlia [215].

This new generation of P2P systems connects peers in an application-level overlay network. In [53], we proposed a generic topological model to capture the essence of this class of overlay networks. This model facilitates the exploitation of new designs of P2P architectures, including the constant-degree lookup-efficient Cycloid system [284]. This chapter presents the topological model and the Cycloid DHT.

9.2 Topological Model of DHT-Based P2P Systems

There are two classes of P2P content-delivery overlay networks: unstructured and structured. Unstructured networks such as Gnutella [133] and Freenet [65] do not assign responsibility for data to specific nodes. Nodes join and leave the network according to some loose rules. Currently, the query method is either flooding [133] where the query is propagated to all neighbors, or random-walkers [207] where the query is forwarded to randomly chosen neighbors until the object is found.

A flooding-based search mechanism brings about heavy traffic in a large-scale system because of exponential increase in messages generated per query. Though random-walkers reduce flooding to some extent, they still create heavy overhead to the network due to the many requesting peers involved. Furthermore, flooding and random-walkers cannot guarantee data location. They do not ensure that querying terminates once the data is located, and they cannot prevent one node from receiving the same query multiple times, thus wasting bandwidth.

Structured networks have strictly controlled topologies. The data placement and lookup algorithms are precisely defined based on a DHT data structure. The node responsible for a key can always be found even if the system is in a continuous state of change. Because of their potential efficiency, robustness, scalability, and deter-

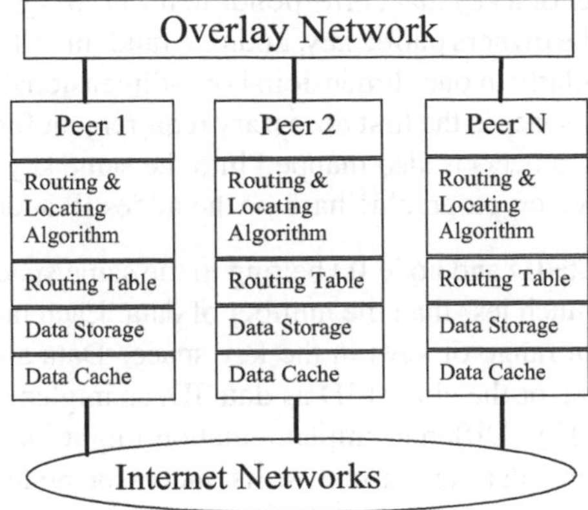

Figure 9.1: A generic topological model of P2P systems.

ministic data location, structured networks have been studied intensively in recent years.

In the following, we develop a generic topological model for structured DHTs and review the existing DHTs by focusing on their topological aspects.

9.2.1 A Generic Topological Model

In developing our abstract model, we made two observations. First, in any P2P system, we can see that there are multiple nodes spreading across the entire Internet and contributing their local resource (e.g., storage and data) to the global P2P system. The Internet can be modeled as a general interconnection network that connects these nodes. Second, we note that the main functions of peer node in a P2P system are message routing and data locating besides data storing and caching. The routing function routes a given request/reply message from the source to the destination node. The data locating function attempts to find out the location or home node of a data item, with the help of a routing table. All peer nodes are connected as an overlay network in a somewhat regular way so that both the routing table size and the routing path arc dramatically reduced. In such systems, entries of the routing table (e.g., IP address of neighbors) serve as logical links to other neighboring nodes and thus are regarded as edges of the overlay networks.

The complexity of an overlay network often determines the size of a P2P system that can be constructed. Likewise, the attainable performance of a P2P system is ultimately limited by the characteristics of the overlay network. Apparently, the selection of the overlay network is the first and also the most crucial step for the design of the new generation of P2P systems. Figure 9.1 displays our simple model. Each component will be clear after we define and describe the following terms.

Data identification (ID) and node ID: Each data (e.g., a file) is assigned a globally unique identifier or a key that corresponds to the cryptographic hash of the files textual name, the owners public key, and/or a random salt. All keys are ordered linearly or circularly in one-dimensional or d-dimensional key space. Note that the ordered key space is the first necessary requirement for an overlay network. Each node (i.e., a peer) is also mapped into the same key space as data. It can be computed as a cryptographic hash of the nodes IP address or public key.

Data placement: Data ID and node ID belong to the same space, but the number of nodes is often much less than the number of data. Each node is responsible for storing a certain range of keys in the key space. Data are stored on the node that has the same or the closest ID as data ID, as implemented in Pastry [266] and Viceroy [210]. Different implementation might have a slightly changed policy. In Chord, data are stored in its successor node; this simplicity can bring about the convenience in self-organization. In CAN [256], all data of one zone are stored on the representative node of that zone.

Hash function: All data or nodes are mapped by hashing into a point in the same key space. The functionality of hash should guarantee a uniform and random distribution of data or nodes within the key space. Consistent hashing [168] was proposed to guarantee the uniformity no matter how the input keys are intentionally chosen. SHA-1 function is used in practice as a good substitute for consistent hash function since producing a set of keys that collide under SHA-1 can be seen as inverting or decrypting the SHA-1 function that is believed to be hard to do [301].

Overlay network: All live nodes are connected logically as an overlay network at application level with the help of routing table. Each entry of the routing table serves as a logical edge, usually containing the IP address of one neighbor. Recent P2P systems have adopted structured topologies such as ring, mesh, and hypercube and showed advantages in scalability and lookup efficiency. Interestingly and importantly, none of them is a simple application of a traditional interconnection network. Each one embodies an insightful anatomy of an existing interconnection network and provides a resilient connection pattern for fault tolerance and self-organization. Seeking for better topologies for an overlay network is becoming a new search topic. Recent work focuses on the constant-degree overlay networks [166, 210, 284].

Routing table: The size (or the number of entries) of the routing table is determined by the outgoing degree of the overlay network. It is the space complexity of an overlay network, i.e., the memory overhead of each node. Moreover, it is not just a measure of the state required to do routing but also a measure of how much state needs to be adjusted when nodes join and leave. For this reason, interconnection networks with small degree (especially with constant degree) should be restudied as possible candidates for P2P overlay network.

Routing and locating algorithm: When a key lookup request is issued or received, it is routed through the overlay network to the home node responsible for that key. Each hop makes the most progress toward resolving the lookup. The notation of progress differs from algorithm to algorithm, but any routing algorithm should be convergent in the sense that each hop makes the distance between the current node ID and the home node ID becomes smaller and smaller. Note that the convergent routing algorithm is the second necessary requirement for an overlay network. The real distance performance is determined not only by the path length in key space but also by the communication delay of each hop in the Internet. Proximity in key space does not mean proximity in geographical space. If the supporting network is represented as a graph in some regular way, the proximity problem can be studied in the form of embedding the overlay network into the supporting network with the smallest possible dilation and congestion. An adjustable hash function might also be helpful if such a function could map nearby IP addresses into close values in key space. Other possible solutions include the neighborhood set [266] and the landmark vector [343].

Self-organizing: To make P2P systems completely distributed, the routing table should be automatically reconfigured when a node joins or departs from the system. A node might leave abruptly by crashing or gracefully by informing its neighbors and submitting its keys. Systems are supposed to be small enough at the beginning so that nodes could just exchange information directly to build the initial routing table. After that, nodes should have an automatic way to detect the node join or node leave to reconfigure their routing table state in such cases. Since not all nodes are alive in the key space and they often join and depart dynamically, the resilient topological connection pattern must ensure the routing table has full or enough numbers of neighbors. When one node fails, another node can substitute. Note that self-organizing is the third necessary requirement for an overlay network.

We remark that a network can be manipulated as an overlay network if and only if the network meets the above three requirements, i.e., an ordered key space which is used to measure the distance between any two nodes, a convergent routing algorithm that sends a message closer to the destination at each step, and self-organizing ability that reconfigures the network topology when nodes join or depart.

9.2.2 Characteristics of Representative DHT Networks

9.2.2.1 Hypercube-Based

Plaxton *et al.* [246] developed perhaps the first routing algorithm that could be scalably used for P2P systems. Tapestry and Pastry use a variant of the algorithm. The approach of routing based on address prefixes, which can be viewed as a generalization of hypercube routing, is common to all theses schemes. The routing algorithm works by correcting a single digit at a time in the left-to-right order: if node number 12345 received a lookup query with key 12456, which matches the first two digits, then the routing algorithm forwards the query to a node which matches the first three digits (e.g., node 12467). To do this, a node needs to have, as neighbors, nodes that match each prefix of its own identifier but differ in the next digit. For each prefix (or dimension), there are many such neighbors (e.g., node 12467 and node 12478 in the above case) since there is no restriction on the suffix, i.e., the rest bits right to the current bit. This is the crucial difference from the traditional hypercube connection pattern and provides the abundance in choosing cubical neighbors and thus a high-fault resilience to node absence or node failure. Besides such cubical neighbors spreading out in the key space, each node in Pastry also contains a leaf set L of neighbors which are the set of $|L|$ numerically closest nodes (half smaller, half larger) to the present node ID and a neighborhood set M which are the set of $|M|$ geographically closest nodes to the present node.

9.2.2.2 Ring-Based

Chord uses a one-dimensional circular key space. The node responsible for the key is the node whose identifier most closely follows the key numerically; that node is called the key's successor. Chord maintains two sets of neighbors. Each node has a successor list of k nodes that immediately follow it in the key space and a finger list of $O(\log n)$ nodes spaced exponentially around the key space. The ith entry of the finger list points to the node that is 2^i away from the present node in the key space, or to that node's successor if that node is not alive. So the finger list is always fully maintained without any null pointer. Routing correctness is achieved with such two lists. A key lookup request is, except at the last step, forwarded to the node closest to, but not past, the key. The path length is $O(\log n)$ since every lookup halves the remaining distance to the home.

9.2.2.3 Mesh-Based

CAN chooses its keys from a d-dimensional toroidal space. Each node is associated with a region of this key space, and its neighbors are the nodes that own the contiguous regions. Routing consists of a sequence of redirections, each forwarding a lookup to a neighbor that is closer to the key. CAN has a different performance profile than the other algorithms; nodes have $O(d)$ neighbors and path lengths are $O(dn^{1/d})$hops. Note that when $d = \log n$, CAN has $O(\log n)$ neighbors and $O(\log n)$ path length like the other algorithms. This actually gives another way to deploy the hypercube as an overlay network.

9.2.2.4 Constant-Degree DHTs

Viceroy [210] and Koorde [166] are two constant-degree DHTs. Both of them feature a time complexity of $O(\log n)$ hops per lookup request with $O(1)$ neighbors per node. But they are different in maintenance of the connectivity between a changing set of nodes and in the routing for efficient key location. Viceroy maintains a connection graph with a constant-degree logarithmic diameter, approximating a butterfly network. Each Viceroy node in butterfly level l has seven links to its neighbors, including pointers to its predecessor and successor pointers in a general ring, pointers to the next and previous nodes in the same level ring, and butterfly pointers to its left, right nodes of level $l+1$, and up node of level $l-1$, depending on the node location.

In Viceroy, every participating node has two associated values: its identity $\in [0,1)$ and a butterfly level index l. The node ID is independently and uniformly generated from a range $[0,1)$ and the level is randomly selected from a range of $[1, \log n_0]$, where n_0 is an estimate of the network size. The node ID of a node is fixed, but its level may need to be adjusted during its lifetime in the system. Viceroy routing involves three steps: ascending to a level 1 node via uplinks, descending along the downlink until a node is reached with no down links, and traversing to the destination via the level ring or ring pointers. Viceroy takes $O(\log n)$ hops per lookup request.

Koorde combines Chord with de Bruijn graphs. Like Viceroy, it looks up a key by contacting $O(\log n)$ nodes with $O(1)$ neighbors per node. As in Chord, a Koorde node and a key have identifiers that are uniformly distributed in a 2^d identifier space. A key k is stored at its successor, the first node whose ID is equal to or follows k in the identifier space. Node $2^d - 1$ is followed by node 0.

Due to the dynamic nature of the P2P systems, they often contain only a few of the possible 2^d nodes. To embed a de Bruijn graph on a sparsely populated identifier ring, each participating node maintains knowledge about its successor on the ring and its first de Bruijn node. To look up a key k, the Koorde routing algorithm must find the successor of k by walking down the de Bruijn graph. Since the de Bruijn graph is usually incomplete, Koorde simulates the path taken through the complete de Bruijn graph, passing through the immediate real predecessor of each imaginary node on the de Bruijn path.

In [284], we proposed another constant-degree DHT, namely, Cycloid. In contrast to Viceroy and Koorde, Cycloid specifies each node by a pair of cyclic and cubic indices. It emulates a cube-connected cycles (CCC) graph by using a routing algorithm similar to the one in Pastry. Although the lookup complexity of all three constant-degree DHTs are of the same order $O(\log n)$, simulation results show that Cycloid has a much shorter path length per lookup request in the average case than Viceroy and Koorde. Cycloid distributes keys and lookup load more evenly between the participating nodes than Viceroy. Also, Cycloid is more robust as it continues to function correctly and efficiently with frequent node joins and leaves.

Table 9.1 summarizes the architectural characteristics of the representative DHTs.

9.3 Cycloid: A Constant-Degree DHT Network

Cycloid combines Pastry with CCC graphs. In a Cycloid system with $n = d \cdot 2^d$ nodes, each lookup takes $O(d)$ hops with $O(1)$ neighbors per node. Like Pastry, it employs consistent hashing to map keys to nodes. A node and a key have identifiers that are uniformly distributed in a $d \cdot 2^d$ identifier space.

Table 9.1: A comparison of some representative P2P DHTs.

Systems	Base Network	Lookup Complexity	Routing Table Size
Chord	Cycle	$O(\log n)$	$O(\log n)$
CAN	Mesh	$O(dn^{1/d})$	$O(d)$
eCAN	Mesh	$O(dn^{1/d})$	$O(d)$
Pastry	Hypercube	$O(\log n)$	$O(\|L\|)+O(\|M\|)+O(\log n)$
Viceroy	Butterfly	$O(\log n)$	7
Koorde	de Bruijn	$O(\log n)$	≥ 2
Cycloid	CCC	$O(d)$	7

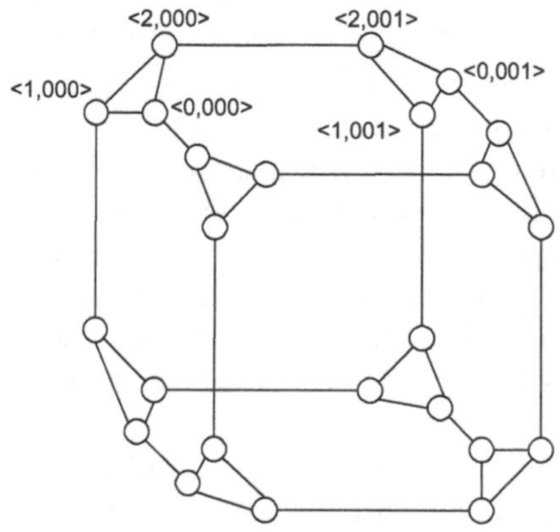

Figure 9.2: A three-dimensional cube-connected-cycles.

9.3.1 CCC and Key Assignment

A d-dimensional CCC graph is a d-dimensional cube with replacement of each vertex by a cycle of d nodes. It contains $d \cdot 2^d$ nodes of degree 3 each. Each node is represented by a pair of indices $(k, a_{d-1}a_{d-2}\ldots a_0)$, where k is a cyclic index and $a_{d-1}a_{d-2}\ldots a_0$ is a cubical index. The cyclic index is an integer, ranging from 0 to $d-1$ and the cubical index is a binary number between 0 and $2^d - 1$. Figure 9.2 shows the three-dimensional CCC.

A P2P system often contains a changing set of nodes. This dynamic nature poses a challenge for DHTs to manage a balanced distribution of keys among the participating nodes and to connect the nodes in an easy-to-maintain network so that a lookup request can be routed toward its target quickly. In a Cycloid system, each node keeps a routing table and two leaf sets with a total of seven entries to maintain its connectivity to the rest of the system. Table 9.2 shows a routing state table for node (4,10111010) in an eight-dimensional Cycloid. Its corresponding links in both cubical and cyclic aspects are shown Figure 9.3.

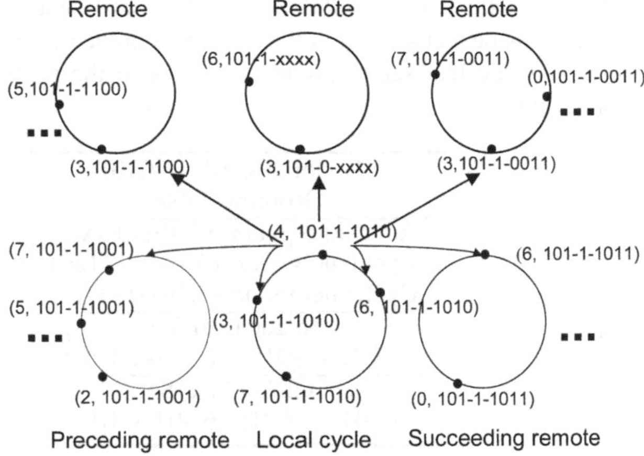

Figure 9.3: Cycloid node routing links state.

In general, a node $(k, a_{d-1}a_{d-2}\ldots a_k \ldots a_0)$ $(k \neq 0)$ has one cubical neighbor $(k-1, a_{d-1}a_{d-2}\ldots \overline{a}_k xx...x)$ where x denotes an arbitrary bit value, and two cyclic neighbors $(k-1, b_{d-1}b_{d-2}\ldots b_0)$ and $(k-1, c_{d-1}c_{d-2}\ldots c_0)$. The cyclic neighbors are the first larger and smaller nodes with cyclic index $k-1$ mod d and their most significant different bit with the current node is no larger than $k-1$. That is,
(k-1, $b_{d-1}\ldots b_1b_0$) = min{$\forall$(k-1, $y_{d-1}\ldots y_1y_0$)$|y_{d-1}\ldots y_0 \geq a_{d-1}\ldots a_1a_0$}
(k-1, $c_{d-1}\ldots c_1c_0$) = max{$\forall$(k-1, $y_{d-1}\ldots y_1y_0$)$|y_{d-1}\ldots y_0 \leq a_{d-1}\ldots a_1a_0$.}
The node with a cyclic index $k = 0$ has no cubical neighbor and cyclic neighbors. The node with cubical index 0 has no small cyclic neighbor, and the node with cubical index $2^d - 1$ has no large cyclic neighbor.

The nodes with the same cubical index are ordered by their cyclic index mod d on a local cycle. The left inside leaf set node points to the node's predecessor and the right inside leaf set node points to the node's successor in the local cycle. The largest cyclic index node in a local cycle is called the primary node of the local cycle. All local cycles are ordered by their cubical index mod 2^d on a large cycle. The left outside leaf set node points to the primary node in the node's preceding remote cycle and the right outside leaf set node points to the primary node in the node's succeeding remote cycle in the large cycle.

The cubical links allow us to change cubical index from left to right, in the same left-to-right order as in Pastry. The cyclic links allow us to change the cyclic index. It is easy to see that the network will be the traditional cube-connected cycles if all nodes are alive. Our connection pattern is resilient in the sense that even if many nodes are absent, the remaining nodes are still capable of being connected. The routing algorithm is heavily assisted by the leaf sets. The leaf sets help improve the routing efficiency, check the termination condition of a lookup, and wrap around the key space to avoid the target overshooting. How the routing table and leaf sets are

Table 9.2: Routing table state of a Cycloid node (4,101-1-1010), with x indicating an arbitrary value, 0 or 1. Inside leaf set maintains the node's predecessor and successor in the local cycle. Outside leaf set maintains the links to the preceding and the succeeding remote cycles.

NodeID(4,101-1-1010) Routing Table
Cubical neighbor: (3,101-0-xxxx)
Cyclic neighbor: (3,101-1-1100)
Cyclic neighbor: (3,101-1-0011)
Inside Leaf Set
(3,101-1-1010), (6,101-1-1010)
Outside Leaf Set
(7,101-1-1001), (6,101-1-1011)

initialized and maintained is the subject of Section 9.3.3.

The Cycloid DHT assigns keys onto its ID space by the use of a consistent hashing function. The key assignment is similar to Pastry, except that the Cycloid associates a pair of cyclic and cubic indices with each node. For a given key, the cyclic index of its mapped node is set to its hash value modulated by d and the cubical index is set to the hash value divided by d. If the target node of a key's ID $(k, a_{d-1} \ldots a_1 a_0)$ is not a participant, the key is assigned to the node whose ID is first numerically closest to $a_{d-1} a_{d-2} \ldots a_0$ and then numerically closest to k.

9.3.2 Cycloid Routing Algorithm

Cycloid routing algorithm emulates the routing algorithm of CCC [247] from source node $(k, a_{d-1} \ldots a_1 a_0)$ to destination $(l, b_{d-1} \ldots b_1 b_0)$, incorporating the resilient connection pattern of Cycloid. The routing algorithm involves three phases, assuming MSDB be the most significant different bit of the current node and the destination.

1. *Ascending*: When a node receives a request, if its $k <$ MSDB, it forwards the request to a node in the outside leaf set sequentially until $k \geq$ MSDB.

2. *Descending*: In the case of $k \geq$ MSDB, when $k =$ MSDB, the request is forwarded to the cubical neighbor; otherwise the request is forwarded to the cyclic neighbor or inside leaf set node, whichever is closer to the target, in order to change the cubical index to the target cubical index.

3. *Traverse cycle*: If the target ID is within the leaf sets, the request is forwarded to the closest node in the leaf sets until it reaches the target.

Figure 9.4 presents an example of routing a request from node (0,0100) to node (2,1111) in a four-dimensional Cycloid DHT. The MSDB of node (0,0100) with

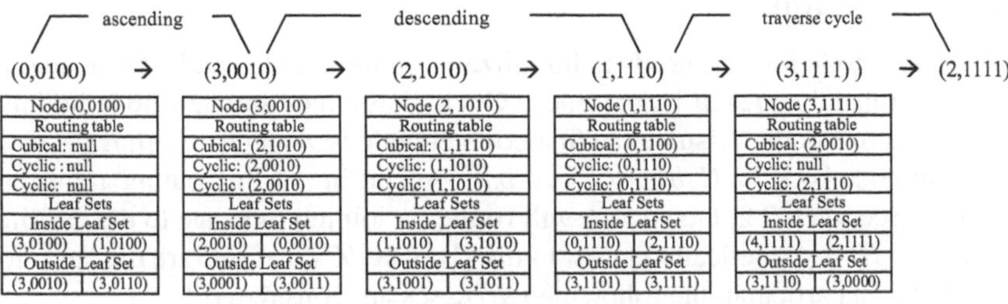

Figure 9.4: An example of routing phases and routing table states in Cycloid.

the destination is 3. As (0,0100) cyclic index $k = 0$ and $k <$ MSDB, it is in the ascending phase. Thus, the node (3,0010) in the outside leaf set is chosen. Node (3,0010)'s cyclic index 3 is equal to its MSDB; then in the descending phase, the request is forwarded to its cubical neighbor (2,1010). After node (2,1010) finds that its cyclic index is equal to its MSDB 2, it forwards the request to its cubical neighbor (1,1110). Because the destination (2,1111) is within its leaf sets, (1,1110) forwards the request to the closest node to the destination (3,1111). Similarly, after (3,1111) finds that the destination is within its leaf sets, it forwards the request to (2,1111) and the destination is reached.

Each of the three phases is bounded by *O(d)* hops and the total path length is *O(d)*. The key idea behind this algorithm is to keep the distance decreasing repeatedly. In [53], we proved the correctness of the routing algorithm by showing its convergence and reachability. By convergence, we mean that each routing step reduces the distance to the destination. By reachability, we mean that each succeeding node can forward the message to the next node. Because each step sends the lookup request to a node that either shares a longer prefix with the destination than the current node, or shares as long a prefix with, but is numerically closer to the destination than the current node, the routing algorithm is convergent. Also, the routing algorithm can be easily augmented to increase fault tolerance. When the cubical or the cyclic link is empty or faulty, the message can be forwarded to a node in the leaf sets. Our discussion is based on a 7-entry Cycloid DHT. It can be extended to include two predecessors and two successors in its inside leaf set and outside leaf set, respectively. We will show in the next section that the 11-entry Cycloid DHT has better performance. Unless otherwise specified, Cycloid refers to the 7-entry DHT.

9.3.3 Self-Organization

P2P systems are dynamic in the sense that nodes are frequently joining in and departing from the network. Cycloid deals with node joining and leaving in a distributed manner, without requiring hash information to be propagated through the entire network. This section describes how Cycloid handles node joining and leaving.

9.3.3.1 Node Join

When a new node joins, it needs to initialize its routing table and leaf sets, and to inform other related nodes of its presence. Cycloid assumes any new node initially knows about a live node. Assume the first contact node is $A = (k, a_{d-1}a_{d-2} \ldots a_0)$ and the new node is $X = (l, b_{d-1}b_{d-2} \ldots b_0)$. According to the routing algorithm discussed in Section 9.3.2, the node A will route the joining message to the existing node Z whose ID is numerically closest to the ID of X. Z's leaf sets are the basis for X's Leaf Sets. In particular, the following two cases are considered:

1. If X and Z are in the same cycle, Z's outside leaf set becomes X's outside leaf set. X's inside leaf set is initiated according to Z's inside leaf set. If Z is X's successor, Z's predecessor and Z are the left node and right node in X's inside leaf set, respectively. Otherwise, Z and Z's successor are the left node and right node.

2. If X is the only node in its local cycle, then Z is not in the same cycle as X. In this case, two nodes in X's inside leaf set are X itself. X's outside leaf set is initiated according to Z's outside leaf set. If Z's cycle is the succeeding remote cycle of the X, Z's left outside leaf set node and the primary node in Z's cycle are the left node and right node in X's outside leaf set. Otherwise, the primary node in Z's cycle and Z's right outside leaf set node are the left node and right node in X's outside leaf set.

We use a local-remote method to initialize the three neighbors in X's routing table. It searches for a neighbor in the local cycle in a decreasing order of the node cyclic index. If the neighbor is not found, then its neighboring remote cycle is searched. The remote cycle search sequence depends on the kth bit in the cubical index. If a_k is 1, the search direction is counterclockwise, otherwise the direction is clockwise. This is done in order to enhance the possibility and the speed of finding the neighbors.

After a node joins the system, it needs to notify the nodes in its inside leaf set. It also needs to notify the nodes in its outside leaf set if it is the primary node of its local cycle. Once the nodes in the inside leaf set receive the joining message, they will update themselves. When the nodes in the outside leaf set receive the joining message, in addition to update themselves, they need to transfer the message to the nodes in their inside leaf set. Thus, the message is passed along in the joining node's neighboring remote cycle until all the nodes in that cycle finish updating.

9.3.3.2 Node Departure

Before a node leaves, it needs to notify its inside leaf set nodes. In Cycloid, a node only has outgoing connections and has no incoming connections. Therefore, a leaving node cannot notify those who take it as their cubical neighbor or cyclic neighbor. The need to notify the nodes in its outside leaf set depends on whether the leaving node is a primary node. Upon receiving a leaving notification, the nodes in the inside and outside leaf sets update themselves. In addition, the nodes in the outside leaf set need to notify other nodes in their local cycle one by one, which will take at most d steps. As a result, only those who take the leaving node as their inside leaf set or outside leaf set are updated. Those nodes who take the leaving node as their cubical neighbor or cyclic neighbor cannot be updated. Updating cubical and cyclic neighbors are the responsibility of system stabilization, as in Chord.

9.3.3.3 Fault Tolerance

Undoubtedly, low degree P2P networks perform poorly in failure-prone environments, where nodes fail or depart without warning. Usually, the system maintains another list of nodes to handle such problems, such as the successor list in Chord [301] and the bucket in Viceroy [210]. In this work, we assume that nodes must notify others before leaving, as the authors of Koorde argued that the fault tolerance issue should be handled separately from routing design.

9.4 Cycloid Performance Evaluation

In [166], Kaashoek and Karger listed five primary performance measures of DHTs: degree in terms of the number of neighbors to be connected, hop count per lookup request, degree of load balance, degree of fault tolerance, and maintenance overhead. In this section, we evaluate Cycloid in terms of these performance measures and compare it with other two constant-degree DHTs: Viceroy and Koorde. Recall that each Cycloid node maintains connectivity to seven neighbors in its routing table. Cycloid can be extended to include more predecessors and successors in its inside and outside leaf sets for a trade-off for lookup hop count. The results due to 11-neighbor Cycloid are included for a demonstration of the trade-off. Similarly, Koorde DHT provides a flexibility to making a trade-off between routing table size and routing hop count. For a fair comparison, in our simulations, we assumed the Koorde DHT maintained connectivity to seven neighbors, including one de Bruijn node, three successors, and three immediate predecessors of the de Bruijn node. Since all of the constant-degree DHTs borrowed ideas from Chord and other DHTs with $O(\log n)$ neighbors, we also include the results of Chord as references. The actual number of participants varied in different experiments.

9.4.1 Key Location Efficiency

It is known that all of the constant-degree DHTs have a complexity of $O(\log n)$ or $O(d)$ hops per lookup request with $O(1)$ neighbors. Although Cycloid contains more nodes than the others for the same network dimension, its average routing performance relative to Viceroy and Koorde is unknown. In this experiment, we simulated networks with $n = d \cdot 2^d$ nodes and varied the dimension d from 3 to 8. Each node made a total of $n/4$ lookup requests to random destinations. Figure 9.5 plots the mean of the measured path lengths of the lookup requests due to various

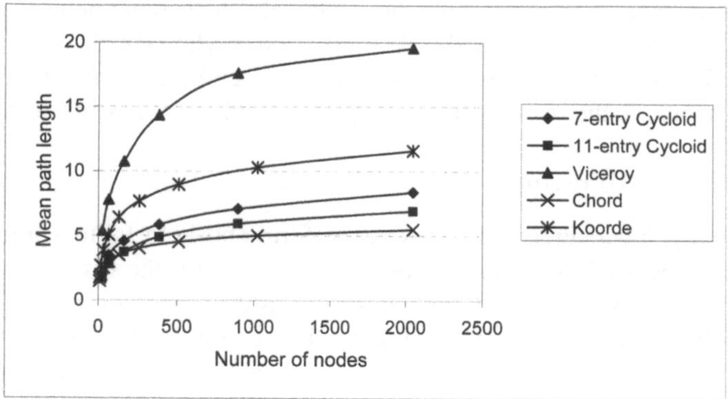

Figure 9.5: Path lengths of lookup requests in various DHTs of different network sizes.

DHT routing algorithms. The path length of each request is measured by the number of hops traversed during its search.

From this figure, we can see that the path lengths of Viceroy are more than 2 times those of Cycloid, although key locations in both Cycloid and Viceroy involve the same ascending, descending, and traverse ring/cycle phases. There are two reasons. First, the ascending phase in Cycloid usually takes only one step because the outside leaf set entry node is the primary node in its cycle. But the ascending phase in Viceroy takes $(\log n)/2$ steps on average because each step decreases the level one at a time. Figure 9.6 presents breakdowns of the lookup cost in different phases in Cycloid. We can see that the ascending phase constitutes less than 15% of the total path length. The descending phase in Cycloid takes d steps because each step redirects the request to a node with longer prefix or is numerically closer to the target. It is followed by another d hops of search in local cycles or cubic neighbor cycles.

In [284], we also measured the lookup cost in different phases in Viceroy and Koorde. We found that the ascending phases in Viceroy took about 30% of the total path length. The distance to the target can be halved each step in the second descending phase. The descending phases constitute around only 20% of the total searching path. That is, more than half of the cost is spent in the third traverse ring phase. In this phase, the lookup request approaches the destination step by step along ring links or level ring links and needs another $(\log n)/2$ steps on average.

In Koorde, each node redirects an incoming lookup request to its first de Bruijn node or a successor. Each selection of a first de Bruijn node would reduce the distance by half. Since the first de Bruijn node may not be the immediate predecessor of the imaginary node of the destination, selection of a successor is to find the immediate predecessor. In [284], we showed that the selection of successors constituted about 30% of the total path length. It implies some nodes might interpose land in between the current node's first de Bruijn node and the imaginary node. In this case, the current node's successors have to be passed in order to reach the immediate pre-

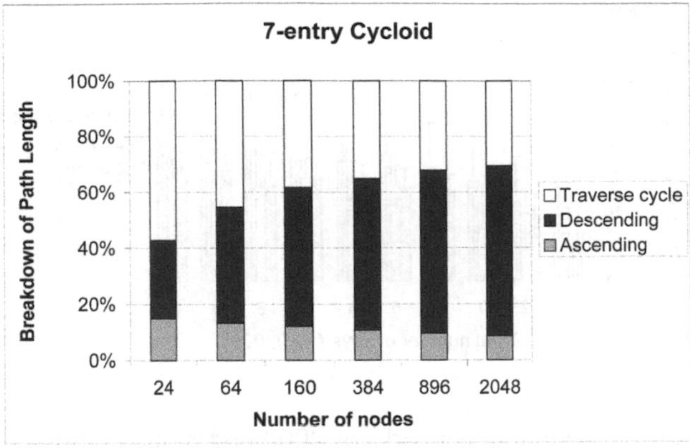

Figure 9.6: Path length breakdown in Cycloid of different sizes.

decessor of the imaginary node. Because of the dense network in which every node is alive, there are only a few nodes at interpose between de Bruijn node and the imaginary node; consequently, the path length of taking successors takes a reasonable percentage of the whole path length. However, Koorde's lookup efficiency is reduced in sparse network. We will discuss this in Section 9.4.3.3.

The principle of Cycloid routing algorithm is almost the same as that of Koorde. In both algorithms, starting from a specific chosen node, the node ID bits are changed one by one until the target node ID is reached. Both of their path lengths are close to d, the dimension of the network in simulation. Since a d-dimensional Cycloid contains more $(d-1) \cdot 2^{\mathrm{d}}$ nodes than Koorde of the same dimension, Cycloid leads to shorter lookup path length in networks of the same size, as shown in Figure 9.5.

From Figure 9.5, we can also see that the path length of Viceroy increases faster than the dimension $\log n$. Its path length increases from 4 in a four-dimensional network to 12.91 in an eight-dimensional network. This means the more nodes a Viceroy network has, the less the key location efficiency.

9.4.2 Load Distribution

A challenge in the design of balanced DHTs is to distribute keys evenly between a changing set of nodes and to ensure each node experiences even load as an intermediate for lookup requests from other nodes. Cycloid deals with the key distribution problem in a similar way to Koorde, except that Cycloid uses a pair of cyclic and cubical indices to represent a node. Viceroy maintains a one-dimensional ID space. Although both Cycloid and Viceroy nodes have two indices to represent their places in the overlay network, the cyclic index is part of the Cycloid node ID but the level is not part of the Viceroy node ID. Also, Viceroy stores keys in the keys' successors. Table 9.3 shows the differences in key assignments between the DHTs.

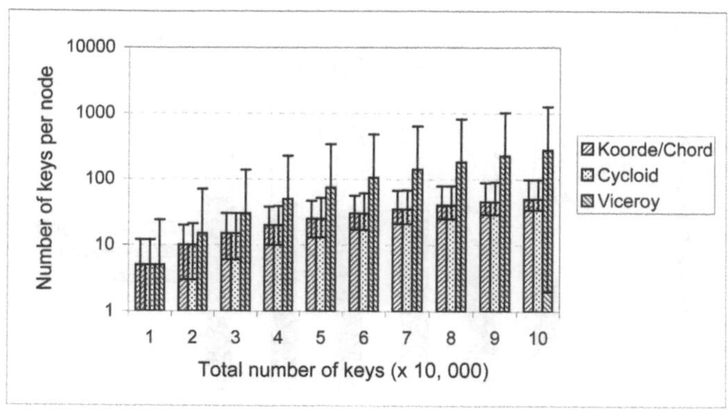

Figure 9.7: Key distribution in networks of 2000 nodes due to different DHTs. Assume the network ID space is of 2048 nodes.

Table 9.3: Node ID space and key assignment in different DHTs.

	Cycloid	Viceroy	Koorde
Base network	CCC	Butterfly	de Bruijn
ID space	$([0, d), [0, d \cdot 2^{\mathrm{d}}))$	$([0, 3\log n), [0, 1))$	$[0, 2^{\mathrm{d}})$
Node identity	$(k, a_{d-1}a_{d-2}\ldots a_0)$ k is static	(level, ID) level is dynamic	ID
Key placement	Numerically closest node	Successor	Successor

In this experiment, we simulated different DHT networks of 2000 nodes each. We varied the total number of keys to be distributed from 10^4 to 10^5 in increments of 10^4. Figure 9.7 plots the mean, the 1st and 99th percentiles of the number of assigned keys per node when the network ID space is of 2048 nodes. The number of keys per node exhibits variations that increase linearly with the number of keys in all DHTs. The key distribution in Cycloid has almost the same degree of load balance as in Koorde and Chord because Cycloid's two-dimensional ID space is reduced to one dimension by the use of a pair of modula and divide operations. By comparison, the number of keys per node in Viceroy has much larger variations. Its poorly balanced distribution is mainly due to the large span of real number ID space in $[0, 1)$. In Viceroy, the key is stored in its successor; that is, a node manages all key-value between its counterclockwise neighbor and itself. Because of Viceroy's large ID span, its node identifiers may not uniformly cover the entire space, and some nodes may manage more keys than the others.

Figure 9.8 plots the mean, the 1st and 99th percentiles of the number of keys per node in Cycloid and Koorde DHTs when there are only 1000 participants in the network. It can be seen that Cycloid leads to a more balanced key distribution for a

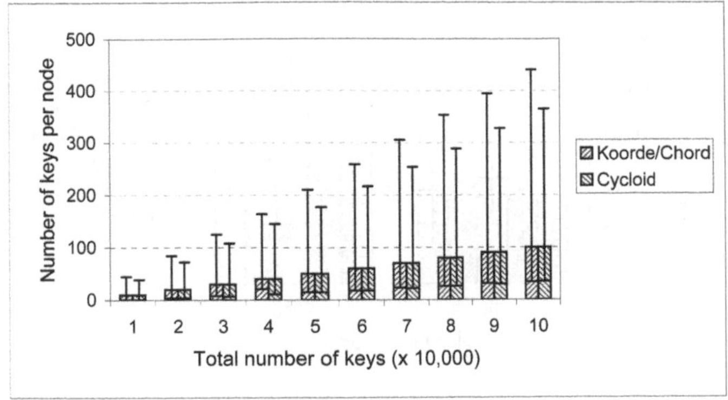

Figure 9.8: Key distribution in networks of 1000 nodes due to different DHTs. Assume the network ID space is of 2048 nodes.

sparse network than Koorde. In Koorde, the node identifiers do not uniformly cover the entire identifier space, leading to unbalanced key allocation as a result of storing the key in its successor. By comparison, using the two-dimensional key allocation method, Cycloid achieves better load balance by storing the key in its numerically closest node; that is, the keys between a node's counter-clockwise neighbor and itself will be allocated to that neighbor or the node itself rather than to itself totally. Chord solved this problem by replicating each node into $O(\log n)$ "virtual nodes," but such replication would destroy the optimality of constant degree in Koorde. In [166], Kaashoek and Karger put forward a question of finding a system that is both degree optimal and load balanced. Cycloid should be an answer.

In summary, when the entire identifier space is mostly occupied, Cycloid's load balance is as good as Chord. When the actual nodes only occupy a small part of the total entire identifier space, Cycloid's load balance is better than Chord.

Key distribution aside, another objective of load balancing is to balance the query load between the participating nodes. The query load is measured as the number of queries received by a node for lookup requests from different nodes.

Figure 9.9 plots the the mean, the 1st and 99th percentiles of query loads of various DHT networks of 64 nodes and 2048 nodes. It is shown that Cycloid exhibits the smallest variation of the query load, in comparison with other constant-degree DHTs. This is partly due to the symmetric routing algorithm of Cycloid.

In Viceroy, the ascending phase consists of a climb using up connections until level 1 is reached and the descending phase routes down the levels of tree using the downlinks until no downlinks remain. As a result, the nodes in the higher levels will be the hot spots; on the other hand, the nodes of the lower levels have smaller workload, which leads to the great workload variation, especially in the large-scale network. In Koorde, the first de Bruijn of a node with ID m is the node immediately precedes $2m$. So, all the first de Bruijn node identifiers are even in a "complete"

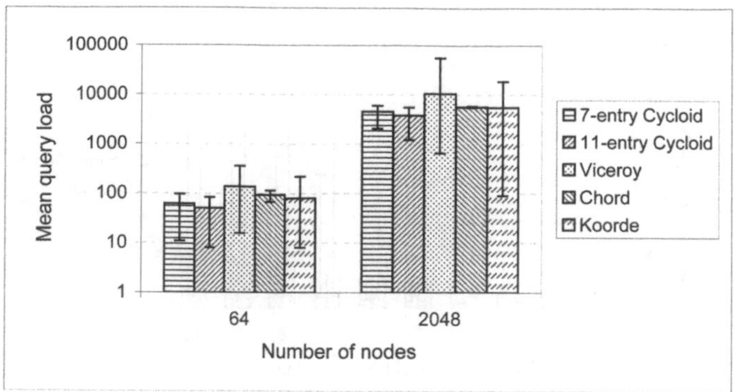

Figure 9.9: Query load variances in various DHTs of different sizes.

(dense) network and with high probability the IDs are even in an incomplete (sparse) network. Consequently, the nodes with even IDs have heavy workload while nodes with odd IDs have light workload according to the lookup algorithm of Koorde. In Cycloid, because of the leaf sets, the nodes with small cyclic index, typically 0 or 1, will be light loaded. However, these nodes constitute only small part of the Cycloid network, in comparison with the hot spots in Viceroy and Koorde. Cycloid performs better than Viceroy and Koorde in the aspect of congestion.

9.4.3 Network Resilience

In this section, we evaluate the Cycloid resilience to node failures and departures. We use the term of departure to refer to both failure and departure. We assume that node departures are graceful; that is, a node informs its relatives before its departure.

9.4.3.1 Impact of Massive Node Failures and Departures

In this experiment, we evaluate the the impact of massive node failures and/or departures on the performance of various DHTs, and on their capability to performing correct lookups without stabilization. Once the network becomes stable, each node is made to fail with probability p ranging from 0.1 to 0.5. After a failure occurs, we performed 10,000 lookups with random sources and destinations. We recorded the number of time-outs occurred in each lookup, the lookup path length, and whether the lookup found the key's correct storing node. A time-out occurs when a node tries to contact a departed node. The number of time-outs experienced by a lookup is equal to the number of departed nodes encountered. Figure 9.10 shows the mean path length of the lookups with the change of departure probability p in different DHTs of size 2048 nodes. The mean, the 1st and 99th percentiles of the number of time-outs of each DHTs are presented in Table 9.4.

In Cycloid, the path length increases due to the increasing of the number of time-

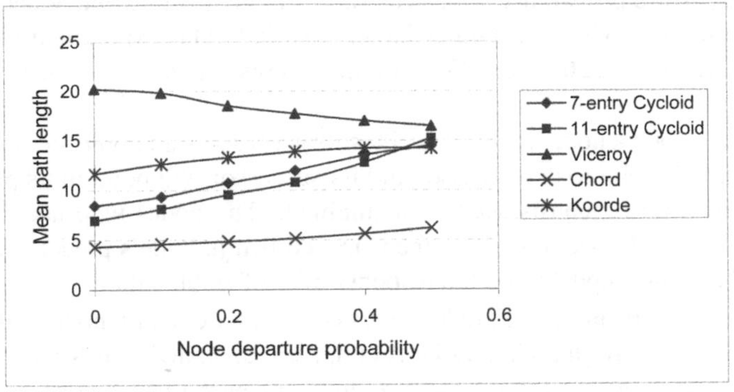

Figure 9.10: Path lengths of different DHTs as more nodes depart/fail.

Table 9.4: Number of time-outs in different DHTs as more nodes depart/fail.

Departure probability	Cycloid	Viceroy	Chord	Koorde
0.1	0.53 (0, 4)	0 (0, 0)	0.62 (0, 6)	0.02 (0, 1)
0.2	1.24 (0, 8)	0 (0, 0)	1.37 (0, 8)	0.04 (0, 1)
0.3	2.46 (0, 11)	0 (0, 0)	2.38 (0, 11)	0.06 (0, 2)
0.4	4.09 (0, 17)	0 (0, 0)	3.91 (0, 16)	0.08 (0, 3)
0.5	5.88 (0, 24)	0 (0, 0)	6.53 (0, 26)	0.09 (0, 4)

outs as the p increases. Recall that when a departed node is met, the leaf sets have to be turned to for the next node. Therefore, the path length increases. All lookups were successfully resolved meaning that the Cycloid is robust and reliable.

We can see from the Figure 9.10 that unlike Cycloid and Koorde, the lookup path length in Viceroy decreases with the increase of p. In Viceroy, a node has both outgoing and incoming connections. A node notifies its outgoing and incoming connections before its departure. Therefore, all related nodes are updated before the node departs. Based on this characteristic, a massive departure has no adverse effect on Viceroy's ability to perform correct lookups. Table 9.4 shows that Viceroy has no time-outs. The decrease of the path length is caused by the decrease of the network size. We can see from Figure 9.10 that when the departure probability is 0.5, the path length is 16.45, which is very close to the 1024 nodes complete network lookup's average path length 16.92 in Figure 9.5.

In order to eliminate the impact of simultaneous node departures in Viceroy, a leaving node would induce $O(\log n)$ hops and require $O(1)$ nodes to change their states. This causes a large amount of overhead. In Cycloid, the path length increased a little with a small fraction of departed nodes. Even though the path length of Cycloid increases slightly, it is still much less than that of Viceroy.

In Figure 9.10, Koorde's path length increased not so much as in Cycloid when the node departure probability p exceeds 0.3. Unlike Cycloid and Viceroy, Koorde has lookup failures when p becomes larger than 0.3. Our experiment results show that there are 791, 1226, and 4259 lookup failures when p = 0.3, 0.4, and 0.5, respectively.

In Koorde, when a node leaves, it notifies its successors and predecessor. Then, its predecessor will point to its successor and its successor will point to its predecessor. By this way, the ring consistency is maintained. The nodes who take the leaving node as their first de Bruijn node or their first de Bruijn node's predecessor will not be notified and their updates are the responsibility of stabilization.

Each Koorde node has three predecessors of its first de Bruijn node as its backups. When the first de Bruijn node and its backups are all failed, the Koorde node fails to find the next node and the lookup is failed. When the failed node percentage is as low as 0.2, all the queries can be solved successfully at a marginal cost of query length with increase path length as shown in Figure 9.10. When p exceeds 0.3, with increasing of time-outs as shown in Table 9.4, the number of failure increases, and the path length increases not so much as before because fewer backups are taken.

From Table 9.4, we can see that although Koorde has much fewer time-outs than Cycloid, it still has a large number of failures. In Koorde, the critical node in routing is the de Bruijn node whose backups cannot always be updated. In contrast, Cycloid relies on updated leaf sets of each node for backup. Therefore, Koorde is not as robust as Cycloid in response to massive node failures/departures. The experiment shows that Cycloid is efficient in handling massive node failures/departures without stabilization.

9.4.3.2 Lookup Efficiency During Node Joining and Leaving

In practice, the network needs to deal with nodes joining the system and with nodes that leave voluntarily. In this work, we assume that multiple join and leave operations do not overlap. We refer the reader to [208] for techniques to achieve concurrency and to handle failures in the system. In this experiment, we compare Cycloid with Viceroy and Koorde when nodes join and leave continuously.

The setting of this experiment is exactly the same as the one in [301]. Key lookups are generated according to a Poisson process at a rate of one per second. Joins and voluntary leaves are modeled by a Poisson process with a mean rate of R, which ranges from 0.05 to 0.40. A rate of R = 0.05 corresponds to one node joining and leaving every 20 seconds on average. In Cycloid, each node invokes the stabilization protocol once every 30 seconds and each node's stabilization routine is at intervals that are uniformly distributed in the 30 seconds interval. Thus, R ranges from a rate of 1.5 joins and 1.5 leaves per one stabilization period to a rate of 12 joins and 12 leaves per one stabilization period. The network starts with 2048 nodes.

Figure 9.11 shows the mean path length of lookup operations in different DHTs as the node join/leave rate R changes. The mean, the 1st and 99th percentiles of the number of time-outs are shown in Table 9.5. There are no failures in all test cases. From the path-length evaluation in Section 9.4.1, we know that the mean path length of Cycloid in steady states is 8.38. From Figure 9.11, we can see that the measured

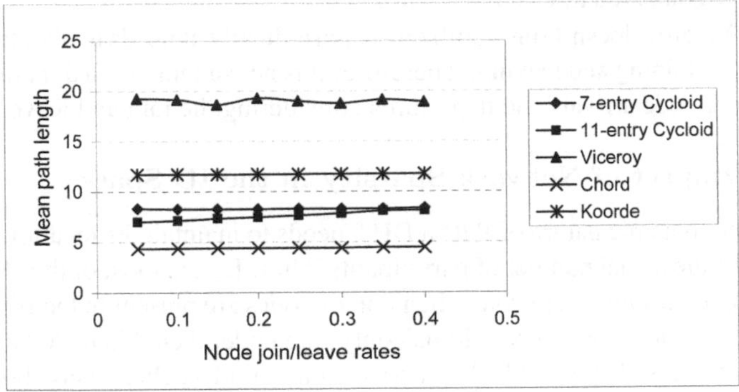

Figure 9.11: Path lengths of lookup requests in different DHTs as the node join/leave rates change.

Table 9.5: Number of time-outs as the node join/leave rate changes.

Node join/leave rate	Cycloid	Viceroy	Chord	Koorde
0.05/1.5	0.005 (0, 0)	0 (0, 0)	0.033 (0, 1)	0.003 (0, 0)
0.10/3.0	0.009 (0, 0)	0 (0, 0)	0.078 (0, 2)	0.013 (0, 1)
0.15/4.5	0.014 (0, 1)	0 (0, 0)	0.130 (0, 2)	0.008 (0, 0)
0.20/6.0	0.031 (0, 1)	0 (0, 0)	0.125 (0, 2)	0.013 (0, 1)
0.25/7.5	0.047 (0, 2)	0 (0, 0)	0.151 (0, 2)	0.016 (0, 1)
0.30/9.0	0.052 (0, 2)	0 (0, 0)	0.191 (0, 3)	0.016 (0, 1)
0.35/10.5	0.058 (0, 2)	0 (0, 0)	0.220 (0, 3)	0.023 (0, 1)
0.40/12.0	0.070 (0, 2)	0 (0, 0)	0.233 (0, 3)	0.023 (0, 1)

path lengths in the presence of node joining and/or leaving are very close to this value and do not change with the rate R. This is because with the help of stabilization, there are less needs for a node to turn to its leaf sets in the case of meeting an absent or departure node. Consequently, a lookup request would experience fewer time-outs and its path length remains unchanged. Compared with the time-out results in Table 9.4, we can see that stabilization removes the majority of the time-outs.

In Koorde, the path lengths changed little compared to 11.59 in a stable network though the time-outs increase with the rate of node joins and leaves. The failure time is reduced to zero compared to the large failure time in Section 9.4.3.1. That's because stabilization updates the first de Bruijn node of each node and the de Bruijn node's predecessors in time. When the first de Bruijn node and its predecessors are all failed, then passed lookups fail with high probability.

The results show that Viceroy's performance is not affected by the node leaving and joining. That's because, before a node leaves and after a node joins, all the related nodes are updated. Although Viceroy has no time-outs, its path length is much longer compared to Cycloid's path.

Though Viceroy doesn't run stabilization periodically, it needs to update all related node for node joining and leaving. Therefore, it is not suitable for a dynamic network with frequent node arrivals and departures considering the join and leave high cost.

9.4.3.3 Impact of Network Sparsity in the ID Space

Due to the dynamic nature of P2P, a DHT needs to maintain its location efficiency, regardless of the actual number of participants it has. But, in most of the DHTs, some node routing table entries are void when not all nodes are present in the ID space. For example, if a local cycle in Cycloid has only one node, then this node has no inside leaf set nodes. It is also possible that a node cannot find a cubical neighbor, or cyclic neighbor. We define the degree of sparsity as the percentage of nonexistent nodes relative to the network size. To examine the impact of sparsity on the performance of other systems, we did an experiment to measure the mean search path length and the number of failures when a certain percentage of nodes is not present.

We tested a total of 10,000 lookups in different DHT networks with an ID space of 2048 nodes. Figure 9.12 shows the results as the degree of network sparsity changes. There are no lookup failures in each test case. We can also see that Cycloid keeps its location efficiency and the mean path length decreases slightly with the decrease of network size. In Viceroy, it's impossible for nodes to fully occupy its ID space because the node ID $\in [0, 1)$. Therefore, it is very likely that some links of a node are void and hence the sparsity imposes no effect on the lookup efficiency. In Koorde, the path length increases with the actual number of participant drops. This is because a sparse Koorde DHT exhibits a large span between two neighboring nodes. Since Koorde routes a lookup request through the immediate real predecessor of each imaginary node on the de Bruijn path, the distance between the imagination node and its immediate predecessor in the sparse DHT leads to a longer distance between the predecessor's first de Bruijn node and the imagination node in the next step. Therefore, more successors need to be taken to reach the immediate predecessor of the imagination, thus larger path length.

9.5 Concluding Remarks

In this chapter, we have introduced a generic topological model to capture the essence of structured DHTs. This model facilitates the design of a new constant-degree DHT, namely, Cycloid. Cycloid is based on Pastry and CCC. Cycloid resembles Viceroy and Koorde in appearance because CCC is a subgraph of butterfly network and de Bruijn is a coset graph of butterfly, as recently proved in graph theories [9, 106]. But they are different in connectivity maintenance of dynamic participants and in routing for key location.

We have evaluated the Cycloid performance in terms of the lookup hop count,

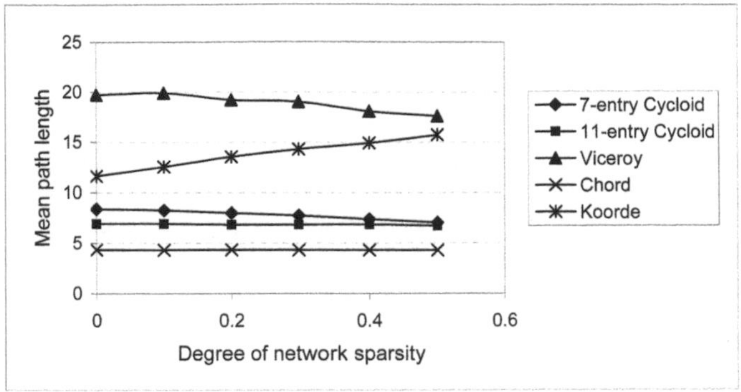

Figure 9.12: Path lengths of lookup requests in different DHTs with the change of the degree of network sparsity.

degree of load balance, degree of fault tolerance, and cost for maintenance, in comparison with Viceroy and Koorde. Experiment results show that Cycloid yields the best average-case location efficiency. Cycloid distributes keys and query load more evenly between the participants than Viceroy. In comparison with Koorde, Cycloid results in higher degrees of load balance for sparse networks and the same degree of load balance for dense networks. Cycloid is more robust because it continues to function correctly when a node's information is only partially correct. Cycloid scales well with the number of nodes, recovers from large numbers of simultaneous node departures and joins, and answers lookups correctly even during recovery.

A common problem with constant-degree DHTs is their weakness in handling node leaving without warning in advance. Keeping more information like a successor list in Chord and Bucket in Viceroy helps resolve the problem, but destroys the optimality of constant degree. Because of this disadvantage, whenever a node joins or leaves, Cycloid needs to notify its inside leaf set. Especially, if the joining or leaving node is the primary node of a cycle in Cycloid, the updates might produce much more overhead. In addition, the initialization and updates of three neighbors in the routing table also might cause a lot of overhead. These issues need to be addressed for an refinement of Cycloid.

Like all other DHTs, Cycloid has an aftermath load balancing problem because a consistent hashing function produces a bound of $O(\log n)$ imbalance of keys between nodes. In [283], we presented a locality-aware randomized load balancing algorithm for the Cycloid network. The algorithm balances the workload of nodes by utilizing the physical proximity information embedded in Cycloid's topological properties. It also features a factor of randomness to deal with the load balancing requirement in P2P networks with churn.

Chapter 10

Semantic Prefetching of Web Contents

Prefetching is an important technique to reduce the average Web access latency. Existing prefetching methods are based mostly on Web link graphs. They use the graphical nature of HTTP links to determine the possible paths through a hypertext system. Although the URL graph-based approaches are effective in prefetching of frequently accessed documents, few of them can prefetch those Web pages that are rarely visited. This chapter presents a semantic prefetching approach to overcome the limitation. It predicts future requests based on semantic preferences of past retrieved Web documents. It is applied to Internet news services.

10.1 Introduction

Bandwidth demands for the Internet are growing rapidly with the increasing popularity of Web services. Although high-speed network upgrading has never stopped, clients are experiencing Web access delays more often than ever. There is a great demand for latency tolerance technologies for fast delivery of media-rich Web contents. Such technologies become even more important for clients who connect to the Internet via low-bandwidth modem or wireless links.

Many latency tolerance techniques have been developed over the years. The two most important ones are caching and prefetching. Web traffic characterization studies have shown that Web access patterns from the perspective of a server often exhibit strong temporal locality. By taking advantage of locality, caching and related content delivery networks are widely used in server and proxy sides to alleviate Web access latency. In the Web client side, caching is often enabled by default in both popular Internet Explorer and Mozilla Firefox. However, recent studies showed that benefits from client-side caching were limited due to the lack of sufficient degrees of temporal locality in the Web references of individual clients [179]. The potential for caching requested files was even declining over the years [6, 24]. As a complement to caching, prefetching technique is to prefetch Web documents (more generally, Web objects) that tend to be accessed in the near future while a client is processing previously retrieved documents. Prefetching can also happen off-line when the client is not surfing the Web at all as long as the client machine is on-line. A recent tracing experiment revealed that prefetching could offer more than twice the improvement due to caching alone [179].

Previous studies on Web prefetching were based mostly on the history of client access patterns [73, 102, 200, 237, 274, 275]. If the history information shows an access pattern of URL address A followed by B with a high probability, then B is to be prefetched once A is accessed. The access pattern is often represented by a URL graph, where each edge between a pair of vertices represents a follow-up relationship between the two corresponding URLs. The weight on the edge from A to B denotes the probability that Web object B will be requested immediately after object A. To make a prefetching prediction, a search algorithm traverses the graph to find the most heavily weighted edges and paths. The URL graph-based approaches are demonstrated effective in prefetching of Web objects that are often accessed in history. However, there is no way for the approaches to prefetch objects that are newly created or never visited before. For example, all anchored URLs of a page are fresh when a client visits a new Web site and none of them is to be prefetched by any URL graph based approach. For dynamic HTML sites where a URL has time-variant contents, few existing approaches can help prefetch desired documents.

In [336, 337], we proposed a semantic prefetching approach to overcome the limitations. In the approach future requests are predicted based on semantic preferences of past retrieved objects, rather than on the temporal relationships between Web objects. It is observed that client surfing is often guided by some keywords in anchor text of Web objects. Anchor text refers to the text that surrounds hyperlink definitions (hrefs) in Web pages. For example, a client with strong interest in soccer may not want to miss any relevant news with a keyword of "World Cup" in anchor texts. An on-line shopper with a favorite auto model would like to see all consumer reports or reviews about the model. We refer to this phenomenon as "semantic locality." Unlike search engines that take keywords as input and responds with a list of semantically related Web documents, semantic prefetching techniques tend to capture the client surfing interest from his/her past access patterns and predict future preferences from a list of possible objects when a new Web site is visited. Based on semantic preferences, this approach is capable of prefetching articles whose URLs have never been accessed. For example, the approach can help soccer fans prefetch World Cup related articles when they enter a new Web site.

Notice that Web HTML format was designed merely for document presentation. A challenge is to automatically extract semantic knowledge of HTML documents and construct adaptive semantic nets between Web documents on-line. Semantics extraction is a key to Web search engines, as well. Our experience with today's search engines leads us to believe that current semantics extraction technology is far away from maturity for a general-purpose semantic prefetching system. Instead, we demonstrated the idea of semantic prefetching in a special application domain of Internet news services. We implemented a keyword-based semantic prefetching approach in a client-side prefetching prototype system, namely, `NewsAgent` [336, 337]. It was coupled with a client-side browser, working on behalf of the client. It monitored the client Web surfing activities and established keyword-based semantic preferences in different news categories. It employed a neural network model over the keyword set to predict future requests based on the preferences. Such a domain-specific prefetching system is not necessary to be running all the time; it could be turned on or off on demand like many other advanced features in popular browsers. This chapter presents the details of the semantic prefetching approach.

Note that semantics-based prefetching services are getting momentum in industry nowadays. For Google (`google.com`), gmail is all about connecting ads to eyeballs. When users log in on gmail, Google knows who they are and what they are interested in, due to their email histories. Google can even build histories of their interests based on the ads they click on. Overtime, Google will show them increasingly relevant ads to them as they read their e-mails. These ads may well be the most personally targeted ads they've ever seen.

10.2 Personalized Semantic Prefetching

10.2.1 Architecture

Figure 10.1 shows the architecture of a generic client-side semantic prefetching scheme. It assumes that a personalized information agent (or proxy) is running behind Web browser and keeping watch on client surfing activities. This agent analyzes the documents that its client is browsing with an attempt to understand their semantic knowledge and find out the semantic relationships between the documents (step 1). The semantic information is represented in an open self-learning capable model that accumulates knowledge about the client preferences (step 2). Based on the knowledge, the agent makes predictions about client future requests (step 3) and "steals" network bandwidth to prefetch related Web documents from the Internet (step 4). The prefetched documents are temporarily stored in the agent's internal cache (step 5). This prefetching agent also serves as a client-side Web proxy and provides its client with requested Web documents transparently from its internal cache, if they are available in the cache (step 6). Otherwise, it fetches the documents from the Internet. Expectedly, the cache would contain some prefetched documents that are never accessed by the client. This information will be used as feedback to improve prediction accuracy.

This personalized semantic prefetching scheme raises a number of key implementation issues. First is Web document semantics extraction. Web documents are often posted in an HTML presentation format. The problem of extracting semantics of general HTML documents on its own is a challenge. It is fundamental in a broader context of information retrieval on the Internet. Although much effort has been devoted to this issue in both academia and industry over the years, recent studies showed that users remained disappointed with performance of today's search engines, partly due to certain misunderstanding of document meanings [175].

Automatic semantics extraction for general HTML documents is beyond the scope of this chapter. In this study, we present a domain-specific semantic prefetching ap-

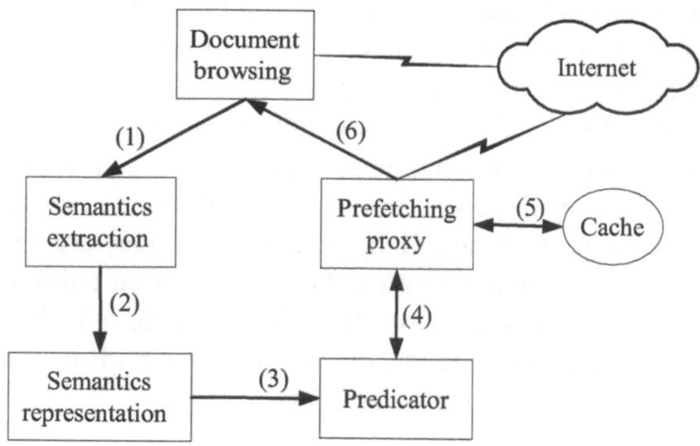

Figure 10.1: Architecture of personalized semantic prefetching scheme.

proach in Internet news services. We are interested in Internet news services because of the following reasons. First, news headlines are often used as their URL anchors and article titles in most news Web sites. The topic and content of a news article can be captured clearly by a group of keywords in its headline. Second, news service sites like `abcnews.com`, `cnn.com`, and `msnbc.com` often have extensive coverage of the same topics during a certain period. Therefore, articles of interest in different sites can be used to cross-examine the performance of the semantic prefetching approach. That is, one site is used to accumulate URL semantic preferences and the others for evaluation of its prediction accuracy. Finally, effects of prefetching in the news services can be isolated from caching and evaluated separately. Prefetching takes effect in the presence of a cache. Since a client rarely accesses the same news article more than once, there is very little temporal locality in news surfing patterns. This leads to a minimal impact of caching on the overall performance of prefetching. We note that a client who is reading news on a specific topic may need to return to its main page repeatedly for other articles of interest. These repeated visits exhibit certain degrees of temporal locality. However, the caching impact due to this type of locality can be eliminated by the separation of a prefetching cache from the browser's built-in cache. The prefetching cache is dedicated to prefetched documents, while the browser cache exploits the temporal locality between repeated visits of the main pages.

10.2.2 Neural Network-Based Semantics Model

Semantics extraction aside, another related implementation issue is to develop a semantics representation model that has the capability of self-learning. There are many approaches available for this issue. An example is neural networks. The neural network approach has been around since the late 1950s and come into practical use for general applications since mid-1980s. Due to their resilience against distortions

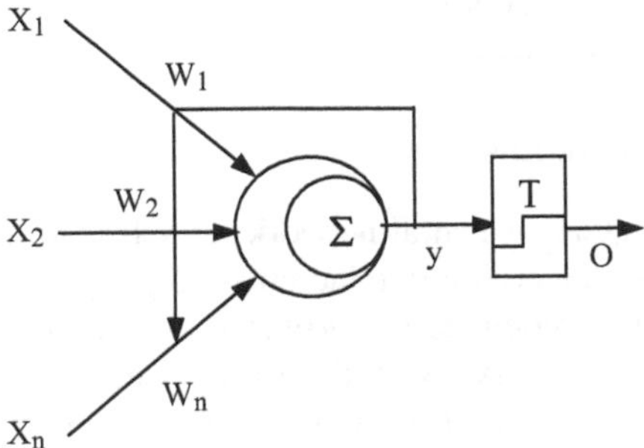

Figure 10.2: A graphical representation of single perceptron.

in the input data and their capability of learning, neural networks are often good at solving problems that do not have algorithmic solutions [151]. In this subsection, we present neural network basics, with an emphasis on prediction accuracy critical parameters and self-learning rules. Details of its application in semantic prefetching in Internet news services will be shown in Section 10.3.

10.2.2.1 Perceptron

A neural network comprises a collection of artificial neurons. Each neuron receives a set of inputs and performs a simple function: computing a weighted sum of the inputs and then performing an activation function for decisions. The activation function can be a threshold operation or a smooth function like hyperbolic operations, depending on the neuron objectives. In order of support predictions in the context of Web document prefetching, we assume a threshold function for the neurons. If the weighted sum of a neuron is less than the predefined threshold, then it yields an output of zero, otherwise one. Specifically, given a set of inputs $x_1, x_2, \ldots, x_n$ associated with weights $w_1, w_2, \ldots, w_n$, respectively, and a net threshold T, the binary output o of a neuron can be modeled as a hard limiting activation function:

$$o = \begin{cases} 1 & \text{if } y = \sum_{i=1}^{n} w_i x_i \geq T \\ 0 & \text{otherwise.} \end{cases} \tag{10.1}$$

The weighted sum y in (10.1) is often referred to as net of the neuron. Figure 10.2 shows a graphical representation of a single perceptron (i.e., neural network with a single neuron). Although it is an oversimplified model of the functions performed by a real biological neuron, it has been shown to be a useful approximation. Neurons can be interconnected together to form powerful networks that are capable of solving problems of various levels of complexity.

In this chapter, we assume single perceptron for a news-prefetching agent. The agent examines the URL anchor texts of a document and identifies a set of keywords for each URL. Using the keywords as inputs, the agent calculates the net output of each URL perceptron. The input x_i, $(1 \leq i \leq n)$ is set to 1 when its corresponding keyword is present in the anchor text, 0 otherwise. Their nets determine the priority of URLs that are to be prefetched.

10.2.2.2 Learning Rules

A distinguishing feature of neural networks is self-learning. It is accomplished through learning rules. A learning rule incrementally adjusts the weights of the neuron inputs to improve a predefined performance measure over time. There are two types of learning rules: supervised and unsupervised. Supervised learning algorithms assume the existence of a supervisor who classifies the training patterns into classes, whereas unsupervised learning does not. Supervised learning algorithms utilize the information about class membership of each training pattern as feedback to the perceptron to reduce future pattern misclassification.

In this study, we deploy supervised learning rules because the client of a prefetching agent classifies each input pattern into desired and undesired categories. By comparing its outputs with the client actual requests in the training phase, the neuron can adjust the input weights accordingly. Usually, the input weights are synthesized gradually and updated at each step of the learning process so that the error between the output and the corresponding desired target is reduced [151].

Many supervised learning rules are available for the training of a single perceptron. A most widely used approach is μ-LMS (least mean square) learning rule. Denote w the vector of the input weights. Denote d_i the desired (target) output associated with an input pattern and y_i the actual output of the input pattern. Let

$$J(w) = \frac{1}{2}\sum_{i=1}^{n}(d_i - y_i)^2, \tag{10.2}$$

be the sum of squared error criterion function. Using the steepest gradient descent to minimize $J(w)$ gives

$$w^{k+1} = w^k + \mu\sum_{i=1}^{n}(d_i - y_i)x_i, \tag{10.3}$$

where μ is often referred to as the learning rate of the net because it determines how fast the neuron can react to the new inputs. Its value normally ranges between 0 and 1. The learning rule governed by (10.3) is called batch LMS rule. The μ-LMS is its incremental version, governed by

$$w^{k+1} = \begin{cases} 0 \text{ or arbitrary} & \text{if k=0} \\ w^k + \mu(d_i - y_i)x_k & \text{otherwise.} \end{cases} \tag{10.4}$$

In this study, we assume the prefetching agent deploys the μ-LMS learning rule to update the input weights of its perceptron.

10.3 NewsAgent: A News Prefetching System

This section presents details of the keyword-based prefetching technique and architecture of a domain-specific news prefetching predictor: `NewsAgent`.

10.3.1 Keywords

A keyword is the most meaningful word within the anchor text of a URL. It is usually a noun referring to some role of an affair or some object of an event. For example, associated with the news "Microsoft is ongoing talks with DOJ" on MSNBC are two keywords "Microsoft" and "DOJ"; the news "Intel speeds up Celeron strategy" contains keywords "Intel" and "Celeron." Note that the predictor identifies the keywords from anchor texts of URLs while users retrieve documents. Since keywords tend to be nouns and often start with capital letters, these simplify the task of keyword identification.

There are two fundamental issues with keyword-based information retrieval: synonymy and polysemy. Synonymy means that there are many ways of referring to the same concept and polysemy refers to the fact that a keyword may have more than one meaning in different contexts. In a broader context of information retrieval, there exist many techniques to address these two issues; see [175] for a comprehensive review. Most notably is latent semantic indexing (LSI) [268]. It takes advantage of the implicit higher order structure of the association of terms with articles to create a multidimensional semantic structure of the information. Through the pattern of co-occurrences of words, LSI is able to infer the structure of relationships between article and words. Due to its high time complexity in inference, its application to real-time Web prefetching requires a trade-off between its time cost and prediction accuracy. For conservativeness in prefetching, we assume a policy of exact keyword matching in the prediction of client future requests. We alleviate the impact of polysemy by taking advantage of the categorical information in news services.

It is known that meaning of a keyword is often determined by its context. An interesting keyword in one context may be of no interest at all to the same client if it occurs in other contexts. For example, a Brazil soccer fan may be interested in all the sports news related with the team. In this sports context, the word "Brazil" will guide the fan to all related news. However, the same word may be of less importance to the same client if it occurs in general or world news categories, unless he/she has a special tie with Brazil. To deal with this polysemy issue in prefetching, Web documents must be categorized into predefined categories. Automated text categorization is a nontrivial task in any sense. It dates back to the early 1960s and has become an active research topic with the popularity of Web services; a comprehensive survey of state-of-the-art techniques can be found in [277]. We note that the existing categorization algorithms are time costly and supposed to be run at the server side. There are few lightweight algorithms that are suitable for on-line document classification at the client side. Fortunately, nearly all news services organize articles into categories (e.g., business, sports, and technology). This server-provided context (categorical) information can be used with keywords to alleviate the impact of keyword polysemy in decision making and hence captures more semantic knowledge than conventional term-document literal matching retrieval methods. A simply strategy is to deploy multiple independent category-specific `NewsAgents`. Since this categorical information is deterministic, different `NewsAgents` can be invoked to analyze the URL links from different categories.

Finally, we note that the significance of a keyword to a client changes with time. For example, news related to "Brazil" hit the headline during the Brazil economic crisis period in 1998. The keyword should be sensitive to any readers of market or financial news during the period. After the crisis was over, its significance decreased unless the reader has special interests in the Brazil market. Due to the self-learning capability of a neural network, the keyword weights can be adjusted to reflect the change.

10.3.2 NewsAgent Architecture

Figure 10.3 shows the architecture of `NewsAgent`. Its kernel is a prediction unit. It monitors the way of its client browsing activities. For each required document, the predictor examines the keywords in its anchor text, calculates their preferences, and prefetches those URL links that contain keywords with higher preferences than the perceptron threshold. Specifically, the system has the following the work-flow: (1) user requests a URL associated with a category; (2) the URL document is retrieved and displayed to the user; (3) the contents of the document are examined to identify the first set of links to evaluate; (4) each link from the set is processed to extract the keywords; (5) keywords are matched to the keyword list; they are added to the list if they occur for the first time; (6) the corresponding preferences are computed using the weights associated with the keywords; (7) links with preferences (i.e., nets) higher than the neuron threshold are sorted and placed in the prefetching queue; (8) links are accessed one by one and the objects are placed in the cache; (9) if the prefetching queue is empty, process the next set of links (step 3); (10) when the user requests a link, the cache list is examined to see if it's there; (11) if it is there, then it is displayed; (12) if not, it is retrieved from the Web and the corresponding keyword's weights are updated.

In most cases, a Web page contains a large number of hyperlinks. The `NewsAgent` examines the URL links by groups. We assume each group contains a fixed number of links and refer to the group size as prefetching breadth. Evaluation of the embedded links by groups is patterned after client browsing behaviors because a client tends to look at a few hyperlinks before he/she decides which to follow. Recall that prefetching needs to be carried out during the time when the client is viewing a page. A recent analysis of client access patterns showed that the average rate that http requests were made was about two requests per minute [24]. In a very short time, the `NewsAgent` may not be able to complete the evaluation of all embedded links and the prefetch of those with high preferences in the current page. The group evaluation strategy provides a quick response to client Web surfing and possibly increases the

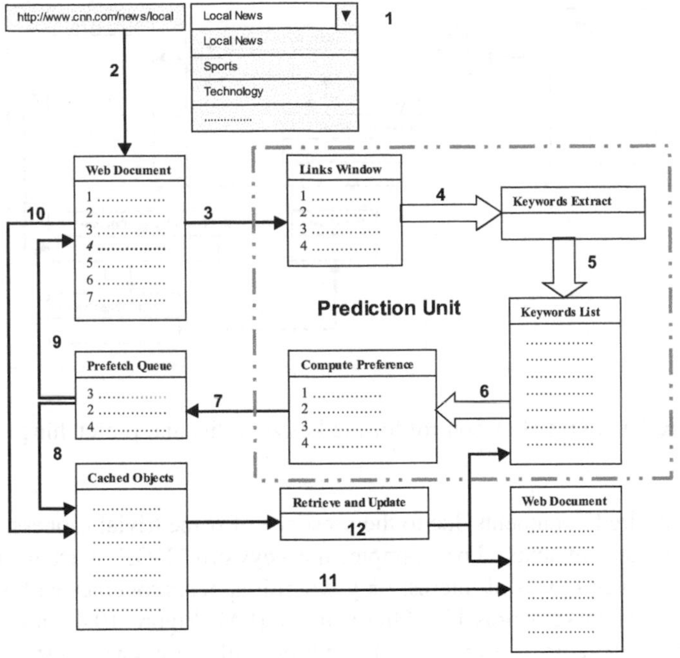

Figure 10.3: Architecture of the NewsAgent.

hit ratio due to prefetching. For example, consider a page containing a large number of URLs. Assume the last URLs in the list have nets of more than 90% and a URL in the beginning has a net of more than 80%. The `NewsAgent` based on the group evaluation strategy will prefetch the moderately preferred link first to meet its client's possible immediate needs. The highest preferred links will be prefetched as the client browsing moves forward.

On receiving a request for an anchored URL, the `NewsAgent` starts to look for the requested pages in the cache. If it has already been available due to prefetching, the page is provided for display. (The weights of its related keywords will remain unchanged by the rule of (10.4)). Otherwise, the `NewsAgent` suspends any ongoing prefetching processes and starts to examine the anchor texts of the requested link for keywords. If a keyword is introduced for the first time, a new weight will be assigned to it. Otherwise its existing weight will be positively updated ($d = 1$) by the rule of (10.4).

10.3.3 Control of Prefetching Cache and Keyword List

Notice that the keyword list could be full with unwanted items when the client loses interest in a long-standing topic. The `NewsAgent` internal cache may also

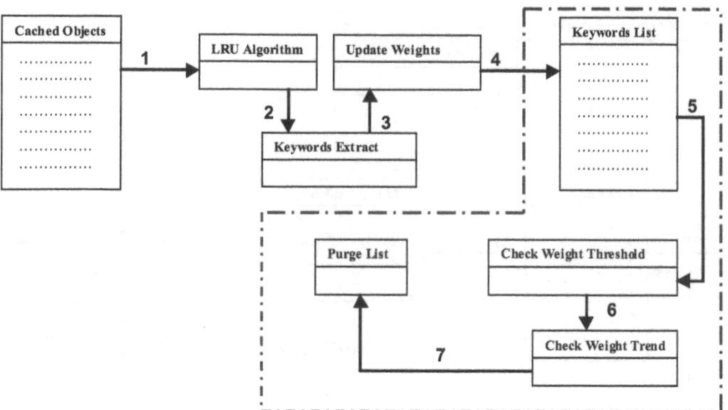

Figure 10.4: Overflow control for the keyword list and prefetching cache.

contain undesired documents due to the presence of some trivial (nonessential) keywords in the anchor texts. For example, the keyword "DOJ" is trivial in a sector of technology news. It is of interest only when it appears together with some other technology-related keywords like Microsoft or IBM. Figure 10.4 shows a mechanism of the `NewsAgent` to control the overflow of the keyword list and the document prefetching cache: (1) when the cache list is full, use LRU algorithm to remove the least recently accessed objects; (2) extract the corresponding keywords; (3) negatively update those weights; (4) compare the updated weights to the minimum weight threshold; (5) if they are larger than the minimum, check the updating trend (decreasing in this case); (6) purge those keywords from the list if they are below the minimum weight threshold; (7) if the keyword list is full, consider steps 5, 6, and 7. However, the trend will not be necessarily decreasing in this case.

By checking the cache, we can identify those prefetched documents that have never been accessed. The `NewsAgent` will then decrement the weights of their related keywords. Also, we set a purging threshold to determine whether a keyword should be eliminated from the keyword list or not. If the weight of a keyword is less than this threshold, we examine its change trend. The weight change trend is a flag that indicates the keyword weight is increasing or decreasing. If it is negative, then the keyword is considered trivial and is eliminated from the list.

The keywords are purged once the keyword list is full. The list size should be set appropriately. A too small list will force the prefetching unit to purge legitimate keywords, while a too large list would be filled with trivial keywords, wasting space and time in evaluation. We set the list size to 100 and the purging threshold value to 0.02 in our experiments. The purging threshold was set based on our preliminary testing, which will be discussed in Section 10.5.4. These settings are by no means optimal choices.

Prefetched documents need to be replaced when the cache has no space to put new documents. The documents to be purged are those that have been in the cache for a long time without being accessed. The URL anchor text (i.e., news captions) that led to their prefetching are first examined. The weights of their related keywords are then negatively adjusted (d = 0) by the rule of (10.4). This assures that these keywords would unlikely trigger the prefetching unit again in the near future.

10.4 Real-Time Simultaneous Evaluation Methodology

The `NewsAgent` system was developed in Java and integrated with a pure Java Web browser, IceBrowser from IceSoft.com. This lightweight browser with its open source code gave us a great flexibility in the evaluation of the `NewsAgent` system. Using such a lightweight browser simplified the process of setting experiment parameters because the browser just did what it was programmed for and there were no hidden functions or add-ins to worry about, as those with full-blown browsers like Netscape and Explorer. Although the `NewsAgent` prototype was integrated with the IceBrowser, it maintained its own cache so as to dedicate this cache to prefetched documents.

Since the Web is essentially a dynamic environment, its varying network traffic and server-load make the design of a repeatable experiment challenging. There were studies dedicated to characterizing user access behaviors [24] and prefetching/caching evaluation methodologies; see [81] for a survey of these evaluation methods. Roughly, two major evaluation approaches are available: simulation based on user access traces and real-time simultaneous evaluation. A trace-based simulation works like an instruction execution trace-based machine simulator. It assumes a parameterized test environment with varying network traffic and workload. It is a common approach to receive repeatable results. There are benchmarks about client access patterns [24]. However, all the benchmarks that are available to the public are about aggregate behaviors of client groups. Due to privacy concerns, there won't be any open traces about individual clients, needless to say, client-side benchmarks with specific URL contents. Moreover, current implementation of the `NewsAgent` is domain specific and it should be turned on only when its client gets to read news. To the best of our knowledge, there are no such domain-specific traces available either.

The real-time simultaneous evaluation approach [80] was initially designed for the evaluation of Web proxy strategies. In this approach, the proxy under test is running in parallel with another reference proxy. Both proxies perform the same sequence of accesses simultaneously so that more time-variant test environment factors like access latency can be considered. Along the lines of this approach, we evaluated the `NewsAgent` performance by accessing multiple news sites simultaneously. One news site was accessed from a prefetching-enabled browser to train the `NewsAgent`. That is, the `NewsAgent` kept accumulating knowledge about its client's access preferences when he/she was reading news from the site. The trained `NewsAgent` then went to work with another browser that has no built-in prefetching functions to retrieve news items from other sites.

All the experimental results were collected while a client was reading news from major on-line news services in different periods. Since this prefetching approach relies upon the reading interest of individual clients accumulated in training phases, results from different clients are incomparable. The test environment was a Windows NT PC (600 MHz Pentium II CPU and 128 MB memory), connected to the Internet through a T-1 connection. Following are the experimental settings:

Topic selection. It is expected that the performance of semantics-based prefetching systems be determined by semantic locality of the access patterns. In order to establish the access patterns systematically, we selected topics of interest to guide the client's daily browsing behaviors. We are interested in those hot topics that have extensive coverage in major news sites for a certain period so that we can complete the experimentation before the media changes its interest. We conducted two experiments in two different periods when the Kosovo crisis and the Clinton affair were at their peak. Both topics lasted long enough in media for prefetching training and subsequent evaluation. Notice that the `NewsAgent` has no intrinsic requirements for the selection of topics in advance. The client interest is expected to change over time and the algorithm works in the same way no matter whether there is a selected topic or not. However, without a preselected topic in evaluation, most of the keywords generated from a random browsing would be too trivial to be purged. As a result, there would be few prefetching activities over a long period due to the lack of semantic locality.

Site selection. We selected `cnn.com` for the `NewsAgent` training, and `abcnews.com` and `msnbc.com` for its evaluation. Each phase took a month (April 1 to May 5, 1999 for training and May 7 to June 14, 1999 for evaluation).

Client access pattern. The client access pattern was monitored and logged daily. Also, the time between two consecutive access requests (i.e., the time allowed to view a page) was set to an average of 1 minute according to a recent study of client-side Web access patterns [24].

10.5 Experimental Results

As indicated earlier in the previous section, the experiment was conducted in two phases: training and evaluation. This section presents the results of each phase.

10.5.1 NewsAgent Training

The `NewsAgent` system relies on a single perceptron to predict its client future request. It follows the μ-LMS learning rule to adapt to the client interest change.

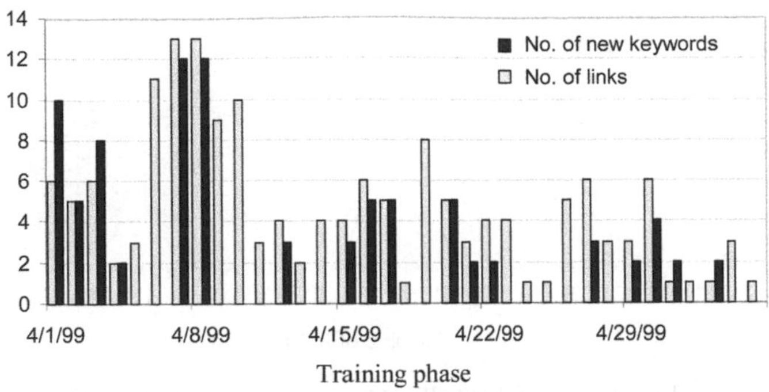

Figure 10.5: Access patterns in the training session.

Before its deployment, the `NewsAgent` must be trained. During its training phase, the `NewsAgent` monitors the client accesses to `cnn.com` server, analyzes the related URL anchor texts, and identifies embedded keywords as inputs to its neural network learning part. It predicts the client access preferences and compares its predictions with the client actual access behaviors. Using the comparative results as a feedback, the `NewsAgent` keeps adjusting the keyword weights recursively. In general, a large set of training data for neural networks with a self-learning capability would lead to highly accurate predictions. However, this may not be the case in news prefetching training. It is because the topics interesting to the client change with time and coverage of a new subject would introduce a set of new keywords. The dynamic nature of keyword inputs makes the `NewsAgent` adaptive to the change of client interests. During this training phase from April 1, 1999 to May 5, 1999, a total of 163 links were accessed. The average link size was 18 KB with a minimum of 9.4 KB and a maximum of 33.4 KB. Figure 10.5 shows the daily access patterns during this period.

It can be observed that in the first four days (4/1–4/4/99), the `NewsAgent` identified many new keywords as more links were accessed. This is expected because the `NewsAgent` started with an initialized neuron with no keyword inputs. The next two days (4/5–4/6) found no new keywords although the media had more coverage about the topic we selected. As time proceeds, the number of new keywords increased sharply in the next two days (4/7–4/8/99) because of new related breaking news. From Figure 10.5, it can also be seen the rise-and-drop pattern occurs repeatedly and demonstrates an important self-similarity characteristic. Due to the dynamics nature of news subjects, whenever a client (or the media) loses interests in a topic, emergence of new host topics would introduce another set of new keywords and the self-similarity pattern repeats.

In the `NewsAgent` training phase, we set its prediction threshold rate T, as defined in (10.1), to a typical median value 0.5. That is, a neuron output of 0.5 or higher was considered a success matching between the neuron prediction and client desire.

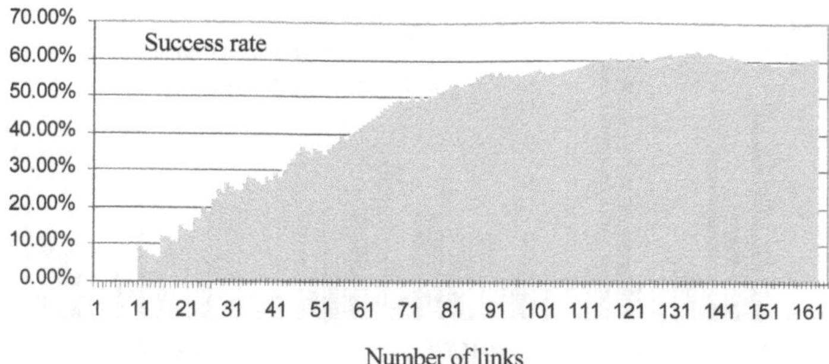

Figure 10.6: Change of the neuron success rate with more links being accessed.

We also set the `NewsAgent` learning rate, as defined in (10.4), to 0.1. This learning rate of 0.1 tends to maintain stability of the neuron and guarantee a smooth weight updating process. In the next section, we will show that high learning rate values can lead to instability and subsequently, the failure of the system. Figure 10.6 plots the success rate of the neuron against the number of links accessed. As we can see, the success rate increases with the number of links accessed until it saturates around 60%. This implies that after a certain training time, the unit could identify six out of ten links as links of interest. Since topic changes tend to generate additional sets of keywords to the neuron, we would not expect an extremely high success rate.

10.5.2 Effectiveness of Semantic Prefetching

The `NewsAgent` was trained while its client was accessing to `cnn.com` news site. After having been trained, it entered into its second evaluation phase in the access of `abcnews.com` new site. During the evaluation phase, the `NewsAgent` kept its keyword inputs and their weights unchanged. Since both the `cnn.com` and `abcnews.com` were covering the similar news topics, we expected the trained `NewsAgent` would prefetch the client's preferred Web documents from `abcnews.com` according to the knowledge accumulated from the training with `cnn.com`. Figure 10.7 shows the access pattern in this phase.

Prefetching systems are often evaluated in terms of four major performance metrics: *hit ratio*, *byte-hit ratio*, *waste ratio*, and *waste byte ratio*. Hit ratio refers to the percentage of requested objects that are found in the prefetching cache; byte-hit ratio refers to the percentage of hit links in size. This pair of metrics, also referred to as *recall*, measures the usefulness of prefetching in terms of Web access latency improvement. Waste ratio refers to the percentage of undesired documents that are prefetched into the cache; byte waste ratio refers to the percentage of undesired documents in size. This pair of metrics reflects the cost (i.e., the extra bandwidth consumption) for prefetching.

Figure 10.8 shows the hit and byte hit ratios in this phase as a function of number of accessed links. While the news in `abcnews.com` had never been accessed be-

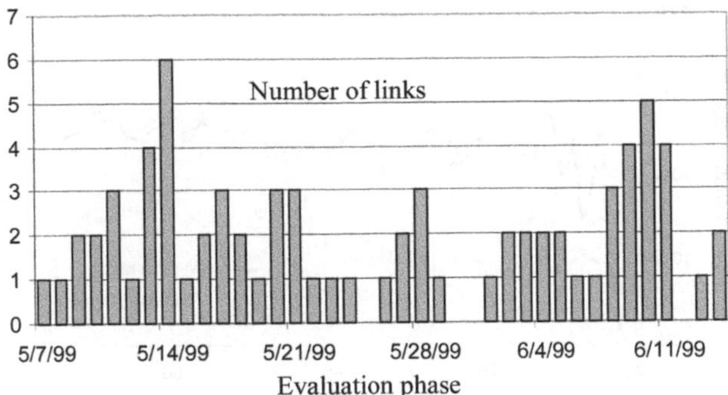

Figure 10.7: Access patterns in the evaluation session.

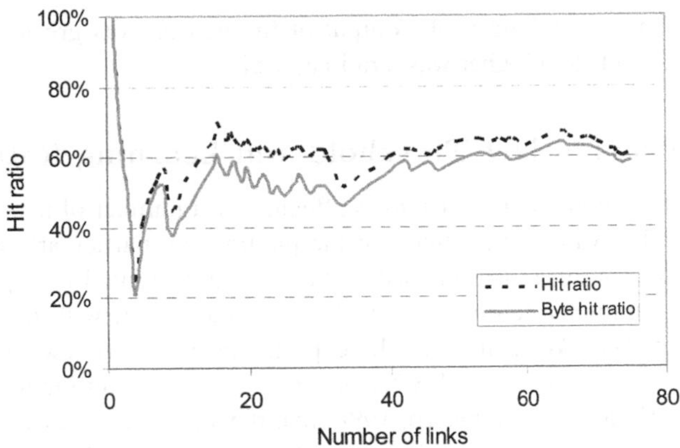

Figure 10.8: Hit ratio and byte hit ratio in the evaluation session.

fore by the client, the `NewsAgent` achieved a hit rate of around 60% based on the semantic knowledge accumulated in the visit of `cnn.com`. Since the news pages have a small difference in size, the plot of the byte hit ratio matches with that of the hit ratio. Since the topic of Kosovo we selected lasted longer than our experiment period, the keywords that were identified in the training phase remained the leading keywords in the evaluation phase. Since the input keywords and their weights in the `NewsAgent` neuron remained unchanged in the evaluation phase, it is expected that accuracy of the prediction will decrease slowly as the client is distracted from the selected topic.

Figure 10.9 shows an average of 24% waste ratio and approximately the same percentages of byte waste ratio during the evaluation phase. This is because many of the links contained similar material and the reader might be satisfied with only one of them. Also, in the evaluation phase, the process of updating weights and purging trivial keywords was not active. However, the neuron kept prefetching those

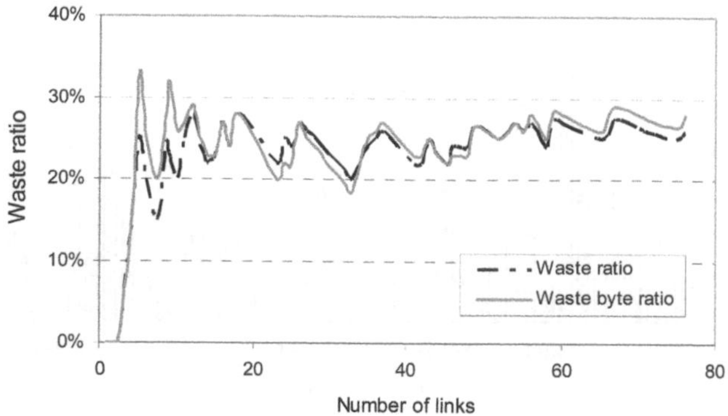

Figure 10.9: Waste ratio and byte waste ratio in the evaluation session.

undesired documents as long as the output of the neuron was greater than the net threshold. This leads to a higher waste ratio as well.

10.5.3 Effects of Net Threshold and Learning Rate

Accuracy of the predictive decisions is affected by a number of parameters in the `NewsAgent`. To evaluate the effects of the parameters, particularly net threshold and learning rate, on hit ratio and waste ratio, we implemented multiple predictors with identical parameter settings, except the one under test, with the same inputs and access sequence. We conducted the experiment over two new sites: Today in America at `msnbc.com` and the US News at `cnn.com/US` during the period from November 18, 1998 to December 16, 1998, five days a week, and two half-an-hour sessions a day. The first contained multimedia documents, while the other was a combination of plain texts and image illustrations. Results from the surfing of these two news sites not only revealed the impact of the parameters, but also helped verify the observations we obtained in the last experiment.

During the experiment time, the topic of "Clinton affair" had much coverage in both sites. As we explained in the preceding section, we selected a guiding hot topic in evaluation because its extensive coverage would help collect more data within a short time period. The `NewsAgent` can identify keywords and establish client access patterns automatically by monitoring his/her browsing behaviors. To reflect the fact that clients may be distracted from their main topic of interest for a while in browsing, we intentionally introduced noise by occasionally accessing headlines that did not belong to the selected topic. We accessed a random link every ten trials. We also assumed each access to be followed by an average of 1 minute viewing, according to a recent analysis of client-side Web access patterns [24].

Figure 10.10 plots the hit ratios due to prefetching with neuron thresholds of 0.25, 0.50, and 0.75, respectively. It shows that a low threshold would yield a high hit ratio for a large cache. This agrees with what (10.1) reveals. Since an extremely small threshold could trigger the prefetching unit frequently, it would lead to prefetching

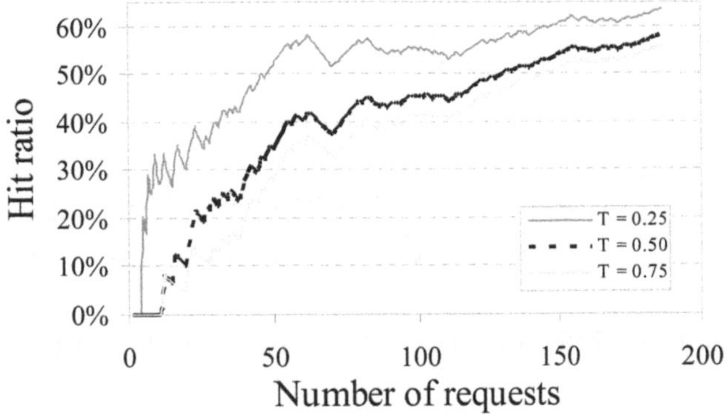

Figure 10.10: Effects of the threshold on cache hit ratio.

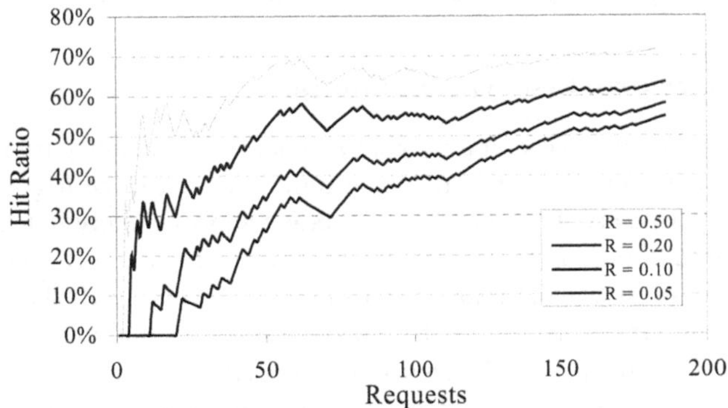

Figure 10.11: Effect of learning rate on cache hit ratio.

of most of the objects embedded in a given Web page. In contrast, the higher the threshold, the smoother the curve is and the more accurate the prediction will be. It is also expected that the lower the threshold value, the more undesired objects get cached. The cache may be filled up with undesired objects quickly. There is a trade-off between the hit ratio and utilization ratio of the cache. The optimum net threshold should strike a good balance between the two objectives.

From the learning rule of (10.4), it is known that the learning rate controls the adaptivity of the `NewsAgent` to keyword inputs. Figure 10.11 shows the hit ratios due to different learning rates. It reveals that a high learning rate leads to a high hit ratio. This is because high learning rates tend to update the weights quickly. Ripples on the curve are due to the fast updating mechanism.

Figure 10.11 also shows that the smaller the learning rate, the smoother the curve is and the more accurate the prediction will be. This is because the weights of trivial keywords are updated with minimal magnitude. This generally affects cache utilization because insignificant weights won't trigger the prediction unit with a lower

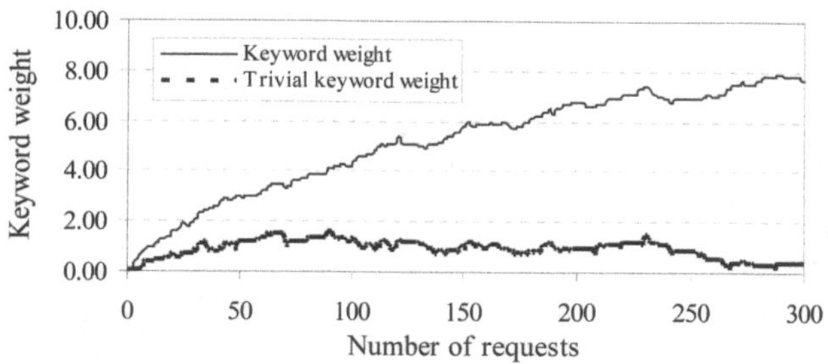

Figure 10.12: Effect of unrelated keywords on prediction accuracy.

learning rate. This may not be the case for a unit with a higher learning rate. Choosing a proper learning rate depends mainly on how fast we want the prediction unit to respond and what percentage of the cache waste we can tolerate.

10.5.4 Keyword List Management

The prefetching unit relies on a purging mechanism to identify trivial keywords and decrease their impact on the prediction unit so as not to be triggered by accident. In the experiments, we found that on average approximately four to five trivial keywords were introduced for each essential keyword. Figure 10.12 plots the distribution of the keyword weights associated with the `NewsAgent`. It shows the system kept the trivial weight minimized as the browsing process continues.

In `NewsAgent`, the keyword weights are reduced whenever their objects in the cache are to be replaced by the prefetching unit. The objects that have never been accessed have the lowest preference value and are to be replaced first. Accordingly, weights of the keywords associated with the replaced object were updated. This guarantees that trivial keywords had minimum impacts on prefetching decisions.

10.6 Related Work

Web prefetching is an active topic in both academia and industry. Algorithms used in today's commercial products are either greedy or user-guided. A greedy prefetching technique does not keep track of any user Web access history. It prefetches documents based solely on the static URL relationships in a hypertext document. For example, WebMirror (`www.maccasoft.com`) retrieves all anchored URLs of a document for a fast mirroring of the whole Web site; NetAccelerator (`www.imsisoft.com`) prefetches the URLs on a link's downward path for a quick copy of HTML formatted documents. The user-guided prefetching technique is based on user-provided preference URL links. There were a number of industry examples in this category, but few succeeded in marketing.

Research in academia is mostly targeted at intelligent predictors based on user access patterns. The predictors can be associated with servers, clients, or proxies. Padmanabhan and Mogul [237] presented a first server-side prefetching approach, in which client access patterns are depicted as a weighted URL dependency graph, where each edge represents a follow-up relationship between a pair of URLs and the edge weight denotes the transfer probability from one URL to the other. The server updates the dependency graph dynamically as requests come from clients and makes predictions of future requests. Schechter *et al.* [275] presented methods of building a sequence prefix tree (path profile) based on HTTP requests in server logs and using the longest matched most-frequent sequence to predict the next requests. Sarukkai proposed a probabilistic sequence generation model based on Markov chains [274]. On receiving a client request, the server uses the HTTP probabilistic link prediction to calculate the probabilities of the next requests based on the history of requests from the same client.

Loon and Bharghavan proposed an integrated approach for proxy servers to perform prefetching, image filtering, and hoarding for a mobile client [200]. They used usage profiles to characterize user access patterns. A usage profile is similar to the URL dependency graph, except that each node is associated with an extra weight indicating the frequency of access of the corresponding URL. They presented heuristic weighting ways to reflect the changing access patterns of users and temporal locality of accesses in the update of usage profiles. Markatos and Chronaki [213] proposed a top-ten prefetching technique for servers to push their most popular documents to proxies regularly according to clients' aggregated access profiles. A concern is that high hit ratio in proxies due to prefetching may give servers a wrong perception of future popular documents.

Fan *et al.* [102] proposed a proxy-initiated prefetching technique between a proxy and clients. It is based on a prediction-by-partial-matching (PPM) algorithm originally developed in a data compressor [74]. PPM is a parametric algorithm that predicts the next l requests based on past m accesses by the user, and limits the candidates by a probability threshold of t. PPM is reduced to the prefetching algorithm of [237] when m is set to 1. Fan *et al.* showed that the PPM prefetching algorithm performs best when $m = 2$ and $l = 4$. A similar idea was recently applied to prefetching of multimedia documents as well by Su *et al.* [303]. The PPM model organizes the URLs that are accessed in a particular server in a Markov predicative tree. Its prediction accuracy comes at the cost of huge memory/storage space for the predicator tree. Chen and Zhang presented a variant of the PPM model that built common surfing patterns and regularities into the predicative tree [59]. By limiting the number of less popular URLs, the model makes it possible to negotiate a trade-off between space and prediction accuracy.

From the viewpoint of clients, prefetching is used to speed up Web access, particularly to those objects that are expected to experience long delays. Klemm [174] suggested prefetching Web documents based on their estimated round-trip retrieval time. Their technique assumes *a priori* knowledge about the sizes of prospective documents for estimation of their round-trip time.

Prefetching is essentially a speculative process and incurs a high cost when its prediction is wrong. Cunha and Jaccoud [73] focused on the decision of when prefetching should be enabled or not. They constructed two user models based on a random-walk approximation and digital signal processing techniques to characterize user behaviors and help predict user next actions. They proposed coupling the models with prefetching modules for customized prefetching services. Crovella and Barford [70] studied the effects of prefetching on network performance. They showed that prefetching might lead to an increase of traffic bursty patterns. As a remedy, they put forward transport rate-controlled mechanisms for reducing smooth traffic caused by prefetching.

Previous studies on intelligent prefetching were based mostly on temporal relationships between accesses to Web objects. Semantics-based prefetching was studied in file prefetching. Kuenning and Popek suggested using semantic knowledge about the relationship between different files to hoard data onto a portable computer before the machine is disconnected [181]. Lei and Duchamp proposed capturing intrinsic corrections between file accesses in tree structures on-line and then matching against existing pattern trees to predict future access [189]. In contrast, we deployed neural nets over keywords to capture the semantic relationships between Web objects and improve prediction accuracy by adjusting the weights of keywords on-line.

Finally, we note that the concept of prefetching is not new. It has long been used in various distributed and parallel systems to hide communication latency. A closely related topic to prefetching in Web surfing is file prefetching and caching in the design of remote file systems. Shriver and Small presented an overview of file system prefetching and explained with an analytical model why file prefetching works [289]. A most notable work in automatic prefetching is due to Griffioen and Appleton [141]. It predicts future file accesses based on past file activities that are characterized by probability graphs. Padmanabhan and Mogul's URL-graph based prefetching work [237] was actually based on this file predictive algorithm.

Prefetching is also used for scalable media delivery. Continuous medias are often bulky in transmission. To reduce clients' perceived access latency, media objects of large sizes can be cached in proxies in segments [332]. Prefetching uncached segments help ensure a timely delivery of the segments of each stream, without causing any proxy jitter [55]. Since the segments of a stream tend to be accessed in order, the determination of the next segment to be accessed is no longer a problem. Segment prefetching focuses on the determination of segment size and the prefetching time for efficient proxy caching.

10.7 Concluding Remarks

This chapter has presented a semantic prefetching approach in the context of Internet news services to hide Web access latency from the perspective of clients. It exploits semantic preferences of client requests by analyzing keywords in the URL anchor texts of previously accessed articles in different news categories. It employs a neural network model over the keyword set to predict future requests. Unlike previous URL graph-based techniques, this approach takes advantage of the "semantics locality" in client access patterns and is capable of prefetching documents that have never been accessed before. It also features a self-learning capability and good adaptability to the change of client surfing interest. This chapter has also presented the details of a personalized `NewsAgent` prefetching prototype.

Like many domain-specific search engines or focused crawling [50, 132, 216], the `NewsAgent` system was targeted at an application domain of Internet news services. The idea of semantic prefetching is applicable to other application domains, such as on-line shopping and e-mail services. Such domain-specific prefetching facilities can be turned on or off completely on user-discretion. This distances the semantic prefetching away from epidemic spyware. Spyware generally refers to programs that can secretly monitor system activity and perform backdoor-type functions to deliver hordes of pop-up ads, redirect people to unfamiliar search engines, or in rare cases, relay confidential information to another machine. Users most often get them by downloading free games or file-sharing software — and consenting to language buried deep within a licensing agreement. Semantic prefeteching systems behave like a spyware, but they can be enabled or disabled as an advanced Web browser feature.

The Web prefetching technologies we have discussed so far are limited to static Web pages. Recent Web traffic studies revealed the increasing popularity of dynamic Web pages that are generated each time they are accessed [238]. This raises new challenges for Web caching and prefetching. There are recent technologies for caching of dynamic Web pages; see [16, 252, 288] for examples. In [257], Ravi *et al.* developed a prefeteching and filtering mechanism for personalized Web e-mails at the network edges and demonstrated the feasibility of prefetching for dynamic Web pages with respect to user-perceived latency.

Chapter 11

Mobile Code and Security

Mobile code, as its name implies, refers to programs that function as they are transferred from one machine to the other. It is in contrast to the conventional client/server execution model. Due to the unique mobility property, mobile code has been the focus of much speculation in the past decade. A primary concern with mobile code is security: protecting mobile code from inspection and tampering by hosts, and protecting hosts from attacks by malicious or malfunctioned mobile code. In this chapter, we survey existing techniques and approaches to enhance the security of mobile code systems. We focus our discussion on mobile agents, a general form of mobile code.

11.1 Introduction

Mobile code is rooted back to the early Internet worm which invaded thousands of networked machines in November 2, 1988 [278, 296]. A worm is a program that propagates itself across a network, using resources on one machine to attack other machines. It spawns copies of itself onto the Internet upon arrival or upon a simple user action. Such worms are examples of malicious mobile codes that tend to break system security measures and gain access to system resources illegitimately.

Mobile codes are not necessarily hostile. In fact, code mobility opens up vast opportunities for the design and implementation of distributed applications. For example, script programs in ActiveX or Javascript are mobile codes that are widely used to realize dynamic and interactive Web pages. One of the most practical uses of such mobile codes is validating on-line forms. That is, a Javascript code embedded in HTML form pages can help check what a user enters into a form, intercept a form being submitted, and ask the user to retype or fill in the required entries of a form before resubmission. The use of the mobile codes not only avoids the transmission of intermediate results back and forth from client to server and reduces the consumption of network bandwidth and server processing power, but also enhances the responsiveness of the user inputs.

Java applets are another form of mobile codes that empower Web browsers to run general-purpose executables that are embedded in HTML pages. When a page that contains an applet is accessed via a Java-enabled browser, the applet's code is down-

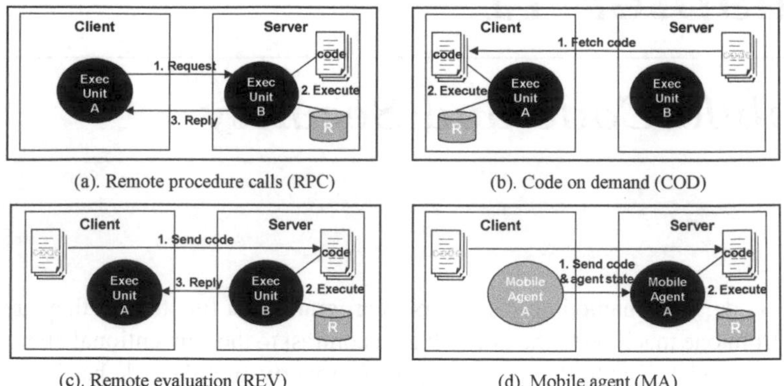

(a). Remote procedure calls (RPC)

(b). Code on demand (COD)

(c). Remote evaluation (REV)

(d). Mobile agent (MA)

Figure 11.1: Illustration of various mobile code paradigms.

loaded on demand and executed by the browser's Java Virtual Machine (JVM). In general, a code-on-demand paradigm allows a client to download an alien code and execute the code in its local executing environment. Java applets of early days were more Web-page-enhancement oriented. In Recent years we have seen the increasing popularity of such code-on-demand technologies in business, scientific, and visualization applications, according to the Java Applet Rating Service (JARS). Another execution model exploiting mobility is remote evaluation (REV) [299]. In REV, a client is allowed to make a request to a server by sending its own procedure code to the server; the server executes the code and returns the results. Unlike pull-based code-on-demand, REV is a push-based execution model.

Both code-and-demand and REV models are limited to one-hop migration between client and server. Like mobile script codes, applet migration deals with the transfer of code only. A more flexible form of migration is mobile agents that have as their defining trait the ability to travel from machine to machine autonomously, carrying its code as well as data and state. An agent is a sort of active object that has autonomy, acting on behalf of its owners. Agent autonomy is derived from artificial intelligence and beyond the scope of this chapter. Figure 11.1 illustrates the three mobile code paradigms, in contrast to the traditional remote procedure call model.

Mobile agents grew out of early code mobility technologies like process migration and mobile objects in distributed systems. Process migration deals with the transfer of code as well as data and running state between machines for the purpose of dynamic load distribution, fault resilience, eased system administration, and data access locality. There are implementations like Sprite [235], Mach [3], and MOSIX [22] in the kernel level and Condor [199] and LSF [355] in the user level; see [225] for a comprehensive review. The literature is also full of application-specific process migration strategies for load balancing and data locality; see [338] for examples. In distributed object systems, such as Emerald [163], Cool [144], and Java RMI, object mobility is realized by passing objects as arguments in remote object invocation. Object migration makes it possible to move objects among address spaces, implementing a finer grained mobility with respect to process-level migration. For example, Emerald [163] provides object migration at any level of granularity ranging from small, atomic data to complex objects. Emerald does not provide complete transparency since the programmer can determine object locations and may request explicitly the migration of an object to a particular node. COOL [144] is an object-oriented extension of the Chorus operating system [21]. It is able to move objects among address spaces without user intervention or knowledge.

Early process and object migration mechanisms are mostly targeted at a "closed" system in which programmers have complete knowledge of the whole system and full control over the disposition of all system components. They assume the process/object authority under the control of a single administrative domain and hence deal with the transfer of authority in a limited way. In contrast, mobile agents are tailored to "open" and distributed environments because of their ability to perceive, reason, and act in their environments. The autonomous requirement makes it possible for the agents to be migrated, with a delegation of resource access privileges, between loosely coupled systems. Mobile agents can also survive intermittent or unreliable connections between the systems. What is more, mobile agents feature a proactive mobility in the sense that the migration can be initiated by the agent itself according to its predefined itinerary.

Lange and Oshima identified seven good reasons for mobile agents [186]: to reduce network load, to overcome network latency, to encapsulate protocols (self-explained data), to support both asynchronous and autonomous execution, to adapt dynamically to the change of environment, to be naturally heterogeneous, and to be robust and fault-tolerant. Although none of the individual advantages represents an overwhelming motivation for their adoption, their aggregate advantages facilitate many network services and application, Harrison *et al.* argued [149]. Because of the unique properties of autonomy and proactive mobility, mobile agents have been the focus of much speculation and hype in the past decade. Overheated research on mobile agents has cooled down as the time of this writing, because there is a lack of real applications to demonstrate their inevitability. However, it remains a technical challenge to design a secure mobile agent system.

In an open environment, mobile codes can be written by anyone and execute on any machine that provides a remote executable hosting capability. In general, the system needs to guarantee the mobile codes to be executed under a controlled manner. Their behaviors should not violate the security policies of the system. For mobile agents that are migrated from one system to another, they may carry private information and perform sensitive operations in the execution environment of the residing hosts. An additional security concern is privacy and integrity. These deal with protecting the agent image from being discovered and tempered by malicious hosts. Since mobile agents offer the highest degree of flexibility for the organization of mobility-aware distributed applications and their deployment raises more security concerns than other forms of mobility, this study focuses on mobile agent systems and related security measures. This chapter draws substantially on our recent review of mobile codes and security [123].

11.2 Design Issues in Mobile Agent Systems

As a special type of mobile code, mobile agents provide a general approach to code mobility. The execution autonomy and proactive multihop migration enable mobile agents to be an appealing paradigm for developing scalable Internet applications. An agent-based system is made up of many components, among which agents and execution environments are two major ones. Agent can move among networked hosts, perform its tasks, and communicate with other agents or its owner. To construct an efficient and secure mobile agent system, several key design issues must be properly tackled. In this section, we discuss the design choices for mobility, communication, naming, and security in mobile agent systems. These techniques are also applicable to other mobile code systems with proper modification.

11.2.1 Migration

The defining trait of mobile agents is their ability to migrate from host to host. Support for agent mobility is a fundamental requirement of the agent infrastructure. An agent can request its host server to transfer it to some remote destination. The agent server must then deactivate the agent, capture its state, and transmit it to the server at the remote host. The destination server must restore the agent state and reactivate it, thus completing the migration.

The image of an agent includes all its code and data, as well as the execution state of its thread, also referred to as its thread-level context. At the lowest level, this is represented by its execution context and call stack. There are two categories of mobility, in terms of the constituents of mobile code that can be migrated. Strong mobility is the ability of a mobile code system that allows migration of both the code and execution state of the mobile code to a different host. Execution state migration makes it possible for the destination server to reactivate the thread at precisely the point where the migration is initiated. This can be useful for transparent load balancing, since it allows the system to migrate processes at any time in order to equalize the load on different servers.

Weak mobility only allows code transfer to a remote host either as a stand-alone code or a code fragment. Stand-alone code is self-contained and will be used to instantiate a new execution entity on the destination host. However, a code fragment must be linked in the context of an already running code and eventually executed. Weak mobility does not support migration of an instruction-level execution state provided by the thread context. However, it is capable of capturing execution state at a higher level, in terms of application-defined agent data, and supporting application-level function migration. It is because the mobile code may be accomplished by various initialization data which direct the control flow appropriately when the state is restored at the destination.

Most current agent systems execute agents using commonly available virtual machines or language environments, which do not usually support thread-level state capture. The agent system developer could modify the virtual machines for this purpose, but this renders the system incompatible with standard installations of those virtual machines. Since mobile agents are autonomous, migration only occurs under explicit programmer control, and thus state capture at arbitrary points is usually unnecessary. Most current systems therefore rely on coarse-grained execution state capture to maintain portability.

Another issue in achieving agent mobility is the transfer of the agent code. One possibility is for the agent to carry all its code as it migrates. This allows the agent to run on any server, which can execute the code. In a second approach, the agent's code is requried to be preinstalled on the destination server. This is advantageous from a security perspective, since no foreign code is allowed to execute. However, it suffers from poor flexibility and limits its use to closed, local area networks. In a third approach, the agent does not carry any code but contains a reference to its code base, which is a server providing its code upon request. During the agent's execution, if it needs to use some code that is not already installed on its current server, the server can contact the code base and download the required code, as in code-on-demand paradigm.

11.2.2 Communication

Agents need to communicate with other agents residing on the same host or on remote ones for cooperation. An agent can invoke a method of another one or send it a message if it is authorized to do so. In general, agent messaging can be peer-to-peer or broadcast. Broadcasting is a one-to-many communication scheme. It allows a single agent to post a message to a group of agents and is a useful mechanism in multiagent systems. Interagent communication can follow three different schemes [185]:

- Now-type messaging. This is the most popular and commonly used communication scheme. A now-type message is synchronous and further execution is blocked until the receiver of the message has completed the handling of the message and replied to it.

- Future-type messaging. A future-type message is asynchronous. The sender retains a handle for future invocation to retrieve the response. Because the sender does not have to wait until the receiver responds, this messaging scheme is flexible and particularly useful when multiple agents communicate with one another.

- One-way-type messaging. A one-way-type message is asynchronous and the current execution is not blocked. The sender will not retain a handle for this message, and the receiver will never have to reply to it. This messaging scheme is similar to the mailbox based approach and it is convenient when two agents are allowed to engage in a loosely connected conversation in which the message-sending agent does not expect any replies from the message-receiving agent.

To support agent communication during migration, one approach is to exploit the mailbox-like asynchronous persistent communication mechanisms to forward messages to the new destination host of an agent. By this way, an agent can send messages to others no matter whether its communication parties are ready. Asynchronous persistent communication is widely supported by existing mobile agent systems; see [327] for a recent comprehensive review of location-independent communication protocols between mobile agents. Another way is to apply the synchronous transient communication mechanisms. Agents communicate with each other only when both parties are ready. This is particularly useful for some parallel applications, which need frequent synchronization in the execution of cooperative agents. Socket over TCP is an example that ensures instantaneous communication in distributed applications. By migrating agent socket transparently, an agent can continue communicating with other agents after movement. Chapter 15 presents the design and implementation of a reliable connection migration scheme.

11.2.3 Naming and Name Resolution

Various entities in a mobile agent system, such as agents, agent servers, resources and users need to be assigned unique names for identification. An agent should be named in such a way that its owner can communicate with or control it while it migrates on its itinerary. For example, a user may need to contact his/her shopping agent to update some preferences it is carrying or simply recall it back to the home host. Agent servers need names so that an agent can specify its desired destination when it migrates. Some namespaces may be common to different entities, e.g., agents and agent servers may share a namespace. This allows agents to uniformly request either migration to a particular server or colocation with another agent with which it needs to communicate.

During the execution of a mobile agent system, there must be a mechanism to find the current location of an entity, given its name. This process is called name resolution. The names assigned to entities may be location dependent, which allows easier implementation of name resolution. Systems like Agent Tcl [139], Aglets [185], and Tacoma [161] use such names, based on host names and port numbers, and resolve them using the Domain Name System (DNS). In such systems, when an agent migrates, its name changes to reflect the new location. This makes agent tracking more cumbersome. Therefore, it is desirable to provide location-transparent names at the application level. This can be done in two ways. The first is to provide local proxies for remote entities, which encapsulate their current location. The system updates the location information in the proxy when the entity moves, thus providing location transparency at the application level. For example, Voyager [258] uses this approach for agent tracking, although its agent servers are identified using location-dependent DNS names. An alternative is to use global, location-independent names that do not change when the entity is relocated. This requires the provision of a name service, which maps a symbolic name to the current location of the named entity. In Ajanta [307], it uses such global names for referring to all types of entities uniformly.

11.2.4 Security

The introduction of a mobile code in a network raises several security issues. In a completely closed local area network, administrated by a single organization, it is possible to trust all hosts and the software installed on them. Users may be willing to allow an arbitrary mobile code to execute on their machines, and dispatch their code to execute on remote hosts. It is because whenever a program attempts some action on a host, the host owner can easily identify a person to whom that action should be attributed. It is safe to assume that the person intends the action to be taken and Trojan Horse attacks are rare.

However, in an open environment like the Internet, it is entirely possible that hosts and users belong to different administrative domains. In such cases, user identity assumption is violated; it becomes hard to identify a responsible person for actions of alien codes. Due to the violation of user identity assumption, user-based security domain in closed distributed systems is no longer true. Authorization based on code sources as in Java applets facilitates fine-grained access control in agent servers to some extent. However, it is insufficient for server protection without identifying the agent owner and understanding his/her access privilege delegation [335].

The inequality between mobile agents and servers in terms of controlling capability complicates the tasks of protecting agents against malicious hosts. An agent executes its program within the environment of a host. The host has full control over the agent action. It may eventually understand the program and therefore change it in any way it wants. This raises another challenge to ensure the correct interpretation and execution by the host [104].

11.3 Agent Host Protections

Agent hosts are the major component of a mobile agent system. They provide an execution environment for mobile agents from different remote hosts. Agents perform their tasks by accessing the host resource. The access requests are issued by agents acting for legitimate users or not. Malicious agents may attempt to exercise unauthorized operations, exhaust host services, or even break down the system. Agent host protection strives to ensure that agent behaviors abide by the security policies specified by the system administrator, and agents cannot execute dangerous operations without the appropriate security mediation.

11.3.1 Security Requirements

Mobile agents on a host may come from different remote hosts. They may traverse multiple hosts and network channels that are secured in different ways, and are not equally trusted. To protect hosts from attacks by malicious or malfunctioned agents in an open environment, the following security requirements should be fulfilled:

- The identity of an agent needs to be determined before allowing it to access sensitive resource. It is realized by the agent authentication mechanisms. With the generated identity information, a host can tell whether an access request is from an agent executing on behalf of a legitimate user.

- An agent cannot be authorized more permissions than those delegated from its owner. Agents roam around networks to perform tasks for their owners. To complete these tasks, agent owners delegate part or all of their privileges to their agents (privilege delegation), whose behaviors are restricted by those permissions. When an agent lands on a remote host, it is granted a set of permissions (agent authorization) according to the delegation from its owner.

- An agent behavior must confirm to the security policy of a visited host. Agents can only access the resource that is allowed by their authorized permissions (agent access control), while other operations will result in security exceptions. A mobile agent may become malfunctioned after being tampered by a malicious host visited in its itinerary. This type of agents also poses a great security threat to the entire system. To control their actions appropriately, we need a coordinated access control scheme, which takes the dynamic behaviors that agents have performed since their creation into account when making access control decisions.

11.3.2 Agent Authentication

Authentication is the process of deducing which principal has made a specific request and whom a mobile agent represents. Lampson *et al.* [184] proposed a theory for authentication in distributed environments. It is based on the notion of principal and a speaks-for relation between principals. A simple principal either has a name or is a communication channel, while a compound principal can express delegated authority. A principal's authority is authenticated by deducing other principals that it can speak for.

Based on this authentication theory, Berkovits *et al.* [32] designed an authentication scheme for mobile agent systems. In this framework, five types of atomic principals are specified: authors, senders, programs, agents, and places. They represent the persons who write the agent programs, those who send the agent on their behalf, the agent programs, the agents themselves, and the servers where agents are executed, respectively. Compound principals can be built from public keys and atomic principals. When an author creates a program, he/she constructs a state appraisal function, which detects whether the program state has been corrupted or not. A sender permission list (SPL) is attached to the program, determining which users are permitted to send the resulting agent. A complementary mechanism is provided by issuing a sender permission certificate (SPC) by the author to allow new users to become the agent senders. In creating a mobile agent, its sender specifies another appraisal function to verify the integrity of the agent's state. The sender also appends a place permission list (PPL), which determines which places are allowed to run the resulting agent on behalf of its sender. Similar to SPC, a place permission certificate (PPC) will be issued if the sender decides to add a new acceptable place.

During the agent migration, a migration request will be sent from the current place to the destination place. The request contains the agent principal, its current state, principal of the current place, and principals on behalf of whom the current and destination places execute the agent. The semantics of a request is determined by the hand-off or delegation relationship between the agent and places. When the destination place receives a request to execute an agent, it will check the author's signature on the agent program and the sender's signature on the agent itself. Then, the place will authenticate the principal on behalf of whom it will execute the agent as indicated in the request, according to the agent's PPL and PPCs. Once the authentication process is completed, the agent is appended with the corresponding principal.

This agent authentication scheme assumes the executing hosts are trustworthy. Otherwise, the migration requests communicated between places may contain false information, which misleads the authentication procedure of the destination place. Besides, the scheme is only applicable to mobile agents with proactive migration, in which the agent sender can predict the sequence of hosts on the agent journey. For places, that an agent reactively migrates to, the authentication procedure cannot succeed according to the agent's PPL and PPCs.

11.3.3 Privilege Delegation and Agent Authorization

A mobile agent can travel across a network to pursue its designated tasks. On each host in its itinerary, an agent needs to be granted certain privileges to execute its program. Privilege delegation specifies the set of permissions that its owner allows it to carry out the tasks. At the same time, the executing host needs to determine which privileges it can grant based on the agent's request and the host security policy.

Farmer *et al.* [103] exploited state appraisal functions to specify permissions delegated from agent author and owner to their agents. State appraisal is to ensure that an agent has not been somehow subverted due to alterations of its state information. A state appraisal function computes a set of privileges to request, as a function of the agent state, when it arrives at a new host. After the state appraisal has determined which permissions to request, the authorization mechanism on the host determines which of the requested permissions will be granted. So, an agent's privileges on an executing host are determined by its current state. An agent whose state violates an invariant will be granted no privileges, while an agent whose state fails to meet some conditional factors may be granted a restricted set of privileges. In this way, a host can be protected from attacks, which alter the agent states maliciously. Besides, the author and owner of an agent can ascertain that it will not be misused in their name by enforcing state invariants.

Before an agent is dispatched from its home host, both its author and owner impose appraisal functions. The author-supplied function max will return a maximum safe set of permits. An owner applies state constraints, req function, to reduce liability and/or control costs. When the author and owner each digitally sign the agent, their respective appraisal functions are protected from undetectable modification. An agent platform uses the functions to verify the correctness of an incoming agent's state and to determine what permissions the agent can possess during execution. Permissions are issued by a platform based on results of the appraisal functions and the platform's security policy. State appraisal provides quite a convenient approach to delegate privileges to an agent from its owner. But, it is not clear how well the theory will hold up in practice, since the state space for an agent could be quite large. While appraisal functions for obvious attacks can be easily formulated, more subtle attacks may be significantly hard to foresee and detect.

11.3.4 Agent-Oriented Access Control

After a host establishes the principal of a mobile agent, the agent may attempt to access system resources, such as files, network sockets, and more. It is not appropriate to grant all permissions to an agent, since the agent may come from a malicious user or through untrusted networks. Access control tries to restrict the actions of an agent to protect a host system.

For a secure system, access control is usually realized by enforcing a security policy, which is a set of access rules, in a reference monitor. However, access control for the mobile agent systems differs from the traditional access control model in many ways. First, there is no fixed set of resources that a host can administer. Different hosts may define different resources. An access control mechanism cannot rely on controlling requests for a specific resource. It should be applicable to any resource that a host may define. Second, the access control model should allow the customization of access control policies from one host to another, one mobile program to another, and one source to another. At the same time, the permission management cannot be too complicated. Third, a mobile agent may have visited multiple hosts before arriving at the current one and its operations may depend on its access history. Therefore it is necessary to integrate spatial information and access history in the secure mobile agent system.

Java has become a popular programming language in the development of mobile agent systems. A practical approach is to utilize the special security mechanisms provided by Java to design an agent-oriented access control scheme. Pandey and Hashii [240] proposed an approach allowing a host to protect and control the local resource that external Java programs could access. In their scheme, a site uses a declarative policy language to specify a set of constraints on accesses to local resources and the conditions under which they will be applied. A set of code transformation tools enforce these constraints on an external Java program by integrating the code, checking access constraints into the program and the host's resource definitions. Execution of the resulting modified mobile program satisfies all access constraints, thereby protecting the host's local resource. Since this approach does not require resource access to make an explicit call to the reference monitor, as implemented in the Java run-time system, the approach does not depend on a particular implementation of the run-time system. However, this approach applies binary editing to modify an agent's program, which leads to violation of the agent integrity. Besides, it is not orthogonal to code obfuscation, by which the agent's program is encrypted to achieve privacy and integrity. In [335], Xu and Fu proposed a fine-grained access control mechanism for mobile agent systems, by using subject-based control in Java. After authentication, an agent is associated with a subject instance, which confines the access permissions authorized to the agent.

In [92], a history-based access control scheme was presented for mobile code security. It maintains a selective history of the access requests made by individual mobile programs and uses this history information to improve the differentiation between safe and potentially dangerous accesses. The permissions granted to a program depend on its identity and behaviors, in addition to some discriminators, such as the location it is loaded from or the identity of its author. When a request is made to a protected resource, a deed security event is prompted. Handlers can be associated with security events. They maintain an event history and check whether it satisfies user-specified constraints. If the checking fails, the handler issues a security-related exception. The access-control policies of the system consist of one or more handlers, which are grouped together. Handlers belonging to a single policy maintain a common history and check common constraints. They ensure the access security.

Although this mechanism considers the dynamic actions of an agent in controlling access requests, it is useful only for applications with one-hop migration, e.g., the Java applets, because the history information and security events are maintained and processed by the local handlers. Besides, the request history was not clearly defined, which results in difficulty in deriving its properties and verifying the soundness of this approach. This access control scheme is essentially a trace-based control in execution monitoring [295]. Like other trace-based approaches, it is difficult for the execution history to record all the security-relevant operations of an agent, when there exist covert channels [183]. A covert channel is a path of communication that is not intended for information transfer at all. For example, an unauthorized agent may manipulate a restricted resource by accessing its attributes or some temporary file. In this way, the history-based access control is circumvented. Recently, Fu and Xu formally defined an access history model for mobile computing, and proposed a coordinated spatio-temporal access control model to specify and reason about the spatial safety property of a mobile code [122]. Its goal is to leverage and enforce the security policies, in which the control of mobile codes access depends not only on the current request but also on the behaviors in the past and maybe on other hosts.

11.3.5 Proof-Carrying Code

Proof-carrying code (PCC) [229] is a software mechanism that allows a host to determine with certainty that it is safe to execute a program supplied by an untrusted source. The basic idea of PCC is that the code producer is required to provide an encoding of a proof that his/her code adheres to the security policy specified by the code consumer. The proof is encoded in a form that can be transmitted digitally. Therefore, the code consumer can quickly validate the code using a simple, automatic, and reliable proof-checking process. PCC shifts the burden of proving that a piece of unknown code is safe to execute from the recipient (code consumer) to the creator (code producer).

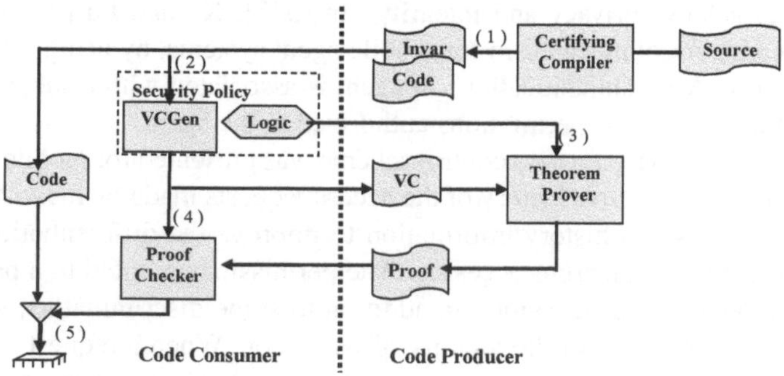

Figure 11.2: An illustration of proof-carrying codes.

A typical PCC session requires five steps to generate and verify the PCC, as shown in Figure 11.2.

1. A PCC session starts with the code producer preparing the untrusted code to be sent to the code consumer. The producer adds annotations to the code, which can be done manually or automatically by a tool such as a certifying compiler. These annotations contain information that helps the code consumer to understand the safety-relevant properties of the code. The code producer then sends the annotated code to the code consumer to execute it.

2. The code consumer performs a fast but detailed inspection of the annotated code. This is accomplished by using a program, called VCGen, which is a component of the consumer-defined safety policy. VCGen performs two tasks. First, it checks simple safety properties of the code. For example, it verifies that all immediate jumps are within the code-segment boundaries. Second, VCGen watches for instructions whose execution might violate the safety policy. When such an instruction is encountered, VCGen emits a predicate that expresses the conditions under which the execution of the instruction is safe. The collection of the verification conditions, together with some control flow information, makes up the safety predicate, and a copy of it is sent to the proof producer.

3. Upon receiving the safety predicate, the proof producer attempts to prove it. In the event of success, it sends an encoding of a formal proof back to the code consumer. Because the code consumer does not have to trust the proof producer, any system can act as a proof producer.

4. Then, the code consumer performs a proof validation. This phase is performed using a proof checker. The proof checker verifies that each inference step in the proof is a valid instance of one of the axioms and inference rules specified as part of the safety policy. In addition, the proof checker verifies that the proof proves the same safety predicate generated in the second step.

5. After the executable code has passed both the VCGen checks and the proof check, it is trusted not to violate the safety policy. It can thus be safely installed for execution, without any further need for run-time checking.

In PCC, a proof-generating compiler emits a proof of security along with the machine code. Also a type checker can verify the code and proof efficiently to allow a safe agent to execute remotely. But the problem is that the proof of security is relative to a given security policy, so in a system where there are many security policies or where the policy is not known beforehand, the proof-generating compiler will not know what to do.

11.4 Mobile Agent Protections

Mobile agents are autonomous entities encapsulating the program code, data, and execution state. Agents are dispatched by their owners for special purposes and they may carry some private information and secrete execution results. When agents are transferred between servers, they are exposed to the threats from both network and server sides.

11.4.1 Security Requirements

A mobile agent is vulnerable to various types of security threats. These include passive attacks, such as eavesdropping, and traffic analysis and active attacks, such as impersonation, message modification, deletion, or forging. Passive attacks are difficult to detect, but can usually be protected against using traditional cryptographic mechanisms. In contrast, active attacks are relatively easy to detect cryptographically, but they are difficult to prevent altogether. The host–host communication often contains sensitive agent data. Therefore, the agent migration process must incorporate confidentiality, integrity, and authentication mechanisms. Most of these requirements can be tackled by various techniques for secure communication [114], which have already been proposed.

11.4.1.1 Protection of Mobile Agents

When an agent performs tasks on a host's execution environment, its internals are in effect exposed to that host. A malicious host can inspect and/or tamper sensitive information stored in the agent, deny agent execution, or replay it multiple times.

- In many applications, algorithms, which the agent program uses to perform tasks, might be proprietary, and the partial results accumulated from visited hosts might be considered private or proprietary. One barrier to achieving such confidentiality of code and state is that the instructions of the agent program must be available for execution on a remote host. Enforcing a security policy that disallows the host's inspection of some parts of the program while allowing it to read and execute others is very difficult, if not impossible.

- Tampering vulnerabilities in mobile agents are similar to inspection ones: a portion of an agent program, including its execution state, may change as the agent moves from host to host, but general security requires that other portions remain immutable. The constant portions include the partial results obtained from visited hosts as well as certain algorithms that might also require immutability to guarantee fair computational results across all remote hosts. Each remote host expects to be able to load a mobile code for execution. Thus, a remote host could potentially read each instruction and modify the agent program before it migrates to a new host. To combat the tampering threats in an untrusted execution environment is a research challenge. Some cryptographic techniques might be used to provide partial solutions.

- Replay attacks occur when a remote host reexecutes a mobile agent in an attempt to infer the semantics of the agent program or to gain extra profits. Effective countermeasures to replay attacks have not been developed. Like any other software system, mobile agents are susceptible to denial-of-execution attacks: a remote host can simply refuse to execute an agent program, thus causing the corresponding application to abnormally terminate. In other cases, a remote host could intercept and deny the mobile agent's requests to databases or other external information sources. Novel methods tackling the denial of execution also need to be proposed for ensuring agent execution.

What is more, the agents' code can also be altered, so that it will perform malicious or malfunctioned actions during the following execution. Multiple remote hosts may even collude to attack a mobile agent.

11.4.1.2 Secure Agent Communication and Navigation

Mobile agents are created by their owners to perform certain tasks. Agents may communicate with each other to synchronize their operations or exchange execution results. When agents roam around an open environment, security of agent communication is a great challenge. The conventional way to keep messages confidential is to use cryptographic techniques. Either the symmetric key or the public key mechanisms is able to encrypt the message contains. However, mobile agents may execute on untrusted hosts. So, it is not secure to let agents carry the private keys that are inherited from their owners or generated for communication sessions. The encryption/decryption algorithms are not effective as a consequence.

Another difficulty posed by multihop migration of agents is how to trace a roaming agent. It is relatively easy to find an agent in a single administrative domain, as certain global information is available. When it comes to a system covering multiple administrative domains, it is hard to let hosts trust each other to provide information of their residing agents. This is particularly difficult when malicious hosts forge false tracing records for private interests.

11.4.1.3 Secure Control of Cloned Agents

Mobile agents are often capable of cloning. Cloned agents may introduce additional security problems. In order to distinguish cloned agents from the original one, they should be assigned different names or identities. So, they will have different message digests. This requires the host, where the original agent resides, to resign the cloned agent with its owner's private key. As we know, it is quite insecure to have an agent carry secret keys in an open, untrusted environment. An alternative is to clone a mobile agent as an exact copy of the original one. The message digest is also copied without change. But the problem of this approach is that we cannot distinguish them, which results in the difficulty of authentication and access control of the cloned versions. Another challenge introduced by agent cloning is how to authorize cloned agents. They should be granted the same privileges as their original or a restricted subset and how to define this subset.

So far, there is no general solution to ensure the secure agent execution on untrusted hosts [104]. There are partial solutions to prevent mobile agents from attacks by malicious hosts or to detect such attacks. Prevention mechanisms strive to make it infeasible or impossible for a host to understand and manipulate the program of an agent during its execution. Static components, such as the program instructions, can be encrypted and signed to ensure their privacy and integrity. Fully protecting a mobile code program is an elusive and open research problem. But we can adopt some techniques that use complexity to make it computationally infeasible, if not impossible, to tamper with or inspect an agent program. Detection mechanisms enable a mobile agent's owner to identify that an attack has occurred on the mobile agent. This can be useful in judging the validity of results that an agent has accumulated, but only after the fact. The most useful detection mechanisms let the owner discover the actual identity of a remote host or aid partially in authenticating the intermediate results produced by the visited hosts.

11.4.2 Integrity Detection

Detection mechanisms aim at identifying illegal modifications of agent code, state, and execution flow. This is useful in judging the validity of execution results as an agent returns to the home host, or in determining the trust of agent information by a host.

Mechanisms for detecting illegitimate mobile code manipulation can be as simple as range checkers, which verify the values of variables within the agent program and state, or certain timing constraints. We can also embed function monitors into the agent programs, and they are called in execution of the program to give assurance that the agent executes correctly. Sophisticated approaches include the appraisal functions in a state appraisal mechanism [103], which check the integrity of agent states by enforcing certain invariants.

Vigna [314] proposed execution tracing to detect unauthorized modifications of an agent through the faithful recording of the agent behavior during its execution on each remote host. This scheme requires each host involved to create and retain a nonrepudiatable trace of operations performed by the agent while residing there, and to submit a cryptographic hash of the trace upon conclusion as a trace summary. A trace is composed of a sequence of statement identifiers and host signature information. It is a partial or complete snapshot of the agent's execution actions. The signature of the platform is needed only for those instructions that depend on interactions with the computational environment maintained by the host. For instructions that rely only on the values of internal variables, a signature is not required and, therefore, is omitted in the trace. The host then forwards the trace and state to the next remote host in the agent itinerary. Vigna also described a protocol for detecting agent tampering by these signed execution traces. Upon the completion of an agent itinerary, the agent's home host verifies the program execution, if it believes that the program has been incorrectly manipulated. The home host simulates the agent execution and asks a remote host for traces from the point when the agent decided to migrate to that host. The remote host will have difficulty in repudiating the request from the owner because it has cryptographically signed and forwarded its trace to the next host. However, this mechanism is not foolproof. A remote host could manipulate the agent program and hide the changes in the traces forwarded to the next host.

11.4.3 Cryptographic Protection of Agents

To protect mobile agents from malicious attacks by untrusted hosts, we can apply cryptographic techniques to mask the semantics of agent program and data, so that it is hard to understand their actual meaning. One possible approach is to adopt tamper-proof hardware. This kind of device executes agents in a physically sealed environment, and the agent integrity and privacy can be easily ensured. Software-based approaches include computing with encrypted functions and code obfuscation. They change agent programs by different methods, and provide agent protection to certain extent.

11.4.3.1 Tamper-Proof Hardware

Tamper-proof hardware, such as secure processors [197], smart cards [36], and secure coprocessors [344] offer an attractive approach to both detection and prevention of attacks on agent execution. It provides a trustworthy computing environment at remote hosts. A mobile agent program can perform cryptographic operations on a host, and in most cases, the crypto keys will be hidden from that host. The tamper-proof hardware can encrypt and digitally sign program code, execution state, and partial results prior to transmitting them to a subsequent host. Alternatively, this hardware can also securely provide critical algorithms to an agent program in hardware form, which would virtually guarantee that the algorithms could not be inspected or modified. For instance, the signature algorithm of an agent can be executed within this hardware using the agent's private key, so that the secrete key is protected by making it available only when decrypted inside the trusted hardware.

Despite these strengths, tamper-proof hardware is not a general solution to secure mobile agent applications. They extend the hardware infrastructure to each remote

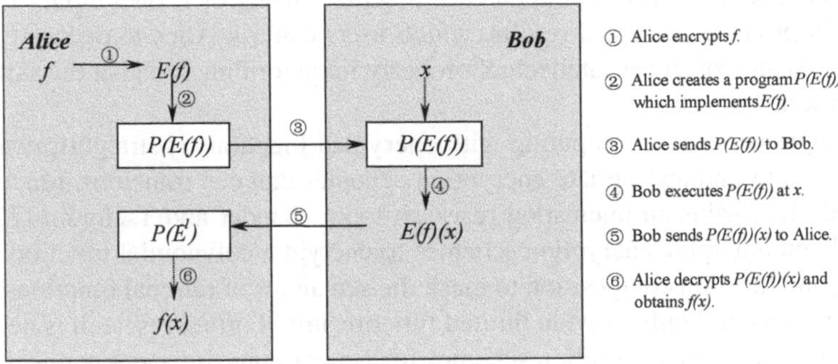

Figure 11.3: Computing with encrypted functions.

host in a way that limits the agent's ability to migrate. This will work only for applications in which tamper-proof hardware can be deployed throughout a controlled network, such as a closed corporate system or secure military network.

11.4.3.2 Computing with Encrypted Functions

Encrypted computing applies cryptographic techniques to provide execution privacy by transforming functions into encrypted forms that become part of an agent program and reveal no information about the original function. Figure 11.3 shows the principle of encrypted computing. Here, Alice has a secrete function f that wants to run on Bob's machine using some input x provided by Bob. To prevent Bob from understanding or modifying function f, Alice encrypts f using her private key and an encryption scheme E to obtain a new program $P(E(f))$ for computing $E(f)$. Then, she sends this program to Bob, who receives the program and runs it on his input x. The program produces an encrypted result $E(f(x))$, which Bob sends back to Alice. Finally, Alice runs a decryption program $P(E-1)$ on result $E(f(x))$ to obtain the desired value $f(x)$. In this way, Bob is able to do useful computation without learning either function f or final result $f(x)$. With this approach, an agent's computation would be kept secret from the executing host, as the sensitive information is carried by the agent. For instance, the means to produce a digital signature can thereby be given to an agent without revealing its private key. This scheme is referred to as computing with encrypted functions (CEF) [270]. However, a malicious host is still able to use the agent to produce a signature on arbitrary data. This leads to a security flaw in agent execution. To tackle this problem, Sander and Tschudin [271] suggested incorporating undetachable signatures to encrypted computing.

CEF can also be extended to function hiding [269], where the result $f(x)$ is returned to Bob after being decrypted by Alice. Function hiding can be used to protect intellectual property from theft and piracy. Suppose function f is some proprietary algorithm that Alice has developed for solving a problem of interest to many people. With function hiding, Alice can embed the encrypted function $P(E(f))$ in an agent program and let users run it on their own machines using their own input. The program will produce encrypted results, which users can ask Alice to decrypt. In this way, Alice can protect her intellectual property for algorithm f, and at the same time charge users for its use.

Although the idea of computing with encrypted functions is straightforward, the challenge is to find appropriate encryption schemes that can transform functions as intended. It remains an interesting research topic. Sander and Tschudin [271] developed homomorphic encryption schemes to encrypt a polynomial function's coefficients, and function composition to mask the semantics of rational functions. However, these schemes only provide limited functionality. Further research is needed to devise more powerful encryption schemes for general-purpose agent applications.

11.4.3.3 Code Obfuscation

To protect the privacy of programs, code obfuscation [196] uses heuristic techniques to modify the program for computing a function. It transforms the program into an equivalent "messy" form, which is quite hard to reverse engineer, but performs the same function as the original one. Collberg and Thomborson [68] have explored the approaches to code obfuscation and classified them into three types: lexical, control, and data transformations.

Lexical transformations change the way that code appears to a malicious attacker. For instance, we can remove the source code formatting information sometimes available from Java bytecode, or scrambling identifier names to make a program look different. In general, lexical transformations are simple to apply, but have relatively low potency because the structure information of the code and data is still mostly preserved and can be eventually understood in spite of the lexical alteration.

Control transformations aim to obscure the control-flow of the original program. They rely on the existence of opaque predicates. A predicate is opaque if its outcome is known at the obfuscation time, but is difficult for the de-obfuscator to deduce. Given such opaque predicates, it is possible to construct obfuscating transformations that break up the control flow of a program. Examples of control transformations include computation transformations, which convert the apparent computation being performed by the program to another equivalent one. For instance, we can change a program by using more complicated but equivalent loop conditions, converting a sequential program into a multithreaded one by using automatic parallelization techniques, etc. Aggregation transformations alter the original code by breaking the abstraction barriers represented by procedures and other control structures, *e.g.* inlining subroutine calls and outlining sequences of statements, and interleaving methods. Ordering transformation modifies the order of statements, blocks, or methods within the code, while maintaining data dependencies.

Data transformations obscure the data structures used by the program. For example, storage and encoding transformations change the usual ways to encode and store data items in a program, such as by splitting variables and converting static to procedural data. Aggregation transformations alter the way data items are grouped together, by restructuring arrays, merging scalar variables, and more. Ordering transformations randomize the order in which data are stored in the program.

Compared with encrypted computing, code obfuscation has the advantage of being more generally applicable, because it can be applied to any computation that can be expressed as a program. However, these schemes are based on heuristic methods and they are not as provably secure as the encrypted computation. In [154], Hohl proposed time-limited blackbox security to tackle this problem by introducing an interval for each obfuscated code. It realizes the protection of mobile agents from malicious hosts by "messing-up" the agent code and data to make it hard to analyze, and then attaching an expiration date to indicate its protection interval. The mess-up algorithm is essentially the same as an obfuscation transformation. Also the approach does not assume it is impossible for an attacker to analyze the agent program, but the analysis simply takes time. So, after creating an obfuscated version of the agent, the agent owner adds the desired protection interval to the current time to get an expiration date. During the execution of the agent, a digital signed certificate including the expiration date information must be presented to any other party that it wants to communicate. If it indicates that the agent has already expired in checking its certificate, communication is refused by others. The expiration date protects the agent even if it migrates to another host. When the agent needs to extend its life for certain reasons, it can get extended by returning to its owner, or some other trusted hosts, which can reobfuscate it using different random parameters for the mess-up algorithms, and then issue a new expiration date.

The time-limited blackbox scheme is an applicable method to ensure agent privacy during its lifetime. It combines the code obfuscation mechanisms with the dynamics of agent behaviors. However, one serious drawback to this technique is the lack of an approach to quantify the protection interval provided by the obfuscation algorithms, thus making it difficult to apply in practice. Furthermore, no techniques are currently known for establishing the lower bounds on the complexity for an attacker to reverse engineer an obfuscated program.

11.5 A Survey of Mobile Agent Systems

In this section, we present an overview for representatives of mobile agent systems, mainly focusing on their architecture, migration and communication mechanisms, and security measures. They are arranged approximately in chronological order of development. Chapter 12 presents the design and implementation of a featured mobile agent system, Naplet.

Telescript. Telescript [324], developed by General Magic, includes an object-oriented, type-safe language for agent programming. Telescript servers, which are called places, offer services by installing stationary agents to interact with visiting agents. Agents can move by using the go primitive, which implements a proactive migration mechanism. A send primitive is also available, which implements remote cloning. Telescript supports strong mobility for agent transfer. In the absolute migration (by the go primitive), the destination host specified by DNS-based host name is required. The name of another colocated agent or resource is needed to provide for a relative migration (by the meet primitive). Colocated agents can invoke each other's methods for communication. An event-signaling facility is also available.

Security has been one of the driving factors in the language design. Telescript provides significant support for security [306], including an access control mechanism similar to capabilities. Each agent and place have an associated authority, which is the principal responsible for it. A place can query an incoming agent's authority, and potentially deny entry of the agent or restrict its access rights. The agent is issued a permit, which encodes its access rights, resource consumption quotas, etc. The system terminates an agent that exceeds its quota, and raises an exception when it attempts an unauthorized operation.

Tacoma. Tacoma [161] is a joint project by University of Troms and Cornell University. Agents in Tacoma are written by Tcl and they can also technically carry scripts written in other languages. The Tcl language is extended to include primitives that support weak mobility. Agents are implemented as Unix processes running the Tcl interpreter. An agent's states must be explicitly stored in folders, which are aggregated into briefcases. Absolute migration to this destination is requested using the meet primitive. The meet command specifies among its parameters, an agent on the destination host that is capable of executing the incoming code. Code shipping of stand-alone code is supported by immediate execution. A briefcase is sent to the destination agent for agent movement. The system does not capture thread-level state during agent migration. Therefore, the Tcl script restarts the agent program after migration. Agents can also use the meet primitive to communicate by colocating and exchanging briefcases. Both synchronous and asynchronous communications are supported. An alternative communication mechanism is to use cabinets, which are immobile repositories for shared state. Agents can store application-specific data in cabinets, which can then be accessed by other agents. No security mechanism is implemented in Tacoma.

Agent Tcl. Agent Tcl [139], developed by Dartmouth College, provides a Tcl interpreter extended with support for strong mobility. Agents are implemented by Unix processes running the language interpreter. Agents can jump to another host, fork a cloned agent at a remote host, or submit code to a remote host. In the first case, an absolute migration enables the movement of a whole Tcl interpreter along with its code and execution state. In the second case, a proactive remote cloning mechanism is implemented. In the third case, code shipping of stand-alone code is supported by asynchronous and immediate execution. Agents have location-dependent identifiers based on DNS hostnames, which therefore change upon migration. Interagent communication is accomplished either by exchanging messages or by setting up a stream connection.

Event signaling primitives are available, but events are currently identical to messages.

Agent Tcl uses the safe Tcl execution environment [193] to provide restricted resource access. It ensures that agents cannot execute dangerous operations without the appropriate security mediation. The system maintains access control lists at a coarse granularity, by which all agents arriving from a particular host are subjected to the same access rules. Agent Tcl calls upon external programs, such as PGP, to perform authentication when necessary, and to encrypt data in transit. However, cryptographic primitives are not available to the agent programmers.

Aglets. Aglets Workbench [185] is a Java-based agent system developed by IBM Tokyo Research Laboratory. Agents, which are called aglets, are threads in a Java interpreter, called aglet contexts, located on different network hosts. Weak mobility is supported by the underlying JVM. Two migration primitives are provided in the system. The primitive dispatch performs code shipping of stand-alone code to the context. The primitive retract exercises code fetching of stand-alone code, and is used to force an aglet to return to the requesting context. Agents are shielded by proxy objects, which provide language-level protection as well as location transparency. Message passing is the only mode of communication and aglets cannot invoke each other's methods. Messages are tagged objects, and can be synchronous, one-way, or future-reply.

A security model for the Java aglets supports architectural definition of the policies. The policy specifies the actions that an aglet can take. In the policy database, the context administrator combines principals that denote aglet groups and privileges into rules. On the other hand, the aglet owner can establish a set of security preferences that will be honored by the visited contexts. They are rules specifying who may access/interact with an aglet on its itinerary. For instance, allowances are preferences dealing with the consumption of resource, such as CPU time or memory. A secure channel established between the source and destination contexts protects aglet migration. The integrity of aglet data is ensured by computing a secure hash value that allows the destination context to perform tampering detection. Cryptograph techniques are applied to make aglet transit confidential.

Concordia. Concordia [76], developed by Mitsubishi Electric, is a framework for developing and executing agent applications. Each node in a Concordia system consists of a Concordia Server that executes on top of a JVM. Like most Java-based systems, it provides agent mobility using Java's serialization and class loading mechanisms, and does not capture execution state at the thread level. An agent object is associated with a separate itinerary object, which specifies the agent's migration path (using DNS host names) and the methods to be executed at each host. The Directory Manager maintains a registry of application services, which enables mobile agents to locate the Concordia Server on each host. Two forms of interagent communication are supported in Concordia: asynchronous event signaling and agent collaboration. It also addresses fault tolerance requirements via proxy objects and an object persistence mechanism that is used for reliable agent transfer. It also can be used by agents or servers to create checkpoints for recovery purposes.

Concordia's security model [318] provides support for three types of protection: agent transmission protection, agent storage protection, and server resource protection. Secure Sockets Layer version 3 (SSLv3) is exploited to create authentication and encryption services for agent migration. When an agent is stored, its information is encrypted and written to persistent stores. The agent's information includes its byte codes, internal states, and travel status. It is encrypted by a symmetric key encryption algorithm. Concordia's resource protection is built upon the standard Java `SecurityManager` class. The security policy for a server is stored in a file called the permissions file. The access requests of an agent are controlled based on the identity of the user on whose behalf the agent is executing.

Ajanta. Ajanta [307] is an object-oriented mobile agent system, developed by University of Minnesota. Each server in Ajanta maintains a domain registry, which keeps track of agents executing on it. Accesses to server resource are realized by proxy interposition between the resource and client agents. Instead of offering direct access to a resource, an agent is given a proxy managed by the server. Ajanta provides various primitives to control an agent. The callback functions, such as arrive and depart methods, define the agent behavior during different phases of its life cycle. They can be overridden by agent programmer for different applications. An agent itinerary records the list of servers to be visited before completing its tasks. Ajanta introduced the concept of abstract migration patterns, which simplifies the construction of complex itineraries by composition of some basic patterns.

In Ajanta security architecture [169], each agent carries a tamper-proof certificate, called its credentials, signed by its owner. It contains the agent's name, and the names of its owner and creator. The agent–server interactions are controlled by an authentication protocol based on challenge–response mechanism. Resource access is granted according to the identity of the agent's owner and an access control list. Agent migration is protected by a secure transfer protocol. An agent image, including its state and code, is selectively hiding and exposing parts of it to different agent servers it visits. Cryptographic techniques are applied to ensure integrity of the agent's read-only states. Some data collected from the visited sites can be stored in an append-only log to prevent any subsequent tampering.

Chapter 12

Naplet: A Mobile Agent Approach

Naplet is an experimental mobile agent system for increasingly important network-centric distributed applications. It provides programmers with constructs to create and launch agents and with mechanisms for controlling agent execution. Its distinct features include a structured navigation facility, reliable agent communication mechanism, open resource management policies, and agent-oriented access control. This chapter presents the design and implementation of the Naplet approach and its application in network management.

12.1 Introduction

An agent is a sort of special object that has autonomy. It behaves like a human agent, working for clients in pursuit of its own agenda. A mobile agent has as its defining trait ability to travel from machine to machine proactively in pursuit of its own agenda on open and distributed systems, carrying its code, data, and running state. The proactive mobility of autonomous agents, particularly their flow of control, leads to a novel distributed processing model on the Internet.

Until recently, mobile agent systems were developed primarily based on script languages like Tcl [137] and Telescript [324]. Recent advance of mobile agent technologies was mostly due to the popularity of Java run-time environment. Java Virtual Machine (JVM) and its dynamic class loading model, coupled with several of other Java features, most importantly serialization, remote method invocation, and reflection, greatly simplify the construction of mobile agent systems. Examples of the Java-based systems include Aglet [185], Ajanta [307], Concordia [76], and D'Agent [137]; see Chapter 11 for a brief review. Readers are also referred to [138, 330] for comprehensive surveys.

Naplet system [334] was started in early 1998, initially designed as an educational package for students to develop understanding of advanced concepts in distributed systems and to gain experience with network-centric mobile computing. Although distributed systems have long been in the core of computer science curriculum, there are few educational software platforms available that are small but full of key concepts in the discipline. A mobile agent system is essentially an agent dock that performs the execution of agents in a confined environment. In addition to mobility support, the dock system not only needs to protect itself from attacks by malicious or misbehaved agents, but also requires isolating the performance of the agents that are concurrently running in the same server and ensuring reliable communication between collaborative agents while they are moving. Systems with such support serve an ideal middleware platform for experiments with related concepts such as naming, process migration, resource management, reliable communication, and security in distributed systems.

An educational system must also be organized in a way that mechanisms are separated from policies so that various algorithms could be implemented without changes in system infrastructure. We reviewed a number of representative mobile agent systems and found none of them met our needs. Most of the systems available in early 1998 were not open-source code yet, because they were mainly targeted at commercial markets. Naplet should be one of the early systems with open-source code. Since its alpha release in 1998, it has been used by hundreds of students at Wayne State University as a platform for programming laboratories and term projects of advanced distributed systems courses. Over the years, it has also evolved into a research testbed for the study of mobility approaches for scalable Internet services. The system was redesigned in the Spring of 2000.

Like other systems, Naplet provides constructs for agent declaration, confined agent execution, and mechanisms for agent monitoring, control, and communication. It has the following distinct features: structured navigation facility, reliable interagent communication mechanism, open resource management policies, and agent-oriented access control. In the following, we present the Naplet architecture and delineate these features. We refer to naplet as an object of `Naplet` class and Naplet server (or server) as an object of `NapletServer`. We conclude this chapter with an application of the Naplet system in network management. The latest release of the software package is available at `www.cic.eng.wayne.edu/software/naplet.html`.

12.2 Design Goals and Naplet Architecture

Code mobility introduces a new dimension to traditional distributed systems and opens vast opportunities for new location-aware network applications. In a networked world, one is obligated to specify not only how to execute designated tasks of an agent, but also where to execute them. A primary goal of the Naplet system is to support the agent-oriented programming paradigm. It is centered around a first-class object: `Naplet`. It is an abstraction of agents, defining hooks for application-specific functions to be performed on visited servers and itineraries to be followed by the agent.

Agent-oriented programming is supported by an object of `NapletServer`. It defines a dock of naplets and provides naplets with a protected run-time environment within a JVM. Naplet server was designed with two goals in mind: *microkernel* and *pluggable*.

- Microkernel: To support the execution of naplets in a confined environment, Naplet server must provide an array of mechanisms for agent migration, agent communication, resource management, security control, etc. The server is designed in a microkernel organization. It is represented by a highly modular collection of powerful abstractions, each dealing with different but specific aspects of the mobility support.

- Pluggable: Each module in the Naplet server is made as an external service to the microkernel. Its default implementation can be easily replaced or enhanced with new implementations. Moreover, application services accessible to naplets should also be installed, configured, reconfigured, and removed dynamically without shutting down the server.

This pluggable and microkernel design not only makes coding easier, but also greatly enhances the service scalability, extensibility, and availability. More importantly, it makes it possible for self-installation of a Naplet server on the network.

12.2.1 Naplet Class

`Naplet` is a template class that defines the generic agent. Its primary attributes include a system wide unique immutable identifier, an immutable codebase URL, and a protected serializable container of application-specific agent running states.

```
public abstract class Naplet implements Serializable, Cloneable {
    private NapletID nid;
    private URL codebase;
    private Credential cred;
    private NapletState state;
    private transient NapletContext context;
    private Itinerary itin;
    private AddressBook aBook;
    private NavigationLog log;
    public abstract onStart();
    public void onInterrupt() {};
    public void onStop() {};
    public void onDestroy() {};
}
```

The naplet identifier (ID) contains the information about who, when, and where the naplet is created. In support for naplet clone, the naplet ID also includes version information to distinguish the cloned naplets from each other. Since a naplet can be recursively cloned, we use a sequence of integers to encode the inheritance information and reserve 0 for the originator in a generation. For example, a naplet ID

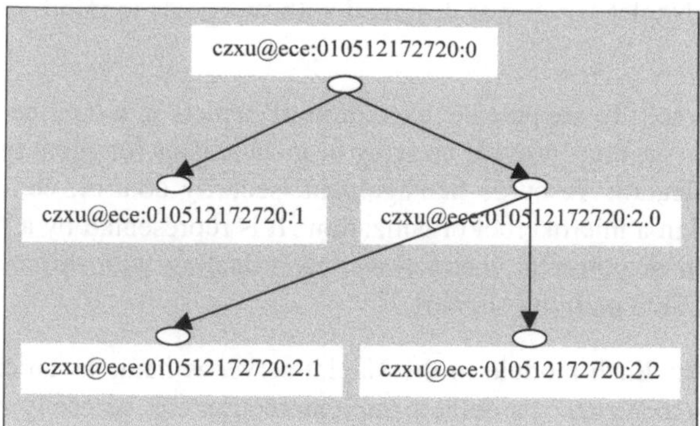

Figure 12.1: Hierarchical naming of naplet ID of a HelloNaplet.

"czxu@ece.wayne.edu:010512172720:2.1/HelloNaplet" represents the hello naplet that was cloned from the original one created by a user "czxu" at 17:27:20 May 12, 2001 in the host "ece.wayne.edu." The inheritance information is shown in Figure 12.1; the naplet name is omitted for brevity.

The Naplet system supports lazy code loading. It allows classes loaded on demand and at the last moment possible. The codebase URL points to the location of the classes required by the naplet. The naplet classes and their associated resources, such as texts and images in the same package can be zipped into an JAR file so that all the classes and resources the naplet needs are transported at a time.

Note that both the naplet ID and code-base URL are immutable attributes. They are set at the creation time and can't be altered in the naplet life cycle. To ensure their integrity, they can be certified and signed by the naplet owner's digital signature. The naplet credential is used by naplet servers to determine naplet-specific security and access control policies.

As a generic class, `Naplet` is to be extended by agent applications. Application-specific agent states are contained in a `NapletState` object. Any object within the container can be in one of the three protected modes: *private*, *public*, and *protected*. They refer to the states accessible to the naplet only, any naplet servers in the itinerary, and some specific servers, respectively. For example, a shopping agent that visits hosts to collect price information about a product would keep the gathered data in a private access state. The gathered information can also be stored in a protected state so that a naplet server can update a returning naplet with new information.

The naplet executes in a confined environment, defined by its `NapletContext` object. The context object provides references to dispatch proxy, messenger, and stationary application services on the server. The context object is a transient attribute and is to be set by a resource manager on the arrival of the naplet. It can't be serialized for migration.

In addition to the attributes, `Naplet` class also provides a number of hooks for application-specific functions to be performed in different stages of the agent life cycle: `onStart`, `onStop`, `onDestroy`, and `onInterrupt`. `onStart` is an abstract method which must be instantiated by extended agent applications. It serves as a single entry point when a naplet arrives at a host. `onStop` and `onDestroy` are event handlers when respective events occur during the execution of the agent. The agent behavior can also be remotely controlled by its owner via the `onInterrupt` method. Details of these will be discussed in Section 12.2.2.

Mobile agents have as their defining trait the ability to travel from server to server. Each naplet is associated with an `Itinerary` object for the way of traveling among the servers. It is noted that many mobile applications can be implemented in different ways by the same agent, associated with different travel plans. We separate the business logic of an agent from its itinerary in `Naplet` class. Each itinerary is constructed based on five primitive patterns: singleton, sequence, parallel, alternative, and loop. Complex patterns can be composed recursively. In addition to the way of traveling, itinerary patterns also allow users to specify a postaction after each visit. The postaction mechanism facilitates interagent communication and synchronization. Details about the itinerary mechanism will be discussed in Section 12.3.

Many mobile applications involve multiple agents and the agents need to communicate with each other. In addition, an agent in travel may need to communicate with its home server from time to time. In support of interagent communication, we associate with each naplet an `AddressBook` object. Each address book contains a group of naplet IDs and their original locations. The locations may not be current, but they provide a starting point for tracing. The address book of a naplet can be altered as the naplet grows. It can also be inherited in naplet cloning. We restrict communications between naplets whose IDs are known to each other.

The last attribute of `Naplet` class is `NavigationLog` for naplet management. It records the arrival and departure time information of the naplet at each server. The navigation log provides the naplet owner with detailed travel information for postanalysis.

12.2.2 NapletServer Architecture

`NapletServer` is a class that implements a dock of naplets within a JVM. It executes naplets in confined environments and makes host resources available to them in a controlled manner. It also provides mechanisms to facilitate resource management, naplet migration, and naplet communication. Each JVM can contain more than one naplet server. The Naplet servers are run autonomously and they collectively form an agent sphere for the naplets.

Naplet servers are run autonomously and cooperatively to form a *naplet sphere*, where naplets live in pursuit of their agenda on behalf of their owners. The naplet sphere can be operating in one of the two modes: with and without a naplet directory. The directory tracks the location of naplets; provisioning of the centralized directory service simplifies the task of naplet management. Figure 12.2 presents the Naplet server architecture. It comprises seven major components: NapletMonitor, NapletSecurityManager, ResourceManager, NapletManager, Messenger, Navigator, and Locator.

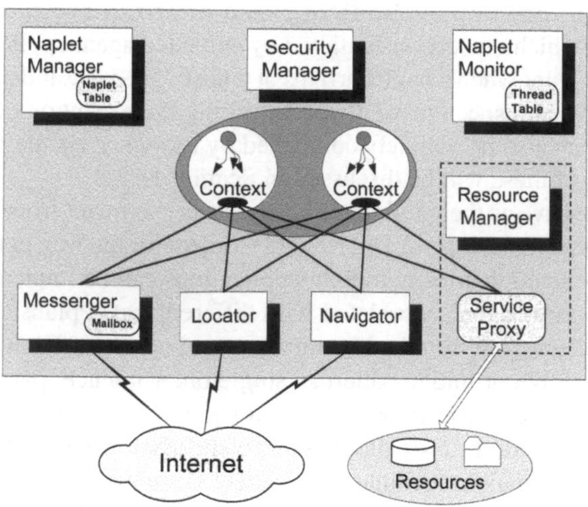

Figure 12.2: The Naplet server architecture.

Each naplet has a home server in the sphere where it is launched through a `Naplet Manager`. The manager provides local users or application programs with an interface to launching naplets, monitoring their execution states, and controlling their behaviors. In addition, the manager maintains the information about its locally launched naplets in a naplet table. Footprints of all past and current alien naplets are also recorded for management purposes.

Naplet launch is actually realized by its home `Navigator`. The launching process is similar to agent migration. On receiving a request for migration from an agent or its local Naplet manager, the navigator consults a local Naplet security manager for a LAUNCH permission. Then, it contacts its counterpart in the destination Naplet server for a LANDING permission. Success of a launch will release all of the resources occupied by the naplet. Finally, the navigator will also report a DEPART event to a `NapletDirectory`, if it exists. The Naplet directory provides a centralized naplet location service. The directory location is provided to naplet servers when they are installed. The directory service is not necessarily required for naplet tracing. In fact, naplet servers can be run without the presence of the directory service.

On receiving a naplet launch request from a remote server, the navigator consults the Naplet security manager and a locally installed `ResourceManager` to determine whether a LANDING permission should be issued. When the naplet arrives, the navigator reports the arrival event to the manager and possibly registers the event with the Naplet directory. It then passes the control over the naplet to a `NapletMonitor`.

A naplet server can be configured or reconfigured with various hardware, software, and data resources available at its host. The hardware resources like CPU cycles, memory space, and network I/O constitute a confined basic execution environment. The software and data resources are largely application dependent and often configured as services. For example, naplets for distributed network management rely on local network management services; naplets for distributed high-performance computing need access to various math libraries. Resource manager provides a resource allocation mechanism, and leaves application-specific allocation policy for dynamic reconfiguration.

Naplets access to local services via a `ServiceProxy`. The proxy provides references to local services to visiting naplets and monitors their resource access operations on-the-fly. Local services available to alien naplets can be run in one of the two protection modes: privileged and nonprivileged. Nonprivileged services, like routines in math libraries, are registered in the resource manager as open services and can be called via their handlers as provided by the service proxy. In contrast, privileged services like getting workload information and system performance must be accessed via a `ServiceChannel`. Each channel is a communication link between alien naplets and local restricted privileged services. It is created by the local resource manager on request. After creation of a channel, the manager passes one endpoint to the privileged service and the other endpoint to the service proxy. It is the service proxy that hands off the service channel endpoint to a requesting naplet. Privileged resources are allocated by the resource manager and the access control is done based on naplet credentials in the allocation of service channels.

Each naplet server contains a `Messenger` for internaplet asynchronous persistent communication. There are two types of messages: system and user. System messages are used for naplet control (e.g., callback, terminate, suspend, and resume); user messages are for communicating data between naplets. On receiving a system message, the messenger casts an interrupt onto the running naplet thread. How the control message should be reacted by the naplet is application dependent and left for programmers to specify. The interrupt handler is given in method `onInterrupt` when a naplet is created. On receiving a user message, the messenger puts it into a mailbox associated with the receiving naplet. The naplet decides when to retrieve the message from its mailbox.

The messenger relies on a `Locator` for naplet tracing and location services and supports location-independent communication. Naplet ID-based message addresses are resolved through a centralized or distributed naplet directory service. Due to the mobility nature of naplets and network communication delay, the location information provided by the directory service may not be current. The messenger provides a postoffice mechanism to handle messages passing between mobile naplets. Section 12.5.1 gives details of the mechanism.

In addition to support for asynchronous communication, each naplet server provides a `NapletSocket` mechanism for a complementary synchronous transient communication between naplets. `NapletSocket` bears much resemblance to Java Socket in APIs, except it is naplet oriented. Conventional TCP has no support for mobility. To guarantee message delivery, an established socket connection must migrate with naplets continuously and transparently. Section 12.5.2 gives `NapletSocket` APIs. Details of the mechanism will be given in Chapter 15.

12.3 Structured Itinerary Mechanism

Mobility is the essence of naplets. A naplet needs to specify functional operations for different stages of its life cycle in each server as well as an itinerary for its way of traveling among the servers. The functional operations are mainly defined in the methods of `onStart()` and `onInterrupt()` in an extended `Naplet` class. The itinerary is defined as an extension of `Itinerary` class. Separation of the itinerary from the naplet's functional operations allows a mobile application to be implemented in different ways following different itineraries. One objective of this study is to design and implement primitive constructs for easy representation of itineraries.

Itinerary representation is a major undertaking of mobile agent systems and various itinerary programming constructs were developed. For examples, Mole provided sequence, set, and alternative constructs, as well as a priority assignment facility in support of flexible travel plans [302]. Ajanta implemented two additional constructs: split, split–join, and loop [307]. They demonstrated the programmability and expressiveness of the constructs mainly by examples. In this section, we define five core structural constructs in the Naplet system: *singleton*, *sequence*, *parallel*, *alternative*, and *loop*. Since each itinerary pattern is associated with a precondition and a postaction, the parallel construct provides flexible support for set, split, and split–join itinerary patterns. In Chapter 13, we extend the core itinerary constructs into a general-purpose mobile agent itinerary language, namely, MAIL. We analyze its expressiveness based on its operational semantics and show MAIL is amenable to formal methods to reason correctness and safety properties regarding mobility.

12.3.1 Primitive Itinerary Constructs

The itinerary of a naplet is mainly concerned about visiting order among the servers. Each visit is defined as the naplet operations from the arrival event through the departure event. The visiting order encoded in the itinerary object is often enforced by departure operations at servers. Correspondingly, we denote a visit as a pair $< S; T >$, where S represents the operations for server-specific business logic and T represents the operations for itinerary-dependent control logic. For example, consider a mobile agent-based information collection application. One or more agents can be used to collect information from a group of servers in sequence or in parallel. At each server, the agents perform information gathering operations (S) (e.g., workload measurement, system configuration diagnosis, etc.), as defined by the application. The operations are followed by agent movement-dependent operations (T) for possible interagent communication and exception handling. Different itineraries would lead to different communication patterns between the naplets. Different itineraries would also have different requirements for handling itinerary-related exceptions. For example, in the case of a parallel search, naplets needs to communicate with each other about their latest search results. Success of the search in a naplet may need to terminate the execution of the others.

We note that servers listed in a journey route may not be necessarily visited in all the cases. Many mobile applications involve conditional visits. For example, in a mobile agent-based sequential search application, the agent will search along its route until the end of its route or the search is completed. That is, all visits except the first one should be conditional visits. We denote a conditional visit as $< C \rightarrow S; T >$, where C represents the guard condition for the visit $< S; T >$.

Based on the concepts of visit and conditional visit, we define visiting order in recursively constructed journey routing pattern. Its base is a singleton pattern, comprising a single visit or conditional visit. Assume P and Q are two itinerary patterns. We define four primitive composite operators seq, alt, and par over the P and Q patterns for constructions of sequential, alternative, and parallel patterns. Specifically,

- `seq(P, Q)` refers to a pattern that the visits of P are followed by the visits of Q by one naplet;
- `par(P,Q)` refers to a pattern that the visits of P and Q are carried out in parallel by a naplet and its clone;
- `alt(P,Q)` refers to a pattern that either the visits of P or Q are carried out by one naplet;
- `loop(C → P)` refers to a pattern that the visits of P are repeated until the guard condition C becomes false.

Formally, the itinerary pattern P is defined in BNF syntax as

$$< V > ::= < S > | < S; T > | < C \rightarrow S; T >$$
$$< P > ::= singleton(V) \mid seq(P, Q) \mid par(P, Q) \mid alt(P, Q) \mid loop(C \rightarrow P)$$

12.3.2 Itinerary Programming Interfaces

The abstraction of itinerary pattern, guard condition, and postaction are expressed in the following public programming interfaces in the Naplet system, respectively.

```
public interface ItineraryPattern extends Serializable, Cloneable {
    public void go(Naplet nap) throws UnableDispatchException;
}
public interface Checkable extends Serializable, Cloneable {
    public boolean check(Naplet nap);
}
public interface Operable extends Serializable, Cloneable {
    public void operate(Naplet nap);
}
```

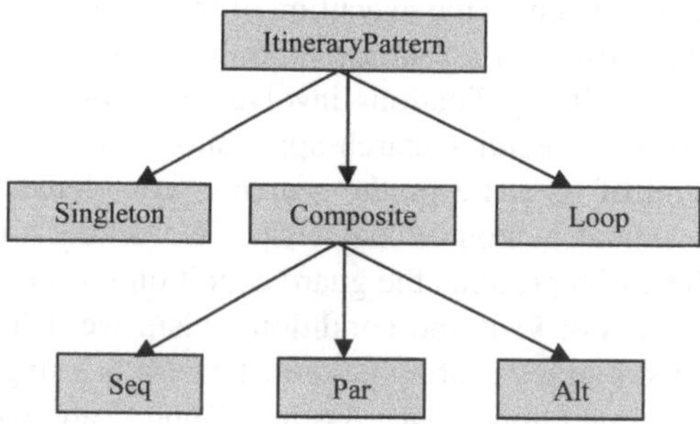

Figure 12.3: Built-in itinerary patterns in the Naplet system.

The Naplet system contains five built-in `ItineraryPattern` implementations: `Singleton`, `SeqPattern`, `AltPattern`, `ParPattern`, and `Loop`. Their class diagrams are shown in Figure 12.3.

In the following, we give two itinerary pattern examples constructed from visits and conditional visits to demonstrate itinerary programming. Consider a mobile agent-based information collection application. One or more agents can be used to collect information from a group of servers $s_1, s_2, \ldots, s_n$ in sequence or in parallel. At each server, the agents perform information-gathering operation (e.g., workload measurement, system configuration diagnosis, etc.), as defined by the application. They are followed by itinerary-dependent operations for possible interagent communication and exception handling. Different itineraries would lead to different communication patterns. In the case of a parallel search, naplets need to communicate with each other about their latest search results. Success of the search by a naplet may need to terminate the execution of the others.

Example 1: The class `MyItinerary1` defines a `SeqPattern` for `MyNaplet`, indicating a sequential information collection. We define `MyNaplet` class as an extension of the base class `Naplet` (line 1). The method `onStart` (line 3) is one of the hooks of the `Naplet` class for application-specific functions to be performed on agent arrival at a server. It contains a location-aware business logic `collectInfo` (line 4). After completion of this function, the agent travels according to its itinerary (line 6). The itinerary is defined in a private class `MyItinerary1` (line 10). It is a simple sequential visiting pattern over an array of servers (line 13). At the end of its itinerary, the agent reports its collected results back to its home by a postaction as defined in the class `ResultReport` (line 9). Since the itinerary class `MyItinerary1` is declared as a private inner class of the naplet, the postaction can be defined on the naplet states. The itinerary is set via a `setItinerary` method of the `Naplet` class (line 13).

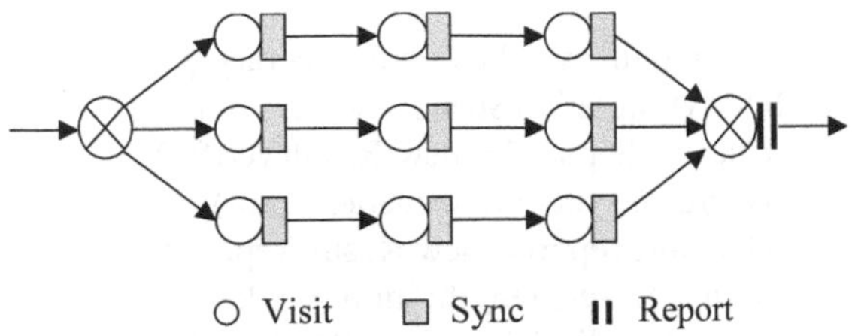

Figure 12.4: An example of parallel search itinerary using three cloned naplets.

```
public MyNaplet extends Naplet {
    ......
    public void onStart() {
        collectInfo();      // Locaion-aware business logic
        try {
            getItinerary().travel( this );
        } catch (UnableDispatchException nde) {};
    }
    private class ResultReport implements Operable {...}
    private class MyItinerary1 extends Itinerary {
        public MyItinerary1(String[] servers) {
            Operable act = new ResultReport();
            setItinerary(new SeqPattern(servers, act));
        }
    }
}
```

Example 2: The class `MyItinerary2` defines a parallel search pattern, as shown in Figure 12.4, by the use of k cloned naplets, each for an equal number of the servers (for simplicity, we assume n can be divided by k). Let $m = n/k$. Totally $m \times k$ visits are defined (lines from 12 through 22). Each visit is of class `Singleton`, comprising a checkable object `ResultVerify` as its guardian precondition (line 12). Whenever a naplet finds the target, it will skip the rest of its servers and meanwhile inform the others. The synchronization is realized by a collective operation defined in the operable object `DataComm` (line 13). A cloned naplet i, $0 \leq i \leq k$, will visit m servers in sequence, as defined in a `SeqPattern` object `journeys[i]` (line 21). The naplets report their results to their home at the end of their journeys via postaction `ResultReport`. All the journeys together form a `ParPattern` itinerary (line 23). The itinerary object is set via a `setItinerary` method of the base class `Naplet`. Naplet cloning is due to a journey of the `ParPattern` itinerary.

```
private class MyItinerary2 extends Itinerary {
    public MyItinerary2(String[] servers, int k) {
        Checkable guard = new ResultVerify();
        Operable sync = new DataComm();
        Operable report = new ResultReport();
        int n = servers.length; int m = n/k;
        Singleton[][] visits = new Singleton[k][m];
        SeqPattern[] journeys = new SeqPattern[m];
        for (int i=0; i < k; i++) {
            for (int j=0; j < m; j++)
                visits[i][j]=new Singleton(guard,servers[i*k+j],sync);
            journeys[i]=new SeqPattern(visits[i],report);
        }
        setItinerary(new ParPattern(journeys));
    }
}
```

12.3.3 Implementations of Itinerary Patterns

In Chapter 13, we will show that the set of itinerary patterns in Figure 12.3 is regular-completeness in the sense that any itinerary in a regular trace can be constructed based on these primitive patterns. However, they are insufficient to express itineraries like "Visiting site s_1 for x times, followed by visiting of s_2 for the same number of times." In the following, we show the implementation details of `Singlton` and `SeqPattern` as programming examples of user-defined itinerary patterns.

In Naplet system, each customized itinerary associated with a naplet is extended from a serializable `Itinerary` class. The itinerary contains a reference to the current pattern and keeps in a stack the naplet trace for recursive traverse.

```
public class Itinerary implements Serializable, Cloneable {
    private ItineraryPattern cur;    // Current itinerary pattern the naplet is on.
    private Stack patterns;          // Stacked itinerary patterns the naplet was on.
    ......
    protected ItineraryPattern popPattern() { return patterns.pop(); }
    protected void pushPattern(ItineraryPattern itin) { return patterns.push(itin); }
    public final void setCurPattern( ItineraryPattern itin ) { cur = itin; }
    public final void travel( Naplet nap ) { current.go( nap ); }
}
```

The `Singleton` class defines the visit of a single server, coupled with precondition and postaction. The `visited` flag is defined as a type-wrapper class `Boolean`, instead of a primitive `boolean` data type. It is because a boolean variable is allocated in a JVM stack and its value will be lost after the migration of the `Singleton` object. The `go()` method shows the details of a naplet migration. Recall that the execution of a naplet is confined to environment defined by a `NapletContext`. The context object contains references to navigator, messenger, and other mobility support services in a naplet server. The migration is accomplished by the navigator.

```
public class Singleton implements ItineraryPattern {
    private URN server;           // Site to be visited
    private Checkable guard;      // precondition of the visit
    private Operable action;      // Post-action of the visit
    private Boolean visited;      // Serializable visit status
        ... ...
    public void go( Naplet nap ) throw UnableDispatchException {
        if ( ! visited.booleanValue() ) {
            visited = Boolean( true );
            if ( guard==null || guard.check(nap) )
                nap.getNapletContext().getNavigator().toDispatch( next, nap );
            else { backtrack(nap); }
        } else {
            if (action ≠ null)
                action.operate( nap );    // Post action
            visited = Boolean( false );
            backtrack(nap);
        }
    }
    private void backtrack(Naplet nap) {
        ItineraryPattern itin = nap.getItinerary().popPattern();
        if (itin ≠ null) {
            nap.getItinerary().setCurPattern( itin );
            itin.go( nap );
        }
    }
}
```

The `SeqPattern` class is defined as an extension of a `CompositePattern` class. A `CompositePattern` object contains a collection of itinerary patterns to be visited, together with a precondition and a postaction. The itinerary patterns are stored in an `ArrayList` data structure, indexed by an `ItineraryIterator` object. The iterator is defined as serializable to replace the Java `ArrayList` iterator. It is because the Java built-in iterator is nonserialiable and its index information will be lost after migration. The `CompositePattern` class is defined as an abstract class because it leaves the visiting order of the servers in the `ArrayList` unspecified.

The `SeqPattern` class defines a sequential visit order in its `go()` method. It recursively traverses each itinerary pattern recorded in the agent itinerary until a `Singleton` is reached.

```
public abstract class CompositePattern implements ItineraryPattern {
    private ArrayList path;              Composite itinerary is stored in an array list
    protected Checkable guard;           Terminate condition for the loop
    protected Operable action;           Post action after the loop
    protected ItineraryIterator iter;    A serializable itierator over the array list
    public abstract void go(Naplet nap) throws UnableDipatchException;
    // Inner object defines a custom iterator over the itinerary array list
    class IteratorImpl implements ItineraryIterator { ... }
}

public class SeqPattern extends CompositePattern {
    ......
    public void go( Naplet nap ) throw UnableDispatchException {
        if (iter.hasNext() ) {
            ItineraryPattern next = iter.next();
            if (guard==null || guard.check(nap)) {
                nap.getItinerary().pushPattern( this );
                nap.getItinerary().setCurPattern( next );
                next.go( nap );  // traverse next pattern recursively
            } else { backtrack( nap ); }
        } else {
            if (action ≠ null) action.operate( nap );
            iter.reset();
            backtrack( nap );
        }
    }
}
```

12.4 Naplet Tracking and Location Finding

12.4.1 Mobile Agent Tracking Overview

Mobility is a defining characteristic of mobile agents. Mobility support poses a basic requirement for tracking agents and finding their current locations dynamically. The agent location information is needed not only for home servers to contact their outstanding agents for agent management purposes, but also for interagent communication.

In general, there are three classes of approaches for the tracking and location finding problem: *broadcast*, *location directory*, and *forward pointer*. A broadcast approach sends a location query message to all servers. It is simple in concept and easy to implement if the system supports broadcast in network and transport layers. In large-scale networks with unreliable communication links, reliable broadcasting is nontrivial by any means. Moreover, any agent location change during the process of broadcasting makes the approach impractical.

A location directory approach is to designate one or more directory servers to keep track of agent locations. The directory service is advertised so that any location query message is directed to the directory. The directory can be organized in a centralized, distributed, or hierarchical way. A centralized organization maintains all location information in a single server. Due to its simplicity in management, this centralized directory approach has been widely used in today's mobile agent prototypes, such as MOA [226], Grasshopper [17], and Aglets [185]. A major drawback of this approach is poor scalability. In highly mobile agent systems, the centralized directory is prone to bottleneck jams.

The hierarchical organization enhances the scalability by deploying a group of directory servers. Each low-level server provides directory services for agents in a region and a high-level server keeps track of the agent regions. This two-level hierarchy can be extended to multilayer hierarchies. Because agents are bound to regional directories dynamically according to their current locations, a challenge with this approach is to ensure location information consistency between the directories when highly mobile agents move across regions. The hierarchical directory approach has never been used in any mobile agent system, because few systems were widely deployed. Stefano and Santoro gave the consistency issue a rigorous treatment [300].

A distributed directory organization maintains the agents' location information in their respective home servers. Whenever an agent migrates, its home server is updated of the new location. This approach was used in OMG's MASIF [75]. Unlike the hierarchical approach that binds agents to different regional directories dynamically, the distributed directory approach binds agents to directories in their home servers statically. This requires that a home server address be retrievable from agent naming.

The third class of agent tracking approach is forward pointer (or path proxies). It relies upon agent footprints left on visited servers to chain them together in a visiting order. In the approach, a location query message will first be sent to the agent home server. The message will then be passed down the agent visiting path. Since the approach requires no updates of agent locations, it incurs no extra overhead in migration. The forward printer approach is used in Mole [29] and Voyager [258].

Finally, we note that the problem of agent tracking bears much resemblance to location finding of a mobile device in a wireless network environment. Both problems are to find mobile entities (logical agents versus physical devices) that are traveling in the network. They differ in a number of aspects. First, physical mobility is often dealt with in network and transport layers, while logical mobility is mainly supported in session layer. Second, physical mobility in wireless networks is characteristic of slow movements in neighbor cells. In contrast, agents can migrate quickly, depending on their business logic to be conducted in each server, and agent migration is lack of locality in service overlay networks. These differences raise different requirements for tracking and location finding algorithms. Broadcasting and distributed directory are widely used in tracking of physical mobility.

12.4.2 Naplet Location Service

The naplet location service interface is defined by `Locator` in the following. It is supported by both location directory and footprint organizations.

```
public interface Locator {
    public URN lookup(NapletID nid) throws NapletLocateException;
    public URN lookup(NapletID nid, long timeout) throws NapletLocateException;
}
```

Recall that `NapletServer` can be running in one of the two modes: with and without naplet directory service. In systems without a directory service, naplets are traced by using naplet footprint information recorded in the naplet manager in each server. `NapletFootPrint`, as defined in the following, contains the source and destination information about each naplet visit.

```
public class NapletFootPrint {
    private URN source;          // Where the naplet comes from
    private Date arrivalTime;
    private URN dest;            // Where the naplet leaves for
    private Date departTime;
}
```

On receiving a location finding request from `Messenger` or other high-level location-independent services, the local `Locator` first checks with its local naplet manager to find out whether the target naplet is in. If not, the `Locator` then retrieves the home server address of the target naplet, as encoded in its `NapletID` object and sends a query message to the home server. The query message is forwarded along the path, starting from the home server, according to the agent footprints left in visited servers until the message catches the target naplet. Note that a query message may arrive at a server before its target because the query message and the naplet may be transferred in different physical routes and the naplet may be blocked in the network. If the query message arrives after the naplet's landing and before its departure, it is responded with the current server location; otherwise, the message needs to be buffered for a certain time period. Readers are referred to Section 12.5.1 for details about in-order message/agent delivery on non-FIFO communication networks.

Since a query message for an agent needs to traverse its whole path, the `lookup` service will be time-out as the agent path stretches out. This is an inherent problem with the forward pointer approach. As a remedy, the Naplet system requires updating the home server with the new location whenever a query is responded. Another solution is to use a forward pointer together with a location directory.

In systems with an installation of `NapletDirectory`, the `Locator` can locate naplets by looking up the directory. Although the location information from the directory may not be current due to the communication delay between a naplet server and the directory, it can be used as a starting point for tracing via the complementary forward pointer approach. Note that we distinguish between two types of naplets:

long-lived and short-lived in terms of their expected lifetime at each server. For stability, the naplet tracing and location service is limited to long-lived naplets.

```
public interface NapletDirectory extends Remote {
    public void register (NapletID nid, URN server, Date time, int event)
                                throw DirectoryAccessException
    public URN lookup( NapletID nid ) throw RemoteException;
}
```

On launching or receiving a naplet, the `Navigator` component of a naplet server registers the ARRIVAL and DEPARTURE events with the directory. The departure event is reported after a naplet is successfully dispatched. However, there is no guarantee of the time when the naplet arrives at the destination. The arrival event is reported after the naplet lands. We postpone the execution of the naplet until the arrival registration is acknowledged. This guarantees that the directory keeps the current location information about the naplets. If the latest registration about a naplet in the directory is a departure from a server, the naplet must be in transmission out of the server. If its latest registration is an arrival at a server, the naplet can be either running in or leaving the server (departure registration may not be needed). The `NapletDirectory` is currently implemented as a component of the naplet server, although its installation is optional. In fact, it can be realized as a stand-alone Lightweight Directory Access Protocol (LDAP) service. One LDAP server can be installed for each naplet sphere (i.e., a collection of naplet servers), independent of naplet servers. To access the LDAP service, each naplet server must authenticate itself to the service.

As we reviewed in Section 12.4.1, a location directory is not necessarily implemented in a centralized manner. The naplet directory services can be provided collaboratively by the naplet manager of each server. Since each naplet has its own home server and the home information is encoded in `NapletID` objects, the naplet location information of can be maintained in their home managers. Correspondingly, any naplet tracing and location requests are directed to respective home managers.

The naplet location service is demanded by `Messenger` for internaplet communication or by `NapletManager` for naplet management. The location service also caches recently inquired locations so as to reduce the response time of subsequent naplet location requests. The buffered naplet location information can be updated on migration either by home naplet managers in systems with distributed naplet directory services, or by remote residing naplet servers in systems with forward pointers.

12.5 Reliable Agent Communication

12.5.1 PostOffice Messaging Service

The messenger service in a Naplet server supports asynchronous message passing between naplets. The messages are naplet ID oriented and location independent.

A naplet can take messages from a specific naplet, any naplet from a naplet server, or any naplet in the sphere if the message sender is not specified. The asynchrony is realized based on a mailbox mechanism. On receiving a naplet, the messenger creates a mailbox for its subsequent correspondences with other naplets or its home naplet manager. Recall that we distinguish messages into two classes: system message for naplet control and user message for data communication. System messages are delivered to their target naplets immediately via interrupts, while user messages are stored in respective mailboxs for the target naplets to retrieve. For flexibility in communication, each messenger also keeps the mailbox open to its naplet so that the naplet can access the mailbox directly, bypassing the send/receive interface.

```
public interface Messenger {
    public void send(NapletID dest, Message msg) throws NapletCommException;
    public void receive(URN server, Message msg) throws NapletCommException;
    public void receive(Message msg) throws NapletCommException;
    public Mailbox getMailbox(NapletID nid);
}
```

The mailbox-based scheme provides a simple and reliable way for asynchronous communication between naplets. Under the hood is a postoffice delivery protocol inside each messenger that implements message forwarding to deal with agent mobility. Each messenger maintains a `MailboxCabinet` to contain all mailboxes of the residing naplets. In addition, it has a special system mailbox, called *s-box*, to temporarily store undelivered messages.

Assume naplet A residing on server Sa is to communicate with naplet B. The naplet A makes a request to Sa's messenger. The messenger checks with its associated Locator to find out naplet B's most recent server or its home server. Due to the mobility nature of naplets and communication delay, this server information is not necessarily current. Without loss of generality, we assume the naplet B used to be in server Sb. Messenger in server Sa sends the message to its counterpart in server Sb. On receiving this message,

1. If naplet B is still running in the server, Sb's messenger replies to Sa with a confirmation and meanwhile inserts the message into naplet B's mailbox. The confirmation message is kept in Sa's messenger only for further possible inquiry from naplet A.

2. If naplet B is no longer in server Sb, Sb's messenger checks with its naplet manager against its naplet trace and forwards the message to the server to which the naplet moved. The forwarding continues until the message catches up to naplet B, say in server Sc. Sc's messenger replies to Sa with a confirmation and inserts the message onto B's mailbox.

3. If naplet B has not arrived in server Sb yet (it is possible because the naplet might be temporarily blocked in the network), Sb's messenger checks with its naplet manager against its naplet trace and finds no record of naplet B. The

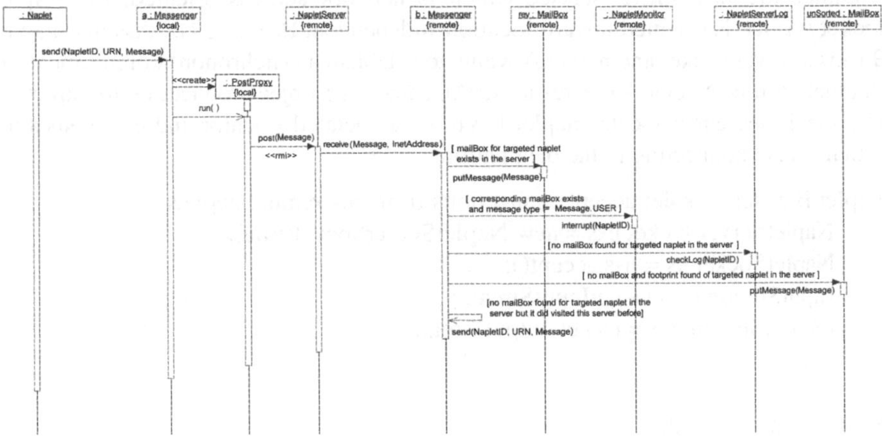

Figure 12.5: Sequence diagram of interagent asynchronous communication.

messenger will insert the message into the *s-box*, waiting for the arrival of naplet B. On receiving the naplet B, Sb's messenger creates a mailbox and transfers the B's messages in the *s-box* to B's mailbox.

12.5.2 NapletSocket for Synchronous Communication

Mailbox-based asynchronous communication aside, Messenger provides a NapletSocket service for synchronous communication between naplets. The NapletSocket service is built on a pair of classes: `NapletSocket` and `NapletServerSocket`. They are implemented as wrappers of Java `Socket` and `ServerSocket`, respectively, providing similar APIs to the Java socket service:

```
public class NapletSocket {
    public NapletSocket(NapletID dest, boolean isPersistent) {}
    public void close() {}
    public InputStream getInputSteam() {}
    public OutputStream getOutputStream() {}
}

public class NapletServerSocket {
    public NapletServerSocket(NapletID dest, boolean isPersistent) {};
    public NapletServerSocket(boolean isPersistent) {};
    public NapletSocket accept() {};
    public void close() {};
}
```

Unlike the Java socket service which is network address oriented, the Naplet socket service is oriented toward location-independent `NapletID`. Assume naplet B runs a server socket and naplet A wants to establish a synchronous communication channel with B. Naplet A creates a `NapletSocket` object connecting to naplet B. The local messenger locates naplet B via its associated Locator and establishes an actual socket connecting to the destination:

Naplet B at server side: accept connections from any remote naplets

```
NapletServerSocket nss = new NapletServerSocket(true);
NapletSocket ns = nss.accept();
InputStream in = ns.getInputStream;
OutputStream out = ns.getOutputStream;
... ...
```

Naplet A at client side:

```
NapletSocket sock = new NapletSocket( B );
InputStream in = sock.getInputStream();
OutputStream out = sock.getOutputStream();
... ...
```

An established socket can be closed by either side. In addition, the NapletSocket service supports connection migration. We distinguish communication channels between *persistent* and *transient*. Persistent channels need to be maintained during migration, while transient channels are not. Consider the socket between naplet A and naplet B. If naplet A is to migrate and naplet B is stationary, the socket is simply suspended before A's migration and resumed as it arrives at the destination. In the case naplet B is about to leave, the Messenger of naplet B needs to suspend all of its outstanding sockets and inform them of its destination for reconnection. This is accomplished by Messenger, transparently to naplet A.

Channel hand-off shouldn't occur until the servers are assured no messages are in transmission. A challenge is how a naplet monitors the status of its naplet sockets. By default the close() method returns immediately, and the system tries to deliver any remaining data. By setting a socket option SO_LINGER, the system is able to set up a zero-linger time. That is, any unsent packets are thrown away when the socket is closed. Details of the mechanism for synchronous persistent Naplet socket services will be presented in Chapter 15.

12.6 Security and Resource Management

A primary concern in the design and implementation of mobile agent systems is security. Most existing computer security systems are based on an identity assumption. It asserts that whenever a program attempts some action, we can easily identify a user to whom that action can be attributed. We can also determine whether the action should be permitted by consulting the details of the action, and the rights that have been granted to the user running the program. Since mobile agent systems violate this important assumption, their deployment involves more security issues than traditional stationary code systems.

12.6.1 Naplet Security Architecture

Over the course of agent execution on a server, server resources are vulnerable to illegitimate access by residing agents. On the other hand, the agents are exposed to the server, their carried confidential information can be breached, and their business logic can even be altered on purpose. The design and implementation of a mobile agent system need to protect agents and servers from any hostile actions from each other. These two security requirements are equally important in an open environment because mobile agents can be authored by anyone and executed on any site that has docking services. However, server protection is more compelling in a coalition environment where the sites are generally cooperative and trustworthy, although mobile codes from different sites may have different levels of trustiness.

The Naplet system assumes naplets are run on trustworthy servers. Security measures focus on protection of servers from any possible naplet attacks. The Naplet Security Architecture (NSA) is based on the standard Java security manager to prevent untrusted mobile codes from performing unwanted actions. Unlike Java's early sandbox security model, which hard coded security policies together with its enforcement mechanism in a `SecurityManager` class, the Java 2 security architecture separates the mechanism from policies so that users can configure their own security policies without having to write special programs.

A security policy is an access-control matrix that says what system resources can be accessed, in what manner, and under what circumstances. Specifically, it maps a set of characteristic features of naplets to a set of access permissions granted to the naplets. It can be configured by a naplet server administrator. It is our belief that any naplet server should be prepared to run an overwhelming number of alien agents from different places. It is cumbersome, if not impossible, to manage the security needs of each individual agent. NSA supports the concept of agent group. A group of agents represent a collection of agents that share certain common properties. For example, cloned agents belong to a group naturally which should be granted similar access permission; agents from thc same owner, organization, or geographical region may form a group that shares the same access control policies. Moreover, agents can also be grouped in terms of their functionalities/responsibilities or particular resources that the agents need to access. Such a group is often referred to an agent role. Administrator, anonymous agents, and normal agents are examples. NSA defines security policies for agents as well as groups and roles. Following is a policy example that grants agents from an Internet domain "`ece.wayne.edu`" to look up a yellow page service.

grant Principal NapletPrincipal "ece.wayne.edu/*" {

```
    permission NapletServicePermission("yellow-page", "lookup");
}
```

Early Java security architecture was targeted at code source. That is, authorization is based on where the code in execution comes from, regardless of the subject of code execution. Subject-based access control is not supported until JDK 1.2. `Naplet` is one of the primary subjects we defined in NSA. Other subjects include `Administrator` and `NapletOwner`. Their authentication is based on a Username/Password LoginModule defined in Java Authentication and Authorization Service (JAAS). In contrast, a naplet is authenticated by the use of its carried certificate. The certificate is issued by the home Naplet manager on behalf of its owner.

The agent-oriented access control is realized via an array of additional security permissions: `NapletServicePermission`, `NapletRuntimePermission`, and `NapletSocketPermission`. They grant access control privileges to system resources as well as application-level services in a flexible and secure manner. Details of the access control model will be presented in Chapter 14.

12.6.2 Resource Management

NSA supports policy-driven and permission-based access control to prevent visiting agents from illegitimate access to local services of a server. It leaves monitoring of the naplet execution and control of resource consumption to naplet monitor and resource manager components of the naplet server.

12.6.2.1 NapletMonitor

A critical type of resource to be monitored is CPU cycles. A mobile agent system without appropriate resource management is vulnerable to denial-of-service attacks. The objective of NapletMonitor is to schedule the execution of multiple agents in a fair-share manner. The Naplet system supports migration and remote execution of multithreaded naplets.

On receiving a naplet, the monitor creates a `NapletThread` and a thread group for the execution of the naplet. The `NapletThread` object assigns a run-time context to each naplet thread and sets traps for its execution exceptions. All the threads created by the naplet are confined to the thread group. The group is set to a limited range of scheduling priorities so as to ensure that the alien threads are running under the control of the monitor. The monitor maintains running states of each thread group and information about consumed system resources including CPU time, memory size, and network bandwidth. It schedules the execution of all residing naplets according to different fair-share resource management policies.

Note that thread scheduling is one of the troublesome problems in Java because it is closely dependent on the underlying scheduling strategies in operation systems. JVM uses fixed-priority scheduling algorithms to decide the thread execution order. If more than one thread exists with the same priority, JVM would switch between the threads in a round robin fashion, if the underlying operating system uses time-slicing

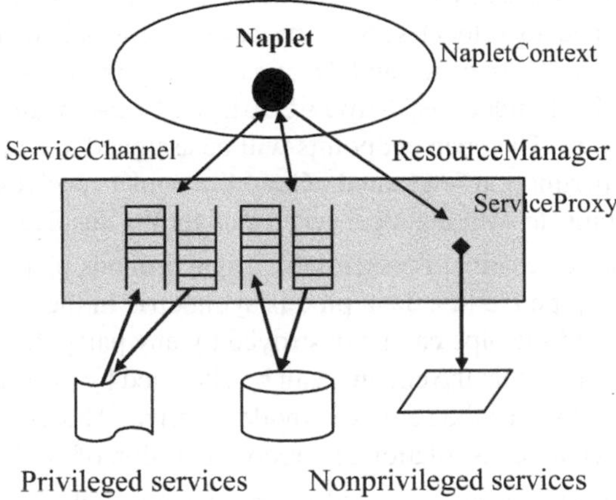

Figure 12.6: Access control over privileged and nonprivileged service.

scheduling. For example, JVM on MS Windows supports fair-share scheduling between threads with the same priority; in contrast, a Java thread on Sun Solaris would continue to run until it terminates, gets blocked, or is preempted by a thread of higher priority. Because of this platform-dependent effect, thread priority is unreliable to support fair-share scheduling. In fact, it is recommended by Java language specification to be used as guides to efficiency only.

Thread priority aside, another performance factor of multithreaded agents is the number of threads. Since JVM cannot distinguish threads from different naplets, a malicious agent can block the execution of other agents by spawning a large number of threads. To ensure fair-share scheduling between naplets, `NapletMonitor` needs to implement a scheduling policy to isolate the performance of naplets, regardless of their priorities and number of threads. The current system release provides the monitoring and control mechanism, leaving scheduling policies undefined.

12.6.2.2 Access to Local Services

Naplets can do few things without access to local services installed on servers and external to the Naplet system. The services include those provided by local operating systems, database management, and other user-level applications. They may be implemented in legacy codes and most likely run in a privileged mode. Although such local services can open to visiting naplets by setting appropriate permissions in `NapletSecurityManager`, visiting naplets should not be allowed to access these services directly. To prevent any threats from misbehaved naplets, resource access operations must be monitored all the time. This is realized by the use of `ServiceProxy` and `ServiceChannel` objects inside the resource manager, as shown in Figure 12.6.

The `ServiceChannel` class defines a bidirectional communication channel between local services and accessing naplets. Each channel is created by the resource manager and attached to a local service by assigning a pair of input/output endpoints: `ServiceInputStream` and `ServiceOutputStream` to the local service. The other pair of endpoints: `NapletOutputStream` and `NapletInput Stream` are left open. The open endpoints will be assigned to a visiting naplet after its service access permission is granted. Once the naplet receives the endpoints, it can start to communicate with the local service under the auspices of the proxy.

Note that the service channel is essentially a synchronous pipe. But it is different from a Java built-in pipe facility. Java pipe is symmetric in the sense that both ends rely on each other and the pipe can be destroyed by any party. In contrast, a service channel is asymmetric in that the channel can be allocated by the service proxy to any authorized naplet as long as the service provider is alive. The asymmetry of service channels enables dynamic installation and reconfiguration of application services.

Local privileged services are accessed via service channels. A naplet server may also be configured with nonprivileged services. Examples are small utility services and math libraries. Nonprivileged services are published with access handlers (e.g., math library function calls). It is also the responsibility of ServiceProxy to allocate the handlers to requesting naplets.

12.7 Programming for Network Management in Naplet

Agent-based mobile computing has been experimented with in various applications, such as distributed information retrieval, high performance distributed computing, network management, and e-commerce; see [37, 43, 236, 164, 176, 340] for examples. It provides a number of advantages over conventional distributed computing paradigms [149, 186]. For example, in the client/server model, clients are limited to services that are predefined in a server. Agent-based mobile computing overcomes this limitation by allowing an agent to migrate carrying its own new service implementation. Due to its unique property of proactive mobility, a featured agent should also be able to find necessary services and information in an open and distributed environment. Other advantages include low network bandwidth due to its on-site computation property, resumed execution after a disconnection from the network is reconnected, ability to clone itself to perform parallel computation, and migration for performance scalability or fault tolerance.

In this section, we illustrate Naplet programming in a network management application. Network management involves monitoring and controlling the devices connected in a network by collecting and analyzing data from the devices. Conventional network management is mostly based on the Simple Network Management Protocol (SNMP). It assumes a centralized management architecture and works in a client/server paradigm. SNMP daemon processes (i.e., SNMP agents) reside on network devices and act like servers. They communicate on request device data to

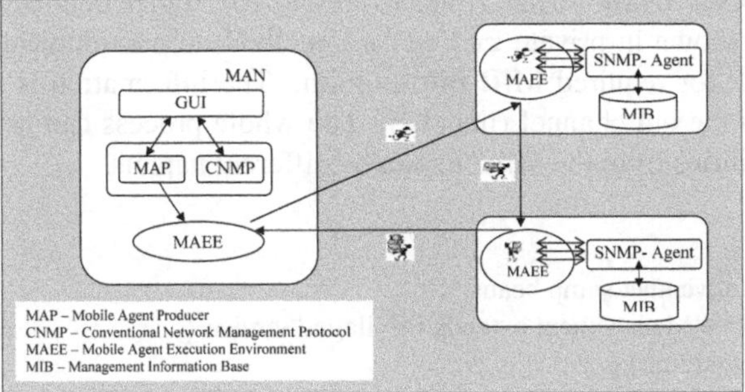

Figure 12.7: Architecture for mobile agent-based network management.

a network management station. The device data are stored in a Management Information Base (MIB) and accessible to local SNMP agents. The management station requests remote MIB information through a pair of fine-grained `get` and `set` operations on primitive parameters in MIBs. This centralized micromanagement approach for large networks tends to generate heavy traffic between the management station and network devices and heavy computational overhead on the management station.

The performance issues in centralized network management architecture can be resolved in many ways. One of the attempts is a mobile agent-based distributed approach. Instead of collecting MIB information from SNMP agents, in this approach, the network management station programs required device statistics or diagnostics functions into an agent and dispatches the agent to the devices for on-site management. Figure 12.7 shows a Naplet-based network management framework. The mobile agent-based framework, namely MAN, is in a hybrid model [176]. It gives the manager the flexibility of using mobile agents or SNMP protocol according to the requirements of management activities.

The MAN management system relies on privileged services provided by the local SNMP agent in each device. In the following, we first present an implementation of the privileged services and then define a naplet class for network management.

12.7.1 Privileged Service for Naplet Access to MIB

In the MAN framework, a network management station creates naplets and dispatches them to target devices. The naplets access to MIB through local SNMP agents of the devices. For communication between Java-compliant naplets and SNMP agents, we deployed an AdvenNet SNMP package on each managed device. The AdventNet SNMP packages provide a Java API for network management.

Following is a `NetManagement` class extended from a `PrivilegedService` base class. It is instantiated by a resource manager and associated with a pair of `ServiceReader` and `ServiceWriter` channels: in and out. Through the input channel, the naplet server gets input parameters from naplets and reorganize them into an AdventNet SNMP format (lines from 6 to 10). It then conducts a sequence of operations, as shown in private `retrieve` method, to communicate with the AdventNet SNMP for required MIB information. The information is returned to the naplet through the out channel (line 12). The whole process can be repeated for a number of inquiries from the same naplet or different naplets.

```
1)  import naplet.*;
2)  import com.adventnet.snmp.beans.*;
3)  public class NetManagement extends PrivilegedService {
4)      public void run() {
5)          for (;;) {
6)              String parms = in.readLine();
7)              Vector values = new Vector();
8)              StringTokenizer paramTokenizer = new StringTokenizer(parms,”;”);
9)              while ( paramTokenizer.hasMoreElements() )
10)                 values.addElement( (String)paramTokenizer.nextToken() );
11)             String result = retrieve( values );
12)             out.writeLine( result );
13)         }
14)     }
15)     private String retrieve( Vector parameters ) {
16)         StringBuffer result = new StringBuffer();
17)         String result = null;
18)         SnmpTarget target = new SnmpTarget();      // Create an enquiry SNMP target
19)         target.loadMibs(”RFC1213-MIB”);            // Load MIB
20)         target.setTargetHost(InetAddress.getLocalHost());    //Set the SNMP target host
21)         target.setCommunity(”public”);
22)         Enumeration enum = parameters.elements();
23)         while(enum.hasMoreElements()) {
24)             target.setObjectID((String)enum.nextElement()+”.0”);
25)             result = target.snmpGet();  // Issue an SNMP get request on managed node
26)         }
27)         return result.toString();
28)     }
29) }
```

12.7.2 Naplet for Network Management

The privileged service defined in `NetManagement` is dynamically configured during the installation of a naplet server. It is accessed by requesting naplets through its registered name "serviceImpl.NetManagement." Following is a naplet example for network management. The `NMNaplet` class is extended from the `Naplet` base class with name, list of servers to be visited, and MIB parameters. It is also instantiated with a `NapletListener` object to receive information retrieved from the servers. All the information will be stored in a reserved `ProtectedNapletState` space. At last, the newly created `NMNaplet` object is associated with a custom designed parallel itinerary pattern shown in lines from 37 to 45.

On arrival at a server, the naplet starts to execute its entry method: `onStart()`. It gets a handler to predefined NetManagement privileged service (lines 16 and 17). It then sends parameters to the server through a `NapletWriter` channel and waits for results from a `NapletReader` channel. Notice that `NapletWriter` and `ServiceReader` are two ends of a data pipe from naplets to servers. Another pipe links a `ServiceWriter` to a `NapletReader`.

When the naplet finishes work on a server, it travels to the next stop (line 27). At the end of its itinerary, the naplet executes an `operate()` method (lines from 30 to 34) to report the results back to its home. Since `NMItinerary` defines a broadcast pattern (lines 40 to 43), the naplet will spawn a child naplet for every server. The spawned naplets will report their results individually.

```
1)  import naplet.*;
2)  import naplet.itinerary.*;
3)  public class NMNaplet extends Naplet {
4)      private String parameters;            // MIB parameters to be accessed
5)      public NMNaplet(String name, String[] servers, String param, NapletListener ch)
                        throws InvalidNapletException, InvalidItineraryException {
6)          super(name, ch);
7)          parameters = param;
8)          setNapletState(new ProtectedNapletState());      // Set space to keep device info.
9)          getNapletState().set("DeviceStatus", new HashTable(servers.length));
10)         setItinerary (new NMItinerary (servers));    // Associate an itinerary with NMNaplet
11)     }
12)     // Entry point of a naplet at each server
13)     public void onStart() throws InterruptedException {
14)         String serverName = getNapletContext().getServerURN().getHostName();
15)         Vector resultVector = new Vector();
16)         HashMap map = getNapletContext().getServiceChannelList();
17)         ServiceChannel channel = map.get("serviceImpl.NetManagement");
18)         NapletWriter out = channel.getNapletWriter();
19)         out.writeLine( parameters );            // Pass parameters to servers
20)         NapletReader in = channel.getNapletReader();
21)         String result = null;
22)         while ( (result = in.readLine()) != EOF ) {
23)             resultVector.addElement( result );
24)         }
25)         Hashtable deviceStatus = (Hashtable) getNapletState().get("DeviceStatus");
26)         deviceStatus.put( serverName, resultVector );
27)         getItinerary().travel( this );
28)     }
29)     private class ResultReport implements Operable {
30)         public void operate( Naplet nap ) {
31)             if ( nap.getListener()!= null ) {
32)                 Hashtable messages = (Hashtable) nap.getNapletState().get("message");
33)                 nap.getListener().report( deviceStatus );
34)             }
35)         }
```

```
    }
    private class NMItinerary extends Itinerary {
        public NMtinerary( String[] servers) throws InvalidItineraryException {
            Operable act = new ResultReport();
            ItineraryPattern[] ip = new ItineraryPattern[servers.length];
            for (int i=0; i<servers.length; i++)
                ip[i] = new SingletonItinerary(servers[i], act);
            setRoute( new ParItinerary(ip) );
        }
    }
}
```

Chapter 13

Itinerary Safety Reasoning and Assurance

This chapter presents an itinerary language, MAIL, which models the mobile behavior of proactive agents. It is an extension of the core itinerary constructs of the Naplet system. The language is structured and compositional so that an itinerary can be constructed recursively from primitive itineraries. We present its operational semantics in terms of a set of inference rules and prove that MAIL is expressive enough for most migration patterns. In particular, it is complete in the sense that it can specify any itineraries of *regular trace models*. Moreover, we show that MAIL is amenable to formal methods to reason about mobility and verify correctness and security properties.

13.1 Introduction

Mobile agent technology supports object migration from one site to another autonomously and *proactively* (that is, not the environment but the agent itself will decide the next site it will visit), performing its designated location-dependent tasks. Although the concept of proactive mobility has recently been demonstrated in several research prototypes, there is a lack of formal treatment of such mobility from the perspective of a distributed programming language. How to specify, model, and reason about travel itineraries of a mobile agent are essential for the development of secure and reliable mobile agent systems.

The proactive mobility of an agent introduces a new dimension to traditional distributed systems and meanwhile raises a new challenge on the specification and modeling of reliable and secure mobile agent systems. In a networked world, one is obligated to specify not only *how* to execute the designated tasks of an agent, but also *where* (its mobility) to execute them. For example, an agent might need to visit more than one site to complete its tasks and these visits must be carried out in a particular order. To model the behavior of an agent, the semantics of an agent migration operation needs to be formalized. For example, for a migration of agent a from site s_1 to site s_2, the semantics should reflect the fact that the location of a is now moved from s_1 to s_2 and a is ready to execute any s_2-dependent tasks. Formal specification and modeling of mobility provide a solid foundation for the analysis of agent security and reliability. For example, given a security policy of "if an agent visits site e, then it must not visit site f." can we analyze the migration pattern of agent a and see whether it will violate such a policy during run time?

From a programming language point of view, although current programming languages such as C++ and Java provide constructs and facilities for the development of traditional parallel and distributed systems, they do not provide facilities for supporting the mobility of agents. For example, they provide neither primitives for moving an agent from one site to another, nor constructs for the specification of different agent migration and communication patterns. Recently, Sewell *et al.* designed a Nomadic Pict language for the representation of location-independent communication between agents [310, 326]. This is achieved by mapping a high-level abstraction of logical channels between two migrating agents to a low-level implementation of physical channels between the two agents based on their current locations. This abstraction greatly facilitates programmers to develop applications in which location-independent communications between agents are required. Although Nomadic Pict supports agent migration from one site to another, it does not provide necessary constructs to specify different migration and communication patterns of proactive agents explicitly. Since *proactive mobility* of agents is at the essence of mobile agent systems, the capability of specifying different migration and communication patterns is fundamentally important and should be an essential component of any agent-oriented programming language.

From a modeling point of view, classic process algebras such as Hoare's CSP [153] and Milner's CCS [223] have been widely used to model parallel and distributed systems. These frameworks provide a solid foundation for mathematical proofs of system properties and a basis for modern model checking technology [64]. However, modeling mobility is not part of their formalisms. Recent π-calculus [224] generalizes the channel-based communication of CCS and its relatives by allowing channels to be passed as data along other channels. This extension introduces a mobility of *communicating channels*, enabling the specification and verification of concurrent systems with dynamically evolving communication topologies. Since π-calculus is mainly designed for communicating systems and there is no formal notation of site (or location), modeling migration patterns in π-calculus is awkward, if not impossible. For example, if one models the visiting of a site by an agent as an interaction between the agent and the site, then one has to model each site as a π-process and introduce a communication link for each site it needs to visit. This is complicated and very inefficient.

There are recent studies focusing on modeling mobile processes; see [83, 280] for surveys. For example, Fournet *et al.* proposed a join calculus as an asynchronous variant of π-calculus [117]. It treats channel and location names as first class values with lexical scopes. This makes it possible to explicitly describe location change of mobile agents. Romann *et al.* developed a Mobile UNITY model [217, 263], as an extension of Chandy and Misra's UNITY model [52], to capture process mobility. It provides a programming notation for process locations and an assertion-style proof logic for verification of the correctness of location changes. Specifically, Mobile UNITY associates each process with a specified location denoted by a program parameter λ. Process migration is modeled by assignments of λ to different location in address space. Later on, Picco *et al.* applied Mobile UNITY to modeling and analysis of mobile agents and fine-grained code mobility [214, 245]. Their focus was on the semantic meaning of mobility constructs such as *move*, *clone*, *activate*, and *deactivate*. They left the semantic model of itinerary constructs unspecified. Most recently, Xu *et al.* provided a predicate/transition model based on Petri net to capture the agent mobility [341]. It models the agents, locations, and their bindings as Petri nets, and agent transfer and location change as transition firing in the Petri nets. These new models provide various support for reasoning of mobile programs that are passive in migration. The predicate/transition net is capable of modeling proactive mobility. However, its focus is on the understanding of dynamic binding of agents with locations.

In [202, 203], we extended the core itinerary constructs in Chapter 12 into a general-purpose mobile agent itinerary language, namely MAIL, for the specification of the mobile behavior of proactive agents. The language is structured and compositional so that an itinerary can be constructed recursively from primitive itineraries. We defined the operational semantics of MAIL in terms of a set of inference rules and prove that MAIL is expressive enough for most migration patterns. Specifically, it is complete in the sense that it can specify any itinerary of a *regular trace model*. In the following, we present the details of MAIL and its semantics. The semantics provides a foundation for mobility reasoning, correctness, and safety verification.

13.2 MAIL: A Mobile Agent Itinerary Language

A mobile agent is a sort of special object that has its own autonomy. It behaves like a human agent, working for clients in pursuit of its own agenda. It has as its defining trait ability to travel from one machine to another proactively in open and distributed systems, carrying its code, data, and running state. This motivates us to define an itinerary language to allow one to specify the itinerary that an agent will follow during its travel.

13.2.1 Syntax of MAIL

Informally, an itinerary specifies the sites that an agent will visit and the ordering of visiting them. Some examples of informal descriptions of itineraries are as follows:

- Visit $s_1, s_2, \ldots, s_n$ in that order.
- Visit s_1 first and then, if $x > 0$, then visit s_2 else visit s_3.
- Visit s_1 and s_2 concurrently (by making a clone of the agent) and then visit s_3.

As a general language for itineraries, we provide constructs to support the specifications of different kinds of itineraries: sequential, conditional, concurrent, and loop itineraries. We formally define MAIL in the sequel.

First, we list the syntactic sets associated with MAIL:

- $\mathcal{S}$: a set of sites. Let $s, s_1, s_2, \ldots$, range over $\mathcal{S}$.
- $\mathcal{Z}$: a set of channels. Let $\alpha, \beta, \gamma, \ldots$, range over $\mathcal{Z}$.
- $\mathcal{V}$: a set of variables. Let $X, Y, Z, \ldots$, range over $\mathcal{V}$.
- $\mathcal{C}$: a set of boolean expressions. Let $c, c_1, c_2, \ldots$, range over $\mathcal{C}$.
- $\mathcal{E}$: a set of arithmetic expressions. Let $e, e_1, e_2, \ldots$, range over $\mathcal{E}$.
- Ξ: a set of signals. Let $\xi, \xi_1, \xi_2, \ldots$, range over ξ.
- $\mathcal{I}$: the set of itineraries definable in MAIL. Let $i, i_1, i_2, \ldots$, range over $\mathcal{I}$.

A *visit* for an agent a is a site $s \in \mathcal{S}$. A visit s for an agent a intuitively specifies "executing a at site s." The capability of specifying *where* to execute agent actions is the motivation of mobile-agent programming paradigm.

Definition 13.1 *(Itinerary) An itinerary is defined as follows:*

- *We define skip and stop are empty and error itineraries, respectively.*
- *We define $\alpha?X$, $\alpha!e$, $signal(\xi)$, and $wait(\xi)$ as primitive communication itineraries.*
- *Every visit $s \in \mathcal{S}$ is a primitive site-visiting itinerary.*
- *If i_1 and i_2 are itineraries and c is a condition, then $i_1; i_2$, $if\ c\ then\ i_1\ else\ i_2$ and $i_1 \parallel i_2$ are itineraries.*
- *If i is an itinerary and c is a condition, then $while\ c\ do\ i$ is an itinerary.*

We summarize it by the following BNF notation:

$$\begin{aligned} i ::= &\ skip \mid stop \mid s \mid \alpha?X \mid \alpha!e \mid signal(\xi) \mid wait(\xi) \mid \\ &\ i_1; i_2 \mid if\ c\ then\ i_1\ else\ i_2 \mid i_1 \parallel i_2 \mid while\ c\ do\ i \end{aligned}$$

where the symbol "::=" should be read as "is defined as" and the symbol "|" as "or." Informally, $i_1; i_2$ and $i_1||i_2$ represent sequential and parallel visit patterns, respectively; $if\ c\ then\ i_1\ else\ i_2$ represents a conditional visit and $while\ c\ do\ i$ a visit loop. $\alpha?X$ and $\alpha!e$ are asynchronous sending and receiving primitives in which channel α is used as a buffer; $signal(\xi)$ and $wait(\xi)$ are synchronous sending and receiving primitives.

An example of itinerary is graphically shown in Figure 13.1, in which $\otimes$ represents concurrent construct, and $\oplus$ with guards represents conditional construct. Using MAIL, the itinerary can be represented by $1; ((if\ x > 5\ then\ 2\ else\ 3) \parallel (4; 5)); 6$.

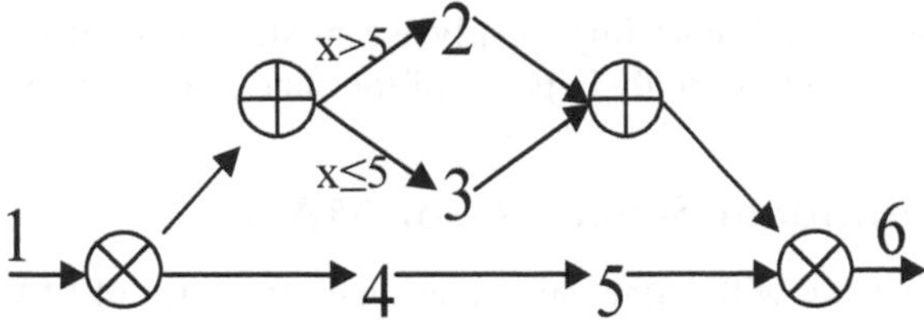

Figure 13.1: An itinerary example.

The constructs used in the above itinerary definition are inspired by the constructs in programming languages. Therefore, an itinerary bears a lot of similarities to a program. However, the following features distinguish an itinerary from a conventional program:

- The primitive building blocks of a program are statements or commands (e.g., an assignment statement) that might change the state of a system when executed. The primitive building blocks of an itinerary are visits, i.e., the sites that an agent can visit during its travel while the corresponding action is specified by the agent's action.
- The semantics of the execution of an itinerary should be considered with respect to a particular agent. Different agents might have different actions and when they execute the same itinerary, the result might be different. Whereas the semantics of a program can be derived from the operational semantics defined for the corresponding programming language.
- During execution, a program itself will not be changed (only the program counter and values of variables are changed), whereas during the execution of an itinerary, in some cases, an itinerary might be transformed into another *residue* itinerary that reflects the remaining sites to be visited. For example, an agent equipped with itinerary $i_1; i_2$ will have a residue itinerary i_2 after the execution of i_1. In other cases, it might be split into a set of itineraries, and a set of agent clones will be made, each of which is responsible for executing exactly one of the resulting itineraries. For example, for itinerary $i_1; (i_2 \parallel i_3)$, after i_1 is executed, a clone of the agent will be made; one of them is responsible for executing i_2, and the other for executing i_3; and during the execution of each, more clones might be made if necessary based on the structures of i_2 and i_3.
- A program is not part of the state space since its content is never changed during execution. An itinerary is part of the state space. In particular, the current location and the acting agent associated with an itinerary are part of the information in the state of a system.

These new situations and complexities motivate us to investigate the semantics of MAIL. Our definition of the operational semantics in the next section reflects these characteristics, and more importantly we model the semantics (meaning) of the mobility of an agent explicitly in terms of the state transitions of the system.

13.2.2 Operational Semantics of MAIL

In the following, we will define the operational semantics of MAIL in terms of a set of inference rules. Using this set of inference rules, one can derive the semantics of the execution of an itinerary i by an agent a. First, we introduce the following notations and assumptions:

- Σ: the set of states of a mobile agent system. The definition of what counts as a state is not important to us, i.e., any notion of a state is allowed here. Let $\sigma, \sigma_1, \sigma_2, \ldots$ range over Σ.

- We assume the evaluation of a condition c in a state σ is defined by function: δ: $\mathcal{C} \times \Sigma \rightarrow \{true, false\}$. Therefore, $\delta(c, \sigma)$ represents the truth value of c in state σ.

- $\mathcal{A}$: a set of agents in the system. Let $a, a_1, a_2, \ldots$ range over $\mathcal{A}$.

- We assume the semantics of an agent a visiting a site s is given by function ρ: $\mathcal{A} \times \mathcal{S} \times \Sigma \rightarrow \Sigma$. Hence, after agent a visits site s starting at state σ, the resulting state will be $\rho(a, s, \sigma)$. We abbreviate it by $\rho(s, \sigma)$ when a is understood from the context. Function ρ models the semantics of an agent's action at a particular site. This semantics can be derived from the definition of conventional programming languages that are used to specify an agent's action.

- $\mathcal{L} ::= 2^{\mathcal{A} \times \mathcal{S} \times \mathcal{I}}$, the set of *layouts* of a system. Each layout is a set of tuples (a_j, s_j, i_j) $(a_j \in \mathcal{A}, s_j \in \mathcal{S}, i_j \in \mathcal{I}, j = 1, \ldots)$ that specifies an agent (or its clone) a_j is located at site s_j with an itinerary i_j. Let $l, l_1, \ldots$ range over $\mathcal{L}$. Note that $\mathcal{L}$ is infinite since $\mathcal{I}$ is infinite.

- $\Omega ::= \Sigma \times \mathcal{L}$, the set of *configurations* of a system. Each configuration is a pair (σ, l) where $\sigma \in \Sigma$ and $l \in \mathcal{L}$. Ω *defines the state space of the whole system.* It is infinite since $\mathcal{L}$ is infinite. However, it will be shown that given an initial configuration, a system can only *reach* a finite set of configurations. Only this set of reachable configurations are our concern for a particular system.

The set of inference rules that defines the operational semantics of MAIL is formalized in Figure 13.2. We explain each of them in the following:

- Rule 1 describes the semantics of executing a *skip* itinerary. Intuitively, it states that "after the execution of itinerary *skip* by agent a, the state component of the configuration will not change, but $(a, s, skip)$ will be deleted from the layout component." The rule has no hypothesis, so it is an axiom. Similar interpretations hold for other rules.

1. $\dfrac{}{(\sigma,\{(a,s,skip),\cdots\}) \rightarrow_{\tau} (\sigma,\{\cdots\})}$

2. $\dfrac{}{(\sigma,\{(a,s,s),\cdots\}) \rightarrow_{s} (\rho(a,s,\sigma),\{\cdots\})}$

3. $\dfrac{s_1 \neq s_2}{(\sigma,\{(a,s_1,s_2),\cdots\}) \rightarrow_{s_1} (\sigma,\{(a,s_2,s_2),\cdots\})}$

4. $\dfrac{}{(\sigma,\{(a,s,skip;i_2),\cdots\}) \rightarrow_{\tau} (\sigma,\{(a,s,i_2),\cdots\})}$

5. $\dfrac{(\sigma,\{(a,s,i_1),\cdots\} \rightarrow_{\lambda} (\sigma',\{a,s',i_1'),\cdots\})}{(\sigma,\{(a,s,i_1;i_2),\cdots\}) \rightarrow_{\lambda} (\sigma',\{(a,s',i_1';i_2),\cdots\})}$

6. $\dfrac{head(\alpha) = n}{(\sigma,\{(a,s,\alpha?X),\cdots\}) \rightarrow_{\alpha?n} (\sigma[n/X][tail(\alpha)/\alpha],\{(a,s,skip),\cdots\})}$

7. $\dfrac{}{(\sigma,\{(a,s,\alpha!e),\cdots\}) \rightarrow_{\alpha!\sigma(e)} (\sigma[\alpha@[\sigma(e)]/\alpha],\{(a,s,skip),\cdots\})}$

8. $\dfrac{}{(\sigma,\{(a,s,signal(\xi)),\cdots\}) \rightarrow_{signal(\xi)} (\sigma[\Gamma\cup\{\xi\}/\Gamma],\{(a,s,skip),\cdots\})}$

9. $\dfrac{\xi \in \Gamma}{(\sigma,\{(a,s,wait(\xi)),\cdots\}) \rightarrow_{wait(\xi)} (\sigma[\Gamma-\{\xi\}/\Gamma],\{(a,s,skip),\cdots\})}$

10. $\dfrac{\delta(c,\sigma) = true}{(\sigma,\{(a,s,if\ c\ i_1\ else\ i_2),\cdots\}) \rightarrow_{\tau} (\sigma,\{(a,s,i_1),\cdots\})}$

11. $\dfrac{\delta(c,\sigma) = false}{(\sigma,\{(a,s,if\ c\ i_1\ else\ i_2),\cdots\}) \rightarrow_{\tau} (\sigma,\{(a,s,i_2),\cdots\})}$

12. $\dfrac{}{(\sigma,\{(a,s,i_1 \| i_2),\cdots\}) \rightarrow_{\tau} (\sigma,\{(a,s,(i_1;!\xi)|||(i_2;?\xi),\cdots\})}$

13. $\dfrac{(\sigma,\{(a,s,i_1),\cdots\}) \rightarrow_{\lambda} (\sigma,\{(a,s,i_{11}\ |||\ i_{12}),\cdots\})}{(\sigma,\{(a,s,i_1;i_2),\cdots\}) \rightarrow_{\lambda} (\sigma,\{(a,s,i_{11}\ |||\ (i_{12};i_2)),\cdots\})}$

14. $\dfrac{}{(\sigma,\{(a,s,i_1\ |||\ i_2),\cdots\}) \rightarrow_{\tau} (\sigma,\{(a,s,i_1),(a,s,i_2),\cdots\})}$

15. $\dfrac{}{(\sigma,\{(a,s_1,!\xi;i_1),(a,s_2,?\xi;i_2),\cdots\}) \rightarrow_{\tau} (\sigma,\{(a,s_1,i_1),(a,s_2,i_2),\cdots\})}$

16. $\dfrac{\delta(c,\sigma) = true}{(\sigma,\{(a,s,while\ c\ do\ i),\cdots\}) \rightarrow_{\tau} (\sigma,\{(a,s,i;while\ c\ do\ i),\cdots\})}$

17. $\dfrac{\delta(c,\sigma) = false}{(\sigma,\{(a,s,while\ c\ do\ i),\cdots\}) \rightarrow_{\tau} (\sigma,\{\cdots\})}$

Figure 13.2: Operational semantics of MAIL.

- Rule 2 describes the semantics of executing a primitive itinerary s at site s by agent a. Intuitively, it states that "after the execution of primitive itinerary s at site s, the state component is transformed into $\rho(a, s, \sigma)$, and (a, s, s) will be deleted from the layout component."

- Rule 3 describes the semantics of migrating agent a from site s_1 to site s_2 for primitive itinerary s_2 ($s_1 \neq s_2$). It models the semantics of mobility of an agent: agent a is located at site s_1 originally, to execute primitive itinerary s_2 where $s_2 \neq s_1$, a has to migrate to site s_2, which results the change of the configuration to one that reflects the new location of agent a while the state component remains unchanged.

- Rule 4 describes the semantics of executing a sequential composition of two itineraries $skip; i_2$ by agent a. Its interpretation is obvious.

- Rule 5 describes the semantics of executing a sequential composition of two itineraries $i_1; i_2$ where $i_1 \neq skip$. It says that "if there exists a configuration transition such that σ is transformed to σ', and layout $\{(a, s, i_1), \ldots\}$ is transformed to $\{(a, s', i_1'), \ldots,\}$ (specified in the hypothesis), then there exists a configuration transition such that σ is transformed to σ', and $\{(a, s, i_1; i_2), \ldots\}$ is transformed to $\{(a, s', i_1'; i_2'), \ldots\}$.

- Rule 6 describes the semantics of receiving a value n from channel α when agent a executes itinerary $\alpha?X$ at site s. The hypothesis specifies that the head of channel α is value n, and the conclusion of the rule specifies that the state of the system will be transformed into one in which X is assigned with value n and the head element n is deleted from channel α. In addition, $(a, s, \alpha?X)$ is transformed into $(a, s, skip)$.

- Rule 7 describes the semantics of sending arithmetic expression e to channel α when agent a executes itinerary $\alpha!e$ at site s. The hypothesis is empty, and the conclusion of the rule specifies that $\sigma(e)$, the evaluation of arithmetic expression e in state σ, is appended to the end of channel α. In addition, $(a, s, \alpha!e)$ is transformed into $(a, s, skip)$.

- Rule 8 describes the semantics of sending signal ξ to signal pool Γ. The hypothesis is empty, and the conclusion of the rule specifies that signal ξ is added into signal pool Γ. In addition, $(a, s, signal(\xi))$ is transformed into $(a, s, skip)$.

- Rule 9 describes the semantics of receiving signal ξ from signal pool Γ. The hypothesis specifies that signal ξ must be present in the signal pool Γ, and the conclusion of the rule specifies that signal ξ is added into signal pool Γ. In addition, $(a, s, wait(\xi))$ is transformed into $(a, s, skip)$.

- Rule 10 and 11 describe the semantics of executing a conditional itinerary $if\ c\ i_1\ else\ i_2$ by agent a when c is evaluated to be true ($\delta(c, \sigma) = true$) or false ($\delta(c, \sigma) = false$), respectively.

- Rule 12 and 13 describe the semantics of operator $\|$ in terms of the semantics of a new operator $|||$ and two new special synchronization itineraries $!\xi$ and $?\xi$, called *send* and *receive* itineraries, respectively. Their semantics will be described later.
  Intuitively, the difference between $|||$ and $\|$ is that synchronization is explicit for $|||$ but implicit for $\|$. Note that $|||$, $!\xi$ and $?\xi$ are not available for itinerary programmers, and are introduced only for the purpose of defining operational semantics of MAIL. Therefore, in an implementation of MAIL, the definition of $\mathcal{I}$ should be extended with these new primitives whereas the original Definition 13.1 should be used for itinerary programmers.

- Rule 14 describes the semantics of parallel operator $|||$. Intuitively, a clone of a is made, and in the resulting configuration, one instance of a is associated with i_1, and the other is associated with i_2.

- Rule 15 describes the semantics of the two special synchronization itineraries $!\xi$ and $?\xi$, which are closely related to the semantics of primitives *send* and *receive* in [39]. Two agent instances need to synchronize between them before either of them can proceed. This semantics is different from $signal\Xi$ and $wait\xi$, which are asynchronous communications.

- Finally, Rule 16 and 17 describe the semantics of executing a loop itinerary $while\ c\ do\ i$: if c is evaluated to be true, then $while\ c\ do\ i$ is reduced to $c; while\ c\ do\ i$; otherwise, the execution of the itinerary terminate, successfully, and it is deleted from the resulting layout. This semantics is closely related to the one for *while statement* in an imperative programming language.

The absence of an inference rule for the *stop* itinerary describes the following semantics of executing a stop itinerary: an agent instance executing a stop itinerary always diverges, i.e., no inference rule is applicable and therefore no progress can be made for that agent instance.

13.3 Regular-Completeness of MAIL

In this section, we study the expressiveness of MAIL in terms of the trace model. Imagine we are observing the execution of a particular itinerary i by an agent a, and we record the sites that are visited by agent a and the order in which this set of sites is visited. When the execution terminates, we get a sequence of visited sites, called a *trace* of itinerary i. Given an itinerary i, it is natural to model i by $traces(i)$ — the set of all traces that itinerary i can perform. We call $traces(i)$ as *the trace model* of i. For example, $traces((a;b)) = \{< a,b >\}$, and $traces((a \parallel b)) = \{< a,b >, < b,a >\}$.

Given two traces s and t, we define the following operators:

- *Concatenation:* the concatenation of s and t is the trace in which s is followed by t and it is denoted by $s \wedge t$. For example $< a, b > \wedge < c, d > = < a, b, c, d >$.

- *Interleaving:* the interleaving of s and t is the trace model that results from all possible interleavings of s and t. It is denoted by $s \bowtie t$ and defined recursively as follows:
  - $s \bowtie <> = \{s\}$.
  - $<> \bowtie t = \{t\}$.
  - If neither s nor t is an empty trace, then

    $$s \bowtie t = \{\text{head}(s) \wedge x \mid x \in \text{tail}(s) \bowtie t\} \cup \{\text{head}(t) \wedge x \mid x \in \text{tail}(t) \bowtie s(t)\},$$

    where head(y) is the first visit of y and tail(y) is the trace consisting of the rest of visits.

- *Kleene closure:* the Kleene closure of s is the trace model consisting of zero or more concatenations of s. It is denoted by s^* and defined as follows:
  - $<> \in s^*$.
  - If $r \in s^*$, then $s \wedge r \in s^*$.

We extend these operators to trace models. Given two trace models P and Q, we define the following operators:

- *Union of P and Q:* $P \cup Q = \{r \mid r \in p \text{ or } r \in Q\}$.
- *Concatenation of P and Q:* $P \wedge Q = \{p \wedge q \mid p \in P \text{and} q \in Q\}$.
- *Interleaving of P and Q:* $P \bowtie Q = \{s \mid s \in (p \bowtie q), p \in P \text{and} q \in Q\}$.
- *Kleene closure:* the Kleene closure of P is denoted by P^* and defined as follows:
  - $<> \subseteq P^*$.
  - If $s \in P$ and $r \in P^*$, then $s \wedge r \in P^*$.

Definition 13.2 *(Trace model of itineraries) The trace model of an itinerary can be characterized by the following rules:*

- $traces(skip) = \{<>\}$, $traces(stop) = \phi$.
- $traces(v) = \{< v >\}$ *where* v *is a visit.*
- $traces(i_1; i_2) = traces(i_1) \wedge traces(i_2)$.
- $traces(if\ c\ i_1\ else\ i_2) = traces(i_1) \cup traces(i_2)$.
- $traces(i_1 \parallel i_2) = traces(i_1) \bowtie traces(i_2)$.
- $traces(while\ c\ do\ i) = traces(i)^*$.

In the following, we define a class of regular trace models and prove that any regular trace model can be expressed by a MAIL itinerary.

Definition 13.3 *(Regular trace model) We define regular trace models over a set of sites $\mathcal{S}$ as follows:*

- *ϕ and $\{v\}$ are regular trace models where $v \in \mathcal{S}$.*
- *If i_1 and i_2 are regular trace models, then $i_1 \cup i_2$, $i_1 \wedge i_2$, $i_1 \bowtie i_2$, i_1^* are regular trace models.*
- *A regular trace model can be obtained only by applying the above two rules in a finite number of times.*

Theorem 13.1 *(Regular-completeness) For each regular trace model m over a set of sites $\mathcal{S}$, there exists an itinerary I such that $traces(I) = m$.*

Proof We prove it by induction. If $m = \phi$, then $traces(stop) = m$. If m = $\{< v >\}$ where $v \in \mathcal{S}$, then $traces(v) = m$. Assume two regular trace models P and Q, for which there exist itinerary I_P and I_Q such that $traces(I_P) = P$ and $traces(I_Q) = Q$. By induction, we have

- $traces(if\ c\ then\ I_P\ else\ I_Q) = P \cup Q$ for some condition c.
- $traces(I_P; I_Q) = P \wedge Q$.
- $traces(I_P \parallel I_Q) = P \bowtie Q$.
- $traces(while\ c\ do\ I_P) = P^*$.

Therefore, given an arbitrary regular trace model m over $\mathcal{S}$, there exists an itinerary I such that $traces(I) = m$. This completes the proof. □

Although MAIL is expressive enough for most applications, there exist some traces which cannot be specified by MAIL. For example, the trace model of site a being visited n times followed by site b being visited n times is not a regular trace model, and cannot be specified by MAIL. However, in practice, this can be achieved in an ad hoc fashion based on the underlying language which is usually Turing-complete.

13.4 Itinerary Safety Reasoning and Assurance

13.4.1 Itinerary Configuration and Safety

Essentially, the set of rules MAIL operational semantics in Figure 13.2 defines a relation between two configurations $\rightarrow \subseteq \Omega \times \Omega$. We denote $c_1 \rightarrow c_2$ if $(c_1, c_2) \in \rightarrow$.

Each relationship $c_1 \rightarrow c_2$ models a possible one-step computation which transforms configuration c_1 to c_2. Therefore, an MAIL language specifies a *transition system* [264], which is a graph in which nodes represent configurations, and edges represent $\rightarrow$ relationships between configurations. However, the corresponding transition system is infinite since the number of configurations Ω in the system is infinite. In the following, we will show that, given an initial configuration, a system can only reach a finite set of configurations. This implies that all systems specified in MAIL are amenable to model checking technique [64, 97] to verify their important correctness and security properties. This is an attractive property of MAIL, in addition to its expressiveness that will be shown in the next section. We first define some notions for the purpose of the proof of Theorem 13.2.

Definition 13.4 *(Reachability) Let* $\rightsquigarrow$ *be a reflexive and transitive closure of* $\rightarrow$. *Given two configurations* c_1 *and* c_2, *we say* c_1 *can reach* c_2 *if* $c_1 \rightsquigarrow c_2$.

Definition 13.5 *(Containment) We define* $\rhd \subseteq \mathcal{I} \times \mathcal{I}$ *as a containment relation between itineraries based on the following rules:*

- $i \rhd i$ *(reflexive).*
- *If* $i_1 \rhd i_2$ *and* $i_2 \rhd i_3$, *then* $i_1 \rhd i_3$ *(transitive).*
- $(i_1; i_2) \rhd i_1$ *and* $(i_1; i_2) \rhd i_2$.
- $(If\ c\ i_1\ else\ i_2) \rhd i_1$ *and* $(if\ c\ i_1\ else\ i_2) \rhd i_2$.
- $(i_1 \parallel i_2) \rhd i_1$ *and* $(i_1 \parallel i_2) \rhd i_2$
- $(While\ c\ do\ i) \rhd i$.
- $(i_1 \parallel i_2) \rhd ((i_1; !\xi) \mid\mid\mid (i_2; ?\xi))$.
- *If* $i_1 \rhd (i_{11} \mid\mid\mid i_{22})$, *then* $(i_1; i_2) \rhd (i_{11} \mid\mid\mid (i_{12}; i_2))$.
- $(i_1 \mid\mid\mid i_2) \rhd i_1$ *and* $(i_1 \mid\mid\mid i_2) \rhd i_2$.

Given an itinerary i, we define $ContainSet(i)$ as the set of itineraries that i contains. By induction on itinerary definition, we can prove that $ContainSet(i)$ is finite for an arbitrary itinerary $i \in \mathcal{I}$. The following theorem tells its reachable configurations are finite, as well.

Theorem 13.2 *(Finite reachable configurations) Let* c *be an arbitrary configuration of a transition system specified in MAIL. The set of configurations that* c *can reach is finite.*

Proof Let $c = (\sigma, l)$ and $IT(c)$ be the set of itineraries in l and $IT(c)'$ be the set of itineraries drawn from IT(c) by substituting $\mathbf{i_1; while\ c\ do\ i_1}$ for $\mathbf{while\ c\ do\ i_1}$ for each itinerary i in $IT(C)$ recursively. Based on the set of rules defined in Figure 13.2, a superset of the configurations that c can reach is SuperReachSet(c) = $\Sigma \times pow(\mathcal{A} \times \mathcal{S} \times ContainSet(IT(c)'))$. Since Σ, $\mathcal{A}$, $\mathcal{S}$, and ContainSet(IT(c)') are finite, SuperReachSet(c) is finite. This proves the theorem. Note that the definition of SuperReachSet(c) is based on $IT(c)'$ instead of on IT(c) because of rule 16 defined in Figure 13.2. □

As a result of Theorem 13.2, numerous verification problems can be defined for mobile agent systems. More importantly, these problems are decidable and can be solved using existing model checking techniques [64]. We give some examples here and leave further development to future work.

Definition 13.6 *(Invariant) Given a condition con, we say con is an invariant of a transition system with respect to initial configuration c if and only if for each $c' = (\sigma', l')$ such that $c \rightsquigarrow c'$, we have $\delta(con, \sigma') = true$.*

Corollary 13.1 *The problem of checking if a condition con is an invariant of a transition system specified in MAIL with respect to initial configuration c is decidable.*

Definition 13.7 *(Safety) A safety partition of Ω, the set of configurations of a system, is a pair (S, IS) such that (1) $S \cup IS \equiv \Omega$ and (2) $S \cap IS \equiv \Phi$. S is called the set of safe configurations, and IS is called the set of unsafe configurations. A system is safe with respect to an initial condition c if and only if for each c' such that $c \rightsquigarrow c'$ we have $c' \in S$, and unsafe otherwise.*

Although for a particular system, the definition of *safe* configurations is application dependent, the following corollary shows that it is decidable to check whether a system is safe or not, due to the fact the initial configuration *c* can only reach a finite number of configurations.

Corollary 13.2 *The problem of checking whether a mobile agent system is safe or not with respect to initial configuration c is decidable.*

13.4.2 Itinerary Safety Reasoning and Assurance

One important aspect of an agent system is the temporal constraint concerning the order of site visits, and the satisfaction checking problem of itineraries associated with temporal constraints. In this section, we propose an itinerary constraint language, *MAILC*, which is based on work flow temporal constraint language *CONSTR* [82]. Using *MAILC*, we can specify that an agent must visit site s_1 before it visits site s_2, and the visit of site *s* causes some other site to be visited or not visited, etc. (more examples are shown below). These constraints are believed to be sufficient for the needs of itinerary reasoning, and they are far beyond the capabilities of current available agent systems. In the following, we define *MAILC* and show that the itinerary satisfaction checking problem is decidable.

Definition 13.8 *(Itinerary constraint language) We define a formula of itinerary constraint language* MAILC *as the following form:*

$$A ::= T \mid F \mid s \mid s_1 \otimes s_2 \mid A_1 \wedge A_2 \mid A_1 \vee A_2 \mid \neg A.$$

We also define the implication connective as $A_1 \rightarrow A_2 ::= \neg A_1 \vee A_2$.

The following examples illustrate some possible constraint expressions and their meanings:

- e: e must be visited eventually.
- $e \wedge f$: both e and f must be visited (in any order) eventually.
- $e \vee f$: at least one of e and f will be visited eventually.
- $\neg e \vee \neg f$: at least one of e and f will not be visited.
- $e \otimes f$: one must first visit e and then visit f (possibly visit other sites in between).
- $\neg(e \otimes f)$: one must not visit e before visiting f.
- $e \rightarrow f$: if e is visited, then f must be visited (either before or after visiting e).
- $e \rightarrow (e \otimes f)$: if e is visited, a visit of f must occur after some visit of e.
- $e \rightarrow (f \otimes e)$: if e is visited, a visit of f must precede some visit of e.

To establish the notion of itinerary satisfaction, we first introduce the notions of trace satisfaction and trace model satisfaction.

Definition 13.9 *(Trace satisfaction) We define the satisfaction relationship* $\models$ *between a trace* t *and a constraint expression* C *as follows by structural induction on* C. *If* $t \models C$ *does not hold, we denote it by* $t \not\models C$.

- $t \models T$ *and* $t \not\models F$.
- *For* $s \in \mathcal{S}$, $t \models s$ *if and only if* $s \in t$.
- *For* $s_1, s_2 \in \mathcal{S}$, $t \models s_1 \otimes s_2$ *if and only if there exist traces* t_1 *and* t_2 *such that* $t_1 \wedge t_2 = t$, $t_1 \models s_1$ *and* $t_2 \models s_2$.
- $t \models A_1 \wedge A_2$ *if and only if* $t \models A_1$ *and* $t \models A_2$.
- $t \models A_1 \vee A_2$ *if and only if* $t \models A_1$ *or* $t \models A_2$.
- $t \models \neg A$ *if and only if* $t \not\models A$.

We extend the notion of satisfaction to those between trace models, itineraries, and constraint expressions.

Definition 13.10 *(Trace model satisfaction) Given a trace model Q and a constraint expression C, we say Q satisfies C, denoted by $Q \models C$, if and only if $Q \neq \Phi$ and for each $t \in Q$, we have $t \models C$.*

Definition 13.11 *(Itinerary satisfaction) Given an itinerary I and a constraint expression C, we say I satisfies C (denoted by $I \models C$) if and only if $traces(I) \models C$.*

Note that given an itinerary I, set $traces(I)$ might be infinite (e.g., I contains a loop construct), and to check if I satisfies a constraint expression C, we have to check for each $t \in traces(I)$ if $t \models C$. This seems to be undecidable when $traces(I)$ is infinite. The following theorem, however, shows that the itinerary satisfaction checking problem is decidable. The proof of the theorem shows itinerary satisfaction checking can be performed in polynomial time.

Theorem 13.3 *(Itinerary satisfaction checking) Given an itinerary I specified in MAIL and a constraint expression C specified in $MAILC$, the problem of checking if $I \models C$ is decidable and the time complexity of checking is $O(m * n)$, where m is the size of itinerary I and n is the size of constraint C.*

Proof We prove it by structural induction on itinerary I in Definition 13.1 and on constraint C in Definition 13.8. Specifically, according to the constraint definition,

- If the constraint **C = T**, then $I \models T$;
- If the constraint **C = F**, then $I \not\models F$;
- If the constraint **C = s** and $sinS$, then $I \models s$ is decidable based on induction of itinerary I:
  - $I \not\models s$ for I = skip, stop, $\alpha?X$, $\alpha!a$, $signal(\xi)$, or $wait(\xi)$.
  - $s \models s$ and $s' \not\models s$ $(s \neq s')$.
  - $i_1; i_2 \models s$ if and only if $i_1 \models s$ or $i_2 \models s$.
  - $if\ b\ then\ i_1\ else\ i_2 \models s$ if and only if $i_1 \models s$ and $i_2 \models s$.
  - $i_1 \parallel i_2 \models s$ if and only if $i_1 \models s$ or $i_2 \models s$.
  - $While\ b\ do\ i \not\models s$.
- If the constraint **C =** $s_1 \otimes s_2$, then $I \models s_1 \otimes s_2$ is decidable based on induction of itinerary I:

- $I \not\models s_1 \otimes s_2$ for I = skip, stop, s, $\alpha?X$, $\alpha!a$, $signal(\xi)$, or $wait(\xi)$.
- $i_1; i_2 \models s_1 \otimes s_2$ iff (1) $i_1 \models s_1 \otimes s_2$, or (2) $i_2 \models s_1 \otimes s_2$, or (3) $i_1 \models s_1$ and $i_2 \models s_2$.
- $If\ b\ then\ i_1\ else\ i_2 \models s_1 \otimes s_2$ iff $i_1 \models s_1 \otimes s_2$ and $i_2 \models s_1 \otimes s_2$.
- $i_1 \parallel i_2 \models s$ iff $i_1 \models s_1 \otimes s_2$ or $i_2 \models s_1 \otimes s_2$.
- $While\ b\ do\ i \not\models s_1 \otimes s_2$.

- If constraint **C** = $A_1 \wedge A_2$, then $I \models C$ iff $I \models A_1$ and $I \models A_2$.
- If constraint **C** = $A_1 \vee A_2$, then $I \models C$ iff $I \models A_1$ or $I \models A_2$.
- If constraint **C** = $\neg A$, then $I \models C$ iff $I \not\models A$.

The theorem is proved. □

13.5 Concluding Remarks

We have defined the syntax and operational semantics of MAIL, a mobile agent itinerary language that allows one to specify the mobile behavior of an agent. MAIL is structured and compositional. It is also expressive in the sense that any regular itinerary traces can be recursively constructed by a set of primitive itineraries. We have also proved that for any initial system configuration, the set of reachable configuration defined by an MAIL itinerary is finite and that the problem of checking whether such a mobile agent system is secure with respect to the initial configuration is decidable. Besides the reachability and decidability, MAIL lays a foundation to assuring more correctness and security properties of mobile agent computing. We are developing a constraint language for the specification of such requirements and integrating the constraints into the MAIL framework.

We note that current version of the MAIL language supports static itineraries only. That is, the initial itinerary associated with an agent cannot be changed during its life cycle, although its actual path can be determined dynamically. Extension of the MAIL to support dynamic itineraries would help further understanding of mobile agent systems in which itineraries and communication might interplay together. For example, two mobile agents might pass from one to another not only primitive values and communicating channels, but also itineraries. It is still not clear what the semantics would be for passing an itinerary from one agent to another since by doing so, the former might affect the navigational behavior of the latter.

Chapter 14

Security Measures for Server Protection

A primary concern in agent-based mobile computing is security. From the perspective of servers, there is nothing more risky than providing docking services to remote general-purpose executables. Since an agent migrates with its code as well as data and running states, its privacy and integrity also need to be protected from attacks by visited servers in the itinerary. These two security requirements are equally important in an open environment because mobile agents can be authored by anyone and execute on any server that has docking services. However, server protection is more compelling in a coalition environment where the servers are generally cooperative and trustworthy, although mobile code from different servers may have different levels of trustiness. This chapter presents the design and implementation of an agent-oriented access control model in the Naplet system and a coordinated spatio-temporal access control model in coalition environments.

14.1 Introduction

Server protection in mobile agents shares the same objective as in mobile codes that deal with the transfer of programs only. However, it raises more challenge issues because of the requirement to support the transfer of agent state and multihop migration as determined by the agent itinerary. Since mobile code systems violate many of the assumptions that underlie security measures in traditional client/server-based distributed systems [62], security in mobile code systems has become a research focus in both academy and industry, with the popularity of Java and mobile code based Internet services.

Java security architecture provides a baseline for secure mobile code systems [135]. It relies upon signatures, authentication, and other cryptographic building blocks to verify trustworthiness of mobile programs or their sources. A limitation of this approach is that mobile codes must be authored by someone of whom the servers have a prior knowledge, because the signer or source of a mobile program can only sign on the code integrity and it cannot attest to the program functions. The Java security measures are limited to support for one-hop code migration. In Chapter 11, we reviewed a number of other security measures for site protection against mobile codes.

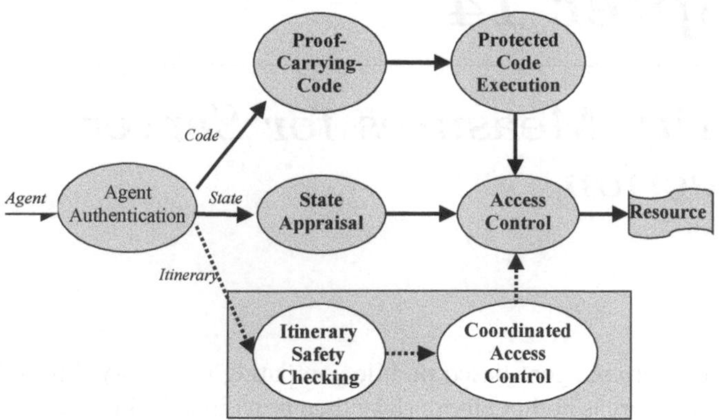

Figure 14.1: Work flow of site protection.

A most noteworthy approach is proof-carrying code (PCC) [10, 228, 229], which allows a host to determine with certainty whether it is safe to execute a remote code. The basic idea of PCC is that the code producer is required to provide an encoding of a proof that his/her code adheres to the security policy specified by the code consumer. The proof is encoded in a form that can be transmitted digitally, so that the code consumer can quickly validate the code using a simple, automatic, and reliable proof-checking process.

PCC is applicable to server protection against untrusted mobile agents. However, agents' multihop mobility complicates the treatment of agent authentication and privilege delegation. Lampson *et al.* [184] proposed a theory for authentication in distributed systems, based on the notion of principal and a speaks-for relation between principals. It can be used to derive one or more principals who are responsible for a request. Based on this theory, Farmer *et al.* [103] and Berkovits *et al.* [32] defined agent authentication and privilege delegation rigorously.

As a mobile agent travels across a network to pursue its designated tasks, it needs to be granted certain privileges for execution at each host. The privileges are determined based on the agent's request and the host security policy. Farmer *et al.* [103] proposed a state appraisal approach to ensure that an agent with corrupted *states* won't be granted any privilege. In the approach, each host computes an agent-carried state appraisal function over the agent state for a set of privileges to a request. The server then determines the granted permissions according to its security policy (access control). In summary, we show the work flow of site protection upon the arrival of an agent in Figure 14.1.

We note that PCC and state appraisal measures together address the needs of server protection by requesting agent authors to provide correctness proofs of agent functions on individual sites. When a server receives an agent, it needs to know not only how the agent would function, but also whether the agent is supposed to visit this server according to its itinerary. This is partially addressed by itinerary reasoning, as

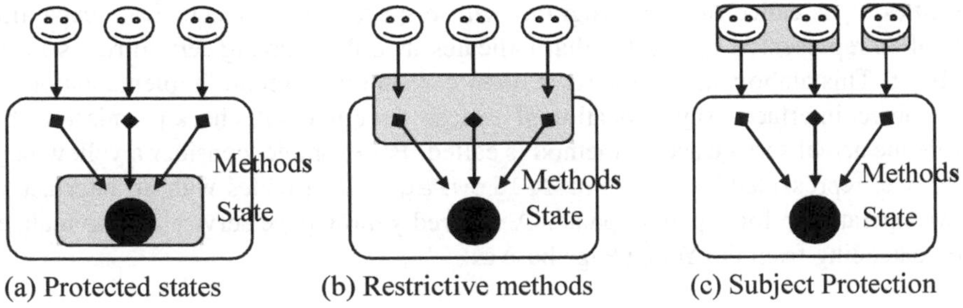

Figure 14.2: Hierarchical access control.

discussed in Chapter 13. Between collaborative servers in a coalition environment, there is also a need for coordinated access control over shared resources in different sites. An example of such requirements is "if an agent accesses a resource r (e.g., a licensed software package or its trial version) on site s_1 for too many times during a certain time period, it is not allowed to access the resource on site s_2 forever." If the resource is private to the server, an agent-oriented access control model suffices to protect the resource. The access control model can be tailored from existing authorization strategies for mobile codes. Synchronized access to shared resources raises unique security requirements for coordinated access control in both time and space domains, because a request originates on a remote host and may traverse multiple hosts and network channels that are secured in different ways and are not equally trusted. In the following, we present the details of two such models we recently developed [122, 335].

14.2 Agent-Oriented Access Control

14.2.1 Access Control in Mobile Codes

Consider a controlled service, represented by an object with an array of internal states and a group of access methods. Access to the object can be protected in different layers. At the bottom layer is protection of invalid access operations, as shown in Figure 14.2(a). It is supported by object-oriented programming languages like Java. Each object defines its internal states as private data and open services as public methods. The compiler guarantees the object to be accessed via public methods only. Note that C++ is not a pure object-oriented language and the internal states of a C++ object can be accessed directly by the use of its reference.

Valid access operations are not necessarily legitimate. Access control at the second layer is to protect the object from unauthorized invocations, as shown in Figure 14.2(b), based on the access control list associated with the object or the capabilities of its callee (user or program) with respect to this object. Each capability is often represented by a ticket that indicates its holder having certain rights to the object. This authorization process is often carried out in secure implementations of the object interface. Any invocation of a secured method will check permissions before the actual service access method is called. For example, consider a yellow page service, represented by a class of `YellowPage`. It provides with an interface of two operations: lookup and update. A secured yellow page service can be realized by extending from the YellowPage base as:

```
Class SecuredYellowPage extends YellowPage {
      void lookup() {
            AccessController.checkPermission()
            super.lookup();
      }
      void update(Address a, Object o) {
            AccessController.checkPermission();
            super.update(a, o);
      }
}
```

Note that the authorization process must be preceded by a callee authentication operation. It is to verify the callee identity. We refer to the authenticated source of a request as an access subject.

At the top layer of access control is protection based on the role of the subject with respect to the object, as shown in Figure 14.2(c). Users in an organization can log in with different roles. According to their responsibilities, the users can be granted different rights with respect to a controlled service. For example, in the yellow page service, administrators are granted permissions to both lookup and update, while normal users are limited to lookup. Accordingly, role-specific service proxies with different confined interfaces to the service object can be created. If the access permission is granted, a reference to the service proxy is returned. With the reference, the subject can access the services via normal method invocations. In contrast to the subject-capability based access control in the second layer, this subject-role-based approach implements access control at a coarse-grained level. Role-based access control simplifies the management of authorization while providing a great flexibility in specifying and enforcing service-specific protection policies [108]. We note that the capability-based and role-based access methods can be combined. In the yellow page service, for example, subjects in the role of normal user can be granted different access permissions according to their capabilities. An extreme of the configuration is subject-specific service proxies that are created on demand when a subject issues the first access request.

In mobile agent systems, an important class of subjects is visiting agents. Since they request services on behalf of their owners on remote machines, agent-oriented access control raises four unique issues. First is delegation of agent owner privilege to the agent. Previous Java-based mobile agent systems assumed a sole subject of server administrator and delegated agent execution to the agent server. The server distinguishes between the agents and exercises access control according to their code sources. The agent privilege delegation avoids the need for switching subjects in access control. However, it is unable to grant access rights to agent owners.

Second is agent naming and authentication. Access control in previous Java-based mobile agent systems was based on code sources, i.e., the code original location as specified by a URL and an optional set of digital signatures of the creator (author). To enable agent-oriented access control, subjects of agent owner and agents must be explicitly named, in addition to the agent creator. Authentication is a process by which the identity of a subject is verified. It typically involves the subject demonstrating some form of evidence (e.g., keywords and signed data using a private encryption key) to prove its identity. The subjects of agent author and owner can be authenticated in a traditional way when they log in. However, there is a need for distributed protocols to verify the identity of mobile agents on remote machines.

Third is agent-oriented access control permission. A security policy is essentially an access control matrix that describes the relationship between agents and the permission they are granted. Previous code source based access control is based on the static characteristics of the agents. To reflect the dynamic nature of mobile agents in access control, new permission structures are needed.

Last is agent-oriented service proxy. For security concerns, the services should not be open to mobile agents directly. Instead, their access should be controlled by a resource manager who creates an agent-specific service proxy (a subset of services) and assigns reference to the service proxy to the agent. For example, a yellow page service contains method for lookup as well as update. For a normal agent, the resource manager can create a service proxy containing an interface to lookup only.

14.2.2 Agent Naming and Authentication in Naplet

A name is a string of bits or characters that uniquely identifies its associated entity in distributed systems. By uniqueness, its means a name refers to at most one identity and the name is immutable. An entity can have more than one name in different formats for different purposes. A naming scheme determines the name format and the identity information to be encoded in the name. For example, each computer on the Internet has a 32-bit IP address name for communication with others on the Internet. The address contains geographical location information for easy location of the machine on the Internet. Each computer on the Internet has another DNS name in an alphanumeric character string for easy reference by human beings.

In mobile agent systems, agent naming plays a key role in agent communication, management, and agent-oriented access control. It is known that a mobile agent generally has a seven-stage life cycle: created by user or applications (author or builder); dispatched by user or applications (owner); landed in agent-enabled servers; executed in servers and access to local resources and services; cloned, suspended and resumed by authorized servers or owner; navigated between server; and terminated or called back by owner.

Each agent needs a unique identifier so that its owner and/or other agents can locate it for communication and management purposes. Their underlying service is agent location. Recent studies on agent naming for location services identified two desirable properties: location transparent and scalable. Due to the dynamic nature of mobile agents, a well-defined name should not contain any current location information of the agent. The name should also be selected autonomously by its owner in an open and distributed environment. A common practice is to name each agent by its function, in combination with its home host address. For example, agent name `ece.wayne.edu:5000/HelloAgent` refers to a hello agent created at agent server `ece.wayne.edu:5000`. The inclusion of the host address not only enables agent owners to name their agent functions independently, but also provides support for home-based agent location. Note that there are systems based on variants of the naming scheme. Voyage supported agent naming based on a reference host [258] and Aglet allows naming an agent with an automatically assigned numerical identifier [185] .

The existing naming schemes were optimized for location services and related agent communication and management. However, they are not suitable for agent-oriented access control. Agent-oriented permissions like "Allow access rights to any naplet from Alice at Wayne State University" requires a naming scheme to include agent owner information in agent names. The naming scheme should also support agents' unique clone operation. More importantly, the naming scheme should give the agent-docking servers a flexibility to specify agent implication-based permissions.

Name implication, as implemented by the Java CodeBase class, is a key in code-sourcer-based access control, as well. For example, any permission granted to a code source `http://ece.wayne.edu/classes/` implies the same permission to code source `http://ece.wayne.edu/classes/foo.jar`. We notice that the CodeBase implication is based on a well-defined way of name resolution of files. For agent name resolution, we need a name resolution in the agent owner space, in addition to a resolution in cloned agents.

A subject in Java refers to an authenticated source of request. Each subject is populated with associated principals, or identities. In the Naplet mobile agent systems, there are three subjects involved in the life cycle of an agent. First is server administrator. When a user wants to install a naplet server, he/she must be authenticated as a server administrator in a normal password-based login context. Only the administrator has the privileges of server setup and administration. It has a principal of `NapletServerAdministrator`. Each naplet server can be configured with one or more application services either statically during the server installation or dynamically when the server is on-line. The service deployment can also be conducted by the server administrator. The services are then run under the privilege of the administrator.

```
LoginContext lc = new LoginContext("ServerAdministrator");
try {
    lc.login();
```

```
} catch (LoginException) {}
Subject admin = lc.getSubject();
Subject.doAsPrivileged( admin, new PrivilegedService() );
```

The second subject in the Naplet system is naplet owner with an associated principal of `NapletOwner`. A naplet acts on behalf of its owner (user or application) who dispatches the naplet. Dispatch of the naplet requires an authentication of its owner. The authenticated owner signs its digital signature on the naplet, indicating his/her responsibilities for the naplet behavior. On arrival at a server, the naplet must be authenticated based on the certificate of its owner issued by an authority or via *a priori* registration. The authenticated naplet needs to access system resources and application services during its execution. The naplet subject has a principal of class `NapletPrincipal`. Once a naplet is landed, the naplet server delegates the naplet execution to the subject of the naplet itself. For example, when a "HelloNaplet" arrives at a naplet server:

```
Subject napSubject = Runtime.authenticate(helloNaplet );
AccessController.checkPermission();
Subject.doAsPrivileged( napSubject, new PrivilegedAction() {
        helloNaplet.init();
        helloNaplet.onStart(); }, null );
```

The naplet can't access controlled services directly. Instead, the service manager creates a proxy with customized secure interface for each request naplet. An important aspect of subjects is principal naming. The server administrator subject is a local object. Since its principals won't be referenced by any remote object, their names can be determined locally by each server. In contrast, the subjects of naplet and naplet owner are global. Their principal names must be in a human-readable and well-defined format so that the server administrator can set up access control permissions based on their names. For example, the administrator can grant permission to an "HelloNaplet" from the owner "czxu of Wayne State University". Recall that each naplet has a system-wide unique naplet identifier. Since this identifier is generated by its dispatcher, it can't be used by remote naplet servers for setting up access control permissions.

NapletOwnerPrincipal. As a user of the Naplet system, the subject of naplet owner can have different distributed naming schemes for its principals. One is X.500 naming format, in which each name is collection of (attribute, value) pairs. The attributes include country (C), locality (L), organization (O), and organization unit (OU). For example, a global unique name `/C=US/O=WSU/OU=ECE` represents the owner analogous to the DNS name ece.wayne.edu. This naming scheme is widely used in distributed applications across administrative domains. However, its implementation in mobile agent systems has to rely on a centralized directory service because it is completely machine location independent. In the Naplet system, we adopt a Kerberos-like naming scheme in a format of user-account-name\hostname. For example, czxu\ece.wayne.edu refers to the user "czxu" at host of ece.wayne.edu. Since the host name is unique on the Internet and the user account information is administrated by the host owner, this e-mail address-like naming scheme guarantees the uniqueness of the naplet owner subject. The Kerberos-like naming scheme defines an implication relationship based on the DNS name resolution of the machines. We also assume a wildcard character '*' for any user on a machine. Consequently, the NapletOwnerPrincipal has implications as shown in this example: a naplet owner of *\wayne.edu implies *\ece.wayne.edu, which in turn implies czxu\ece.wayne.edu.

NapletPrincipal. Naplet is a first class subject that acts on behalf of the naplet owner. The naming scheme for its principals must be scalable and location transparent. The scheme must also be expandable so as to support cloned naplets. Existing mobile agent naming schemes in the format of "code-source/agent-name" are scalable and location transparent. But they are neither expandable for cloned agents nor inclusive of agent owner information. In the Naplet system, we define a naming scheme in a format of "naplet-owner/naplet-name: naplet-version." The naplet name part is a case-insensitive alphanumeric string, selected by its owner. The naplet version part is a sequence of integers with dot delimiters, representing the naplet generation in its family. The original naplet has a version sequence of "0". For example, a name of czxu\ece.wayne.edu/HelloNaplet:0 refers to the original "Hello Naplet" from the naplet owner of czxu\wayne.edu. Its first clone is represented as czxu\ece.wayne.edu/HelloNaplet:1.0 and the second one is czxu\ece.wayne.edu/HelloNaplet:2.0. The cloned naplet can be further cloned. A name of czxu\ece.wayne.edu/HelloNaplet:2.1.0 represents the third generation of the original naplet. The principal naming scheme defines an implication relationship based on the name resolution of NapletOwner and naplet clone generations. For example, a naplet czxu\ece.wayne.edu/? (i.e., any naplet from owner czxu\ece.wayne.edu) implies czxu\ece.wayne.edu/HelloNaplet:? (i.e., any version of the HelloNaplet from czxu\ece.wayne.edu), which in turn implies the cloned naplet czxu\ece.wayne.edu/HelloNaplet:2.1.0.

14.2.3 Naplet Access Control Mechanism

The security behavior of a Java run-time system is specified by its security policies. Represented in an access-control matrix, each security policy describes a set of permissions granted to a subject in the access of resources such as network sockets and files. The Naplet system defines a set of additional permission types to control access rights of naplet subjects over restricted services on the naplet servers.

NapletRuntimePermission(target). This is a subclass of java.lang.BasicPermission. It has a target but no actions. The target for the permission can be any string of characters. This type of permission essentially offers a simple naming conversion. For example, `NapletRuntimePermission`("land") denotes the permission for naplet landing and `NapletRuntimePermission`("clone") for naplet clone. Other possible target names include:

- "Shutdown" for shutting down the local naplet server,
- "Execute" for execution permission of a naplet,
- "Suspend" for suspension permission of a naplet.

As an example, the following policy grants the server administrator to shut down the server and the hello naplet from `czxu\ece.wayne.edu` to land.

```
grant Principal NapletAdministrator "administrator" {
    permission NapletRuntimePermission("shutdown");
}
grant Principal NapletPrincipal "czxu\wayne.edu/HelloNaplet.*" {
    permission NapletRuntimePermission("land");
}
```

NapletServicePermission(service, actions). This permission class represents a permission of access to the target service. This can be granted to either naplets or the local server administrator. For example, the following security policy grants the server administrator to look up and update a yellow page service and a "query naplet" from `czxu\wayne.edu` to look up the service only.

```
grant Principal ServerAdministrator "administrator" {
    permission NapletServicePermission( "yellow-page", "lookup, update");
}
grant Principal NapletPrincipal "czxu\wayne.edu/QueryNaplet" {
    permission NapletServicePermission("yellow-page", "lookup");
}
```

NapletSocketPermission(server, actions). This permission class represents permissions granted to naplet subjects to access a target server. The target server can be represented as a string in the format of "hostname:port/ napletservername." For example, `ece.wayne.edu:2400/YellowPage` represents a "YellowPage" server running on port 2400 of the `ece.wayne.edu` machine. Unlike the Java built-in `SocketPermission` that controls access to machines, this `NapletSocketPermission` object provides a high-level abstract for naplet communication and migration between naplet servers. The following policy grants "hello naplet" from `czxu\wayne.edu` a permission to talk to naplets running on a remote server and a permission to migrate to the remote server.

```
grant Principal NapletPrincipal "czxu\wayne.edu/HelloNaplet" {
    permission NapletSocketPermission(
            "ece.wayne.edu:2400/YellowPage", "talk, travel");
}
```

The base `Permission` class has an abstract method, `boolean implies (Permission perm)`, and it must be implemented by each subclass above. Basically, "Permission p implies permission q" means if a subject is granted permission p, the subject is guaranteed permission q. The implication relationship between two permissions p and q of the same class can be determined easily based on the name resolution of `NapletPrincipal`.

The permission classes represent access to system resources and application services. Currently, all Java permission classes, including those defined above, are positive in that they represent access approvals, rather than denials. Past practice and experience demonstrated that the Java security architecture with positive permissions worked very well for code-source-based access control. However, there are many occasions in need of subject-oriented access denials. A not unusual example is to allow any hello naplets from `wayne.edu` to land for service "ABC," except those from `xyz\wayne.edu`. To enforce such a subject-oriented permission, the Naplet system provides the server administrator a handler to define his/her own customized permission check. The Naplet system utilizes a `checkPermission()` method of the JRE SecurityManager to achieve this end. It separates security policies from a permission-handling mechanism for flexible access control. The following permission grant and customized permission check codes realize the above negative permission example.

```
grant Principal NapletPrincipal "?\wayne.edu/HelloNaplet" {
    permission NapletRuntimePermission("land");
}
public class NapletSecurityManager extends SecurityManager {
    public void checkPermission( Permission perm) {
        if (perm instanceof NapletRuntimePermission) {
            Subject sbj = Subject.getSubject(AccessController.getContext());
            String service = perm.getName();
            String actions = perm.getActions();
            if (sbj.getPrincipals() contains "xyz\wayne.edu/HelloNaplet"
                        and service is "ABC" and actons contains "land" )
                throw new SecurityException("HelloNaplet from xyz is denied");
            else
                super.checkPermission( perm );
        }
    }
}
```

14.3 Coordinated Spatio-Temporal Access Control

In this section, we present a coordinated access control model in support of synchronized access to shared resources in both time and space domains. For timewise synchronization, permissions granted to an agent must be associated with durations of validity. For spacewise synchronization, a grant of the temporal permissions must also be subject to certain spatial constraints related to agent access history. We extend the role-based access control (RBAC) model [108] to support the specification of temporal and spatial constraints for the coordinated access control in mobile computing. It also provides formalisms for the constraint satisfaction reasoning.

There were recent temporal access control models for monitoring time-sensitive activities. For example, Bertino *et al.* [34] presented a time-based scheme by using periodic authorizations and derivation rules. Each authorization is assigned a periodicity constraint, which specifies a set of intervals when the authorization is enabled. Later on, the authors integrated such interval-based temporal constraints into RBAC and developed a temporal RBAC (TRBAC) model to efficiently manage permissions of temporal roles [35]. TRBAC was recently generalized by Joshi *et al.* [162] by incorporating a set of language constructs for the specification of various temporal constraints on roles, user-role assignments, and role-permission assignments. TRBAC models can be tailored to support temporal constraints in agent-based mobile computing, but they can't accommodate spatial constraints because the concept of access history is not defined in the model.

In [92], Edjlali *et al.* presented a mechanism to use a selective history of the access requests of a mobile code in access control. Their history-based access control model is limited to one-hop mobile codes. Because it has no concept of itinerary and shared resource (objects), this model is hardly extensible to include the access history of a mobile agent on different sites.

14.3.1 Temporal Constraints

Timing constraints are essential for controlling time-sensitive activities in various applications. For example, in a work flow management system tasks usually have timing constraints and need to be executed in certain order. The existing timing constraints in TRABC [35] and GTRABC [162] are based on intervals of periodic events that are imposed on roles. Each interval has explicit beginning and ending time points, indicating the time when a role becomes enabled or disabled. Since the periodic constraints are imposed on roles in TRBAC, a disabling event of a role would revoke all of its granted privileges. Considering the fact that different permissions authorized to a role often have different temporal constraints, more roles need to be defined in TRBAC. Moreover, because there is no global clock in distributed systems and the arrival time of an agent on a server is unpredictable, the interval timing models are not appropriate to characterize the time-sensitive activities of mobile agents on different servers. To tackle these problems, we propose an extended RBAC control to coordinate the agent's activities by the use of time durations (i.e., intervals with no beginning and ending time points). Instead of associating periodic events with roles, we define temporal constraints as additional attributes of permissions so as to avoid complicating permission management.

As in [35], we denote $perm = (op, obj)$ as a permission associated with a subject through an active role, indicating that the subject is authorized to execute the operation *op* on object obj. We define a temporal permission, denoted by $tperm$, as a tuple $< \Delta t, perm >$, where Δt is a time duration expression specifying the validity span of the permission $perm$. It represents that the subject, being granted the permission, is allowed to exercise operation *op* on object *obj* only for a time duration of Δt since the corresponding role is activated. A permission without any temporal constraint can be expressed as a temporal permission whose validity span is infinity.

With respect to the validity of time duration, each temporal permission can be in one of three different states for an agent: *inactive*, *active-but-invalid*, and *valid*. A temporal permission is *inactive* if it is not assigned to any role of the requesting subject or it is assigned to a role of the subject but that role has not be activated. An active permission becomes invalid when the validity duration of the temporal permission expires. That is, as we apply the discrete time model as used in [35].

Definition 14.1 *(Valid permission) A permission is valid at time* t *on a site* s*, when it is active at* t *and* t *is within its validity duration. That is,*

$$(\forall perm \in Permission)(\forall t \in \mathcal{N})\ valid(t, s, perm) \Leftrightarrow$$
$$(\exists t_a \in \mathcal{N}) activate(t_a, s, perm) \wedge t_a \leq t \leq t_a + dur(perm),$$

where the type permission Permission *contains all the access permissions specified on a site, the predicate $activate(t_a, s, perm)$ denoting permission* perm *is activated since the time point t_a on site s, and the function $dur(\cdot)$ returning the validity duration attribute associated with a permission.*

As a result, the *object access authorization* of the RBAC model [108] is adjusted to incorporate the validity of a permission as a necessary condition for authorizing an access request. The temporal constraints associated with permissions require the access control mechanism on each site not only to check the availability of related permissions, but also to verify their validity. This validity duration requirement is referred to as *validity duration obligation* property, which can be derived directly by the definition of valid permissions and the semantics of the related temporal operators.

Proposition 14.1 *(Validity duration obligation) A permission becomes invalid after the validity duration of the corresponding temporal permission expires, since its activation. That is,*

$$(\forall perm \in Permission)\ \Box(activate(s, perm) \Rightarrow \Diamond dur(perm) \neg valid(s, perm)), \tag{14.1}$$

where the modality $\Diamond np$, which means "It will be the case the interval n hence that p," is equivalent to Fnp *in [248], and* activate(s, perm) *for a permission* perm *has the same semantics as* activate(t_a, s, perm) *in the temporal logic context.*

The validity duration is a predefined default value for each temporal permission, no matter which agent activates it. To accommodate differentiated services for different agents, the validity duration can be adjusted by the use of a validity duration adjustment certificate of an agent for that permission, after its successful authentication. A duration adjustment certificate can be abstracted as a triple $(pn, \Delta t', perm)$, in which $pn \in \{+, -\}$ indicates whether the adjustment duration $\Delta t'$ is positive or negative to the base duration of the temporal permission for *perm*. Therefore, the right-hand side of (14.1) becomes $\Diamond dur(perm) \Diamond \Delta t' \neg valid(s, perm))$. According to the *FF Axiom* in [248], it derives $\Diamond(dur(perm) + \Delta t')\neg valid(s, perm))$, indicating an extension of the validity duration when the adjustment is positive.

14.3.2 Spatial Constraints

A mobile agent roams across a network following its itinerary. Its future travel route may be branching and depend on the execution results on visited sites. However, at each time point in its lifetime, the past travel trace, a sequence of visited sites, is unique. As in Section 13.3, we denote the trace of a mobile agent by $< s_0, s_1, \ldots, s_n >$. At each site, the agent performs operations with access to resources of different types. Since private resources in a server can be accessed under local control, we focus on access to shared resources between the sites.

Spatial constraints are defined over agent access actions. The constraint specification language must be expressive so that most security requirements related to synchronized access to shared resources can be represented. Meanwhile, the language must be simple and intuitive so that the constraints can be reasoned easily. A permission grant requires constraint satisfaction checking at run time right after an agent is authenticated and its role is activated. There are many general-purpose constraint languages, such as object constraint language [319]. They are expressive, but too complex for spatial constraint reasoning and run-time constraint satisfaction checking. In the following, we present a simple language called SRAC based on propositional logic for synchronized resource access.

A key concept of the language is the agent *access history*. It is essentially a list of accesses performed by an agent in its itinerary. We represent an access by a tuple (s, o, r), where s is a variable of type site $S = (s_0, s_1, s_2, \ldots)$, o is a variable of type operation $o = (read, write, execute, \ldots)$, and r is a variable of type shared resource $R = (r_0, r_1, \ldots)$. We define a set of all accesses by an agent as $A = \{(s, r, o)\}$. Over the set A, we define a operation selection (σ) to select a subset of accesses that meet certain conditions. For example, selection $\sigma_{o=update}(A)$ contains all accesses in the history that perform update operations. We refer to the set of accesses on all shared objects in a site as a *synchronized access record* (SAR). That is, $SAR_i = \sigma_{s=s_i}(A)$.

Definition 14.2 *(Access history) The access history ($\mathcal{H}$) of a mobile agent is a sequence of SARs on the visited sites according to the agent itinerary. That is, $\mathcal{H} = (SAR_0, SAR_1, \ldots, SAR_n)$.*

After an agent's execution on a site, the corresponding SAR is appended to its access history. It is also possible that there are several SARs for a site in the access history of a mobile agent, due to its repeated visits to the site. In that case, the selection operation for the site, $\sigma_{s=s_i}(\mathcal{H})$ will return a set of SARs. The access history maintains all the access information of an agent needed by cooperative sites for security control. The definition of access history is independent of its implementation. It can be centralized or distributed, carried by the agent, or maintained by sites. We only require that each site in the coalition can access the history information of its executing agents.

Over the agent access history $\mathcal{H}$, spatial constraints can be defined. We define the constraints recursively based on two primitive constraints. One is the existence of access (s, o, r) in the history $\mathcal{H}$. The other is a cardinality constraint over set A or its subsets, denoted by $\#(m, n, A)$. It states the number of elements in set A must be between two predefined integers m and n. For example, a constraint $\#(0, 5, \sigma_{r=RSW}(A))$ indicates a restricted software (RSW) package, either licensed or trial version, cannot be accessed by more than 5 times, no matter where the agent is run. Let C_1 and C_2 be constraints; a compound constraint is a conjunction of C_1 and C_2 by the use of logical operators AND, OR, NOT. Like the itinerary constraint language in Chapter 13, we define a shared resource access constraint language SARC as follows:

Definition 14.3 *Shared resource access constraint language*

$$C ::= T \mid F \mid (s, o, r) \mid \#(m, n, A) \mid C_1 \wedge C_2 \mid C_1 \vee C_2 \mid \neg C$$

We also define the implication connective as $C_1 \rightarrow C_2 ::= \neg C_1 \vee C_2$.

The semantics of the spatial constraint satisfaction for an agent subject is as follows.

- For a permission *perm* with a (s_i, o_a, r_a) constraint, only when the agent has exercised operation o_a to resource r_a on site s_i, the constraint for permission *perm* is satisfied on the current site, denoted by s. That is,

  $$satisfied(s, perm) \Rightarrow \sigma_{s=s_i, o=o_a, r=r_a}(\mathcal{H}) \neq \varnothing,$$

  where the predicate $satisfied(\cdot)$ states that the spatial constraint associated with a permission can be satisfied according to the access history of the agent.

- For a permission *perm* with a $\#(m, n, A)$ constraint on resource r_a, the constraint checking for the permission *perm* cannot be satisfied if the agent has accessed the corresponding resource more than the stipulated times. That is,

  $$satisfied(s, perm) \Rightarrow \#(\sigma_{r=r_a}(\mathcal{H})) \leq n, \text{ for } m = 0.$$

The spatial constraints for a permission state the necessary conditions for the authorization of that permission. We define the *permission activation obligation* to express the property of controlling the permission state by the spatial constraints.

Definition 14.4 *(Permission activation obligation) A permission cannot be activated on a site s, if its related spatial constraints are not satisfied. That is,*

$$(\forall perm \in Permission)\ \neg satisfied(s, perm) \Rightarrow \neg activate(s, perm).$$

This obligation indicates the execution behavior of an agent must meet the requirement of spatial constraints before the desired permission becomes active. According to the definition of valid permissions (14.1), the state of a permission is invalid unless it is activated. Together with the semantics of the spatial constraint satisfaction, we can derive the following proposition.

Proposition 14.2 *(Spatial authorization obligation) A permission* perm *for an agent subject* a *on site* s_i *is valid with regard to a primitive spatial constraint* (s_j, o, r)*, only if the required access action has been exercised on the specific site. That is,*

$$\begin{aligned}&\square(valid(s_i, perm) \Rightarrow \\ &\quad \circ_p \Diamond_p((\exists a' \in Subject_j)\ SU_i(a) = SU_j(a')\ \wedge\ access(s_j, a', o, r))),\end{aligned}$$

where $Subject_j$ *refers to the set subject on site* s_j*,* SU_i *and* SU_j *being the* subject/user *assignment functions on* s_i *and* s_j*, respectively, and the predicate* $access()$ *denoting the access action on a site. They are defined in the RBAC model [108]. Operators* $\circ_p$ *and* $\Diamond_p$ *represent the past-tense temporal counterparts to* $\circ$ *(reading as "next will be") and* $\Diamond$*, as in [96].*

The obligation property for a primitive constraint $\#(m, n, A)$ can be similarly derived, and a compound spatial constraint is checked recursively following the semantics of the connectives. Thus, an agent is authorized to access certain resource only when it has (not) exercised required operations on a site visited before. Also, the access control mechanism checks these constraints based on the agent's access history.

14.4 Concluding Remarks

From the discussion above, we can see the model for itinerary safety reasoning or coordinated access control consists of a set of target objects, events on this object set

Table 14.1: Relationship between the itinerary safety reasoning and the coordinated access control models.

Components	Itinerary safety reasoning	Coordinated access control
Target objects	Set of sites	Set of shared resources
Events	Site visit	Access to a shared resource
Event ordering	Itinerary by MAIL	Time/space-wise access pattern
Execution instance	Trace	Access history
Constraint	Itinerary constraint	Shared resource access constraint

and their ordering, the agent execution instances, and the constraints on the agents' behaviors. We note that there is a one-to-one association between components in the itinerary safety reasoning and those in the coordinated access control scheme; Table 14.1 shows the association. The latter controls the actions of an agent according to the secure access policies. As we are interested in the accesses to the shared resources in coalition environments, the agent access pattern can be expressed, by a resource access language, as a set of access operations on the shared resources. The resource access language is similar to the itinerary language in Chapter 13 with sequential, parallel, conditional, and repetitional access actions. Besides, the syntax and semantics of their constraint languages are similar, and the goals of their constraint checking procedures are both to determine whether a sequence of events satisfies some predefined policies. As a consequence, the methodology used in the itinerary safety checking can also be applied to tackle the constraint satisfaction reasoning for the coordinated access control scheme.

Chapter 15

Connection Migration in Mobile Agents

In multiagent systems, agents often need to communicate with each other repeatedly during their life cycles. In such a situation, persistent connection allows a pair of agents to keep exchanging information on an established channel without frequent disconnection and reconnection. Persistent connection provides an alternative communication protocol to the message-oriented mailbox scheme. Connection persistence for mobile agents poses a challenge. This chapter presents a reliable connection migration mechanism that supports concurrent migration of both endpoints of a connection and guarantees exactly once delivery for all transmitted data. In addition, a mobile code access control model is integrated to ensure secure connection migration.

15.1 Introduction

In agent-based mobile computing, it is often necessary for agents to communicate with each other for collaboration. Communication between mobile agents has long been an active research topic. Agent communication languages like KQML [110] and FIPA's ACL [111] were designed for message-oriented interactions between autonomous agents that originate in different places. Communication between a pair of sending and receiving agents may be either synchronous or asynchronous. In synchronous communication, both the sending and receiving agents synchronize at every message. The sending agent is blocked until its message is actually delivered to the recipient; the receiving agent is blocked until a message arrives. In asynchronous communication, a message send operation is nonblocking in the sense that the sending agent is allowed to proceed as soon as the message is copied to a local system buffer.

In Section 12.5.1, we described a post office protocol for asynchronous communication between agents. In the approach, a message is sent to an intermediate message proxy (i.e., messenger) in the local naplet server. The messenger transmits the message to its counterpart at the receiving side and buffers it in a mailbox in the remote messenger, no matter whether the receiving naplet is running or not. The message is delivered to the receiving agent if the agent is running and ready to take it or waits for retrieval by the agent. In the case that the receiving agent migrates to a new location before the message is delivered, the message is forwarded following the agent footprints left behind.

Asynchronous communication plays a key role in many distributed applications and is widely supported by existing mobile agent systems; see [327] for a review of location-independent communication protocols between mobile agents. Since it is hard for a sending agent to determine whether and when the receiver gets the message, it is not always appropriate and sufficient for applications that require agents to closely cooperate. Synchronous communication requires both the sending and receiving agents to be synchronized at each message. But there are situations where the agents need to communicate with each other repeatedly during their life cycles. For example, in the use of mobile agents for parallel computing [340], cooperative agents need to exchange data frequently as the computation proceeds. In such a situation, it is desirable for the agents to establish a connection and keep the connection alive until it is closed explicitly by one of the agents. A persistent connection would keep the agents working more closely and efficiently.

Socket is a solution for synchronous communication in distributed applications. Conceptually, it is an abstraction of communication endpoint at the transport layer, to which an application can write data to and read data from. The traditional TCP or UDP has no support for mobility because it was designed with the assumption that the communication peers are stationary. To guarantee message delivery in case of agent migration, a connection migration scheme is desirable so that an established socket connection would migrate with its endpoints continuously and transparently.

15.1.1 Related Work

There were recent studies on mobile TCP/IP in both network and transport layers to support the mobility of physical devices in mobile computing. We refer to this type of mobility as physical mobility, in contrast to logical mobility of codes. A network-layer implementation allows a user to use the same IP address even when he/she changes network attachment point. Mobile IP [156] is such an example, which works on a home agent concept associated with the mobile host. Every package destined to a mobile host by its home address is intercepted by its home agent and forwarded to it. Representatives of transport-layer protocols include MSOCKS [211],TCP-R [127], M-TCP [304], and Migrate [294]. They use an end-to-end mechanism to handle host mobility, by extending the TCP protocol with a TCP migrate option while preserving the semantics of TCP. Although these protocols provide feasible ways to link moving devices to network, they have no control over the logical mobility. Moreover, they require a change of TCP stacks. This hinders the protocols from wide deployment.

Mobile agent systems are often implemented as a middleware. Agent connection migration requires support of session-layer implementations in the middleware. In the past, a few session-layer connection migration mechanisms were proposed. Zhang and Dao presented a persistent connections model [350], while relying on a centralized service to update all hosts of any location change of a host. A similar approach is due to Qu *et al.* [250]. It preserves upper-layer connections by using some kernel interfaces to access the system buffer for undelivered data. Okoshi *et al.* presented a MobileSocket library on top of Java Socket [232]. It uses a dynamic socket switch to update connection and application layer window so as to guarantee the delivery of all in-flight data at a user level.

Similar mechanisms were used for robust TCP connections for the purpose of masking server crash or communication failures. Robust TCP connection mechansim [94] addresses reliable communication problems for fault tolerant distributed computing. Reliable Sockets (Rocks) [348] allows TCP connections to support changes in attachment points with emphasis on reliability over mobility. It has support for automatic failure detection and a protocol for interoperates with endpoints that do not support Rocks.

We note that agent-related connection migration involves two unique reliability and security problems. Since both endpoints of a connection would move around, a reliable connection migration needs to do more for exactly once delivery for all transmitted data. Security is a major concern in agent-oriented programming. Socket is one of the critical resources and its access must be fully controlled by agent servers. Agent-oriented access control must be enforced during the setup of connections. Connection migration is vulnerable to eavesdropper attacks. Additional security measures are needed to protect transactions over an established connection from any malicious attacks due to migration.

In [353], we presented an integrated mechanism that deals with reliability and security in connection migration. It provides an agent-oriented socket programming interface for location-independent socket communication. The interface is implemented by a socket controller that guarantees exactly once delivery of data, including data in transmission when the communicating agents are moving. To assure secure connection migration, each connection migration is associated with a secret session key created during connection setup. We implemented a prototype of the mechanism as a `NapletSocket` component in the Naplet mobile agent system. In the following, we present the details of the design and implementation of the `NapletSocket`.

15.2 NapletSocket: A Connection Migration Mechanism

15.2.1 NapletSocket Architecture

NapletSocket API is provided by means of classes: `NapletSocket(agent-id)` and `NapletServerSocket(agent-id)`. They resemble Java Socket and ServerSocket in semantics, except that the NapletSocket connection is agent oriented. It is known that Java Socket/ServerSocket establishes a connection between a pair of endpoints in the form of (host IP, port). For security reasons, an agent is not allowed to specify a port number for its pending connection. Instead, the underlying Naplet-

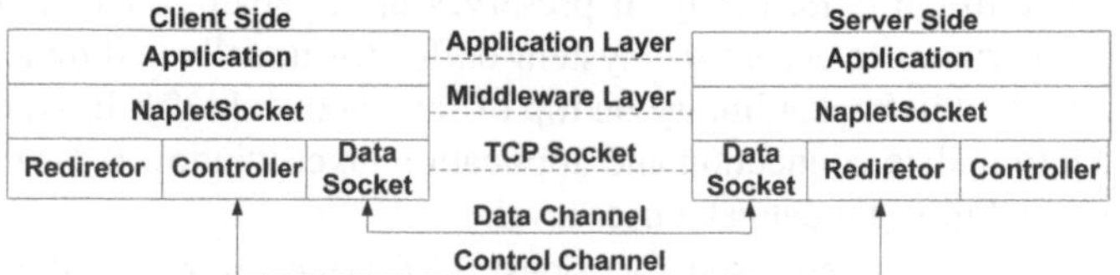

Figure 15.1: NapletSocket architecture.

Socket system allocates ports to connection requests based on resource availability and access permissions. The Naplet system contains an agent location service that maps an agent identifier to its physical location. This ensures location transparent communication between agents. Once the connection is established, all communication is through the connection and no more location service is needed.

To support connection migration, the NapletSocket system provides two new methods *suspend()* and *resume()*. They can be called either by agents for explicit control over connection migration, or by a Naplet docking system for transparent migration.

Figure 15.1 shows the NapletSocket architecture. It comprises three main components: data socket, controller, and redirector. The component of data socket is the actual channel for data transfer. It is associated with a pair of send/receiver buffers to keep undelivered data. The controller is used for management of connections and operations that need access rights to socket resources. The redirector is used to redirect socket connection from a remote agent to a local resident agent. Both controller and redirector can be shared by all NapletSockets so that only one pair is necessary.

During connection establishment, the controller of the client agent sends a request to the counterpart at the server. After the request is acknowledged, the client connects to the redirector at the server side and the connection is then handed to the desired agent. After a connection is established, the two agents communicate with each other through accessing the data socket, no matter where their communication parties are located. Under the hood is a sequence of operations by the NapletSocket library. The underlying data socket is first closed, when the NapletSocket takes a suspend action before agent migration. During agent migration, no data can be exchanged since the connection is suspended. After the agent lands on the destination, the NapletSocket system will resume the connection by connecting to the redirector at the other side. The data sockets of both client and server are then updated and new input/output streams are re-created atop of the socket.

Since there is no need for message forwarding, communication over NapletSocket is efficient. In Section 15.4, we will show suspend and resume operations incur marginal overheads to keep connections persistent.

15.2.2 State Transitions

The design of NapletSocket can be described as a finite state machine, extended from the TCP protocol. It contains 14 states, as listed in Table 15.1. The states in ital-

Table 15.1: States in NapletSocket transitions.

State	Description
Closed	Not connected
Listen	Ready to accept connections
Connect_Send	Sent a CONNECT request
Connect_Acked	Confirmed a CONNECT request
Established	Normal state for data transfer
Sus_Sent	Sent a SUSPEND request
Sus_Acked	Confirmed a SUSPEND request
Suspend_Wait	Wait in a SUSPEND operation
Suspended	The connection is suspended
Res_Sent	Sent a RESUME request
Res_Acked	Confirmed a RESUME request
Resume_Wait	Wait in a resume operation
Close_Sent	Sent a CLOSE request
Close_Acked	Confirmed a CLOSE request

ics are newly added to the standard TCP state transitions. The states SUSPEND_WAIT and RESUME_WAIT are used only for concurrent connection migrations, which will be covered in later sections. A NapletSocket connection is in one of these states. Certain action will be taken when an appropriate event occurs, according to the current state of the connection. There are two types of events: calls from local agents and messages from remote agents. Actions include sending messages to remote agents and calling local functions.

Figure 15.2 shows the state transitions of a NapletSocket connection. The solid lines show the transitions of clients connecting to servers and the dotted lines are for the servers. Details of the open, suspend, resume, and close transactions are as follows:

Open a connection. Both client and server are initially at the CLOSED state. When an agent does an active open, a CONNECT request is sent to the server and the state of the connection changes to CONNECT_SENT. If the request is accepted, the client side NapletSocket receives an ACK and a socket ID to identify the connection. Then it sends back its own ID and the state changes to ESTABLISHED.

Connection in server side switches to the LISTEN state once an agent does a listen. When a CONNECT request comes from a client, the server acknowledges it by sending back an ACK and a socket ID. It then changes to the CONNECT_ACKED state. After the socket ID of the client side is received, it switches to ESTABLISHED and the NapletSocket connection is established. Now data can be transferred between the two peers as normal socket connection.

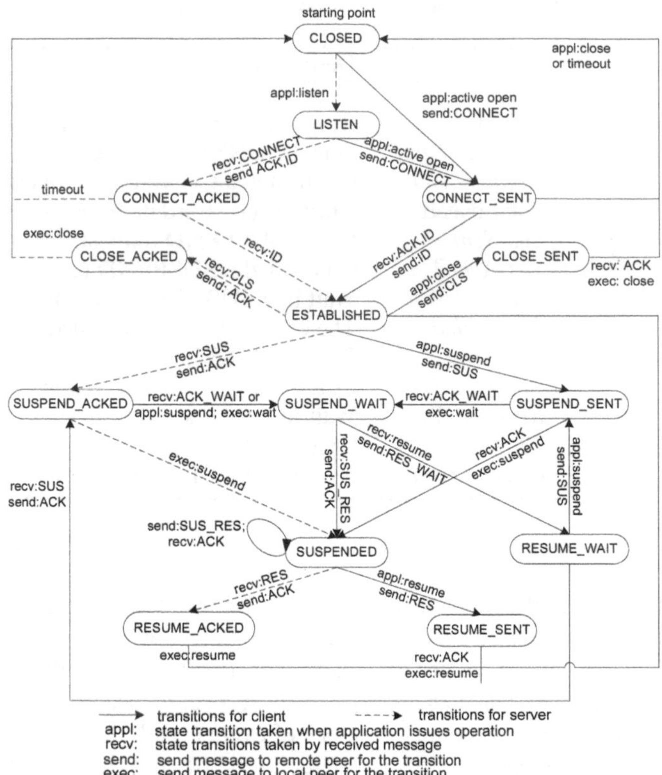

Figure 15.2: NapletSocket state transitions diagram.

Suspend/resume a connection. After a connection is established, either of the two parts may suspend it. The one who wants to suspend a connection invokes the suspend interface and a SUS is sent to the peer. If the request is acknowledged, an ACK is sent back and triggers the action of closing underlying input/output streams and data socket. The connection state then switches to SUSPENDED.

When the other side of NapletSocket receives the SUS message, it sends back an ACK if it agrees to suspend. Then it closes the underlying connection. After that, the state for this peer also changes to SUSPENDED. Now connections at both peers are suspended. No data can be exchanged in this state.

At the SUSPENDED state, when either of the agents decides to resume the connection, it invokes the resume interface. The resume process first sets up a new connection to the remote redirector and sends a RES message. If an ACK is received, it then resumes the connection and the state switches back to ESTABLISHED. Once the remote peer in the SUSPENDED state receives a resume request, it first sends back an ACK. Then the redirection server hands its connection to the desired NapletSocket and new input/output streams are created. After that, both peers change back to the ESTABLISHED state.

Close a connection. In either an ESTABLISHED or SUSPENDED state, if an agent decides to close the current connection, it requests the `NapletSocket` to send a CLS(CLOSE) request to the remote peer. After the request is acknowledged, the local data socket is closed. The other side of the connection closes passively after receiving a CLS request. It first acknowledges the request and then closes the underlying socket and streams. At the time, data socket at both sides is closed and the state changes to CLOSED.

15.3 Design Issues in NapletSocket

15.3.1 Transparency and Reliability

If there is any established connection before its agents' migration, the connections should be migrated transparently. We first discuss the case for only one connection. Then we extend it to multiple connections. The main approach for connection migration is to use a data socket under NapletSocket. Each time the agent migrates, the underlying data socket is closed before migration and updated to a new data socket after migration. When two agents migrate freely, it is possible that migration happens at the same time when there are data being transferred and in this case, the data may fail to reach their destinations. The presence of mobility causes a problem for reliable communication. Furthermore, two agents may migrate simultaneously, which makes it more difficult to achieve reliability.

To guarantee that messages can be finally delivered, we added an input buffer to each input stream and wrapped them together as a `NapletInputStream`. To suspend a connection, the operation retrieves all currently undelivered data into the buffer before it closes the socket. The data in the `NapletInputStream` migrate with the agent. When migration finishes and the connection is resumed at the remote server, a read operation first reads data from the input buffer of its `NapletInputStream`. It doesn't read data from socket stream until all data from the buffer have been retrieved.

In case both agents of a connection want to move simultaneously, there is a problem since the resume operation from one agent only remembers the previous location of the other agent. It is not difficult for NapletSocket to know where the next host is to support two migrations at the same time. The problem is an agent may migrate frequently in the network. A resume operation may have to chase a mobile agent if the two agents are allowed to migrate concurrently. In order to avoid the chasing problem and provide a simplified solution, we delay the migration of one agent if two are issued at the same time, which means only one can migrate at a time and the other cannot leave until the first one finishes. Therefore agent migrations occur sequentially although they are issued simultaneously. But from the viewpoint of high-level applications, the underlying sequential migration is transparent and there is no restriction for agents migration.

To delay one operation when two suspend operations are issued around the same time, two states SUSPEND_WAIT and RESUME_WAIT are introduced. One of them is delayed and the state of the connection is put into the SUSPEND_WAIT state. At this state, the operation is blocked until the agent finishes migration and sends a notification message. Then the connection is switched to the SUSPENDED state and the second agent can migrate to another host.

RESUME_WAIT is the state when a resume operation is blocked. In the case of concurrent migration, when one of the peers finishes migrating, it invokes the resume operation to update underlying connection and I/O streams. If we approve the resume operation finish as usual, then a connection is to be established between the agent who has just finished migration and the one who is to migrate. The connection switches to the ESTABLISHED state. But the connection has to be suspended again due to the second agent migration and the switches of states from SUSPENDED to ESTABLISHED and back are not necessary. The purpose of RESUME_WAIT is to prevent the state transition from SUSPEND to ESTABLISHED. During the resume operation, instead of establishing a new socket connection, we change the state of the connection to RESUME_WAIT and block the resume operation. The resume operation will be signaled after the other agent finishes migration. Since no new connection is established after the first agent migration, there is no need to suspend the connection before the second agent migration. By using this RESUME_WAIT state, we save time for a suspend operation and part of a resume operation.

During concurrent agent migration, it is possible for a socket controller to issue a suspend operation and at the same time receive a SUSPEND request from the other side of the connection. Depending on when a suspend operation is issued, we classify the problem into two types. One is overlapped concurrent connection migration when the suspend operation is issued before an acknowledgement for the SUSPEND request from the other side is sent; the other is nonoverlapped when the operation is issued after an acknowledgment for SUSPEND has been sent and response for the SUSPEND is still in progress.

Overlapped concurrent connection migration. In the first case, both sides of a connection issue SUSPEND requests at about the same time and neither receives an acknowledgment before the SUSPEND request from the other side arrives. Each side has to decide whether to grant the request. To approve one and only one request, we give priority to one of the agents to let it migrate first. The other one is delayed until the first finishes. A time sequence for this case is in Figure 15.3(a). At the beginning of the sequence, there is a connection between sides A and B. Both of them issue a SUSPEND request at about the same time. We assume side B has a high priority. Side B receives the request and since it has sent a SUSPEND request, this is treated as concurrent connection migration. Side B sends back an ACK_WAIT to delay the migration. The situation is the same for side A. But side A always acknowledges a SUSPEND request since it has a low priority. Then agent in side B migrates and the state of the connection in side A is switched to SUSPEND_WAIT. After finishing migration, side B informs side A with a SUS_RES to continue the blocked suspend operation. Side A confirms with an acknowledgment and finishes the issued suspend operation. Then the connection is kept in a SUSPENDED state for both sides A and B. At the time, agent in side A may migrate to another host. After that, it resumes the connection by normal resume operation. Then the connection in both sides returns to the ESTABLISHED state.

Nonoverlapped concurrent connection migration. In the second case, one side of a connection issues a suspend operation after it acknowledges a SUSPEND request from the other side and the previous request hasn't finished. The second suspend operation shouldn't continue until the first one finishes. This is another case for concurrent connection migration and we call it nonoverlapped migration since the processing of the two suspend operations are not overlapped. In this case, it doesn't matter which side of the connection has a high priority since the acknowledged request has to finish first. A time sequence is presented in Figure 15.3(b) for this case. At first, side A issues a SUSPEND request and side B replies with an acknowledgment. Then side A suspends the connection and starts to migrate. Before it finishes, agent in side B decides to migrate so a suspend operation is invoked from socket controller in side B. Since this request couldn't get approved, instead of sending a SUSPEND request to side A, we switch the connection state in side B to SUSPEND_WAIT. After the agent in side A finishes migration, it sends a RESUME request to resume the connection. Because side B has a blocked suspend operation to finish, it replies with a RESUME_WAIT message to block the resume operation and continues with its suspend operation. After that, connection in side B is in the SUSPENDED state and in side A it is kept in the RESUME_WAIT state. Then agent in side B can migrate and a normal resume operation is used to switch the state to ESTABLISHED after the migration.

We give priorities to one of the agents when both agents want to migrate at the same time. A question is how to determine which one is assigned a high priority. A simple solution is to determine the priorities according to their roles in the connection. For example, we can give a high priority to all connections established from ServerSocket. But this solution is prone to deadlock in an environment with multiple agents connected with each other. For example, in a configuration of three agents X, Y, and Z, which are clients of Y, Z, and X, respectively. This forms a circular waiting list if all of the agents want to move at the same time. To prevent deadlock in simultaneous migration, we determine the migration priority of each agent based on its unique agent ID. During connection setup, a hash function is applied to each agent ID and generates hash values for both agents. We assign their priorities according to their ordered hash values.

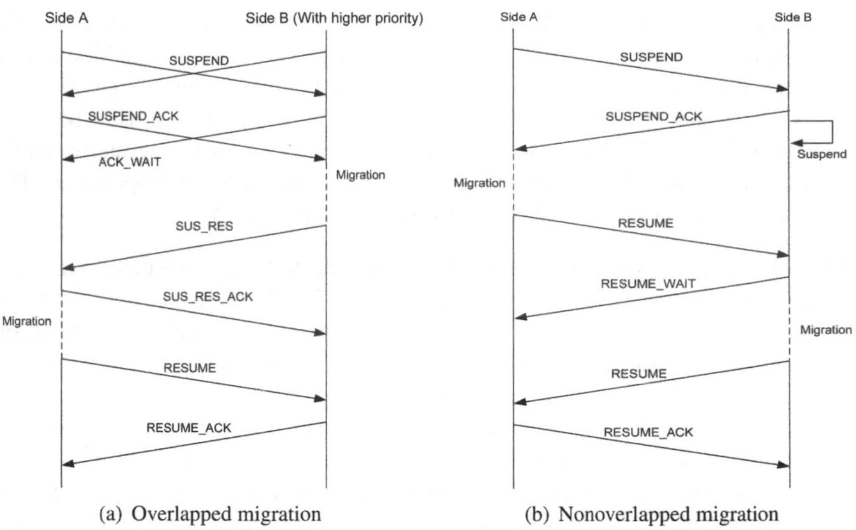

(a) Overlapped migration (b) Nonoverlapped migration

Figure 15.3: Time sequence of concurrent connection migration.

15.3.2 Multiple Connections

In preceding discussions, we focused on concurrent agents migration for one connection. In case multiple connections are established before agent migration, all connections should be suspended. When suspending all these connections, it is possible for them to be suspended in different orders. For example, suppose there are two connections for agents in sides A and B, represented as #1 and #2. Side A may suspend connections in the order of #2 #1 and side B in the order of #1 #2. When the two sides issue suspend operations simultaneously, it is possible for side A to suspend connection #2 while at the same time side B is suspending #1. Either one of these concurrent suspend operations needs to be delayed because each is operating on different connections. Thus both of the connections are successfully suspended. When suspending the second connection, side A on #1 while side B on #2, both sides will find the connection has already been suspended. By default a suspend operation needs to do nothing for a suspended connection. Therefore both sides successfully suspend their connections and agent migrations happen at the same time. As a result, either of the connections knows where the other peer is after landing on a new host.

To have only one agent migrate at a time, suspend operations for one of the connections must be delayed. We achieve this by giving different responses to a suspend operation when it is operating on a connection that is in the SUSPENDED state. Suspend operations are distinguished by whether they are issued locally or invoked by remote messages. When controller issues a suspend operation due to agent migration, it is local suspending. When the connection is suspended on receiving a SUSPEND request, it is remote suspending. During a suspend operation, if the connection has already been suspended remotely which means agent migration is happening in the other side, we decide whether to continue or block depending on the priority of the peer. If it has a low priority, we block the suspend operation; if it has high priority, we finish the operation without further actions since it is already in the SUSPENDED state. The blocked suspend operation will get notified when the one with high priority finishes migration.

Figure 15.4 gives an illustration of the protocol. In this example, we assume side B has a high priority. Initially, side A issues a SUSPEND request for connection #2 and side B issues a SUSPEND request for connection #1 at about the same time. Because the operations are on different connections, they both get approved. Then connection #2 in side A and connection #1 in side B become locally suspended. Connection #1 in side A and connection #2 in side B become remotely suspended. After that, side A issues a suspend operation for connection #1. It is blocked because the connection is suspended remotely and side A is in a low priority. The same thing happens with connection #2 for side B. But because side B has a high priority, the suspend operation returns without further actions. Then the agent in side B proceeds with migration while the connection in side A is in the SUSPEND_WAIT state. After the agent in side B lands on a new host, side B issues a resume operation for the blocked suspend in side A for connection #1 and resume connection #2. Side A sends back RESUME_WAIT for the resume connections because it is to migrate. After receiving the message, connection in side B switches to the state of RESUME_WAIT. After side A finishes agent migration, it sends RESUME messages to side B for both connection #1 and # 2; the two connections then switch to the ESTABLISHED state.

15.3.3 Security

Security is a primary concern in mobile agent systems. NapletSocket addresses security issues in two aspects. First, the agent should not be able to cause any security problems to the host where it resides, either at the original host or at the host it migrates to. Second, the connection itself should be secure from possible attacks like eavesdropping. More specifically, a connection can only be suspended/resumed by the one who initially creates it.

Regarding the first issue, any explicit request to create a Socket or ServerSocket from an agent is denied. Permissions are granted only to requests from the NapletSocket system. Thus the only way for an agent to use socket resources is through the service provided by the mobile agent system. Now the problem becomes whether we can deny permission if a request is from an agent and grant it if it is from the system.

This problem can be solved by an agent-oriented access control [335]. It allows permissions to be granted according to who is executing the piece of code (subject), rather than where the code comes from (code base). A subject represents the source of a request such as a mobile agent or NapletSocket controller. By denying access requests from the subject of agents for socket resources and granting them to local users such as administrators, we achieve the security goal in a simple manner. In mobile agent applications, an agent subject has no permissions to access local socket resources by default. When it needs access to a socket resource, it submits a request

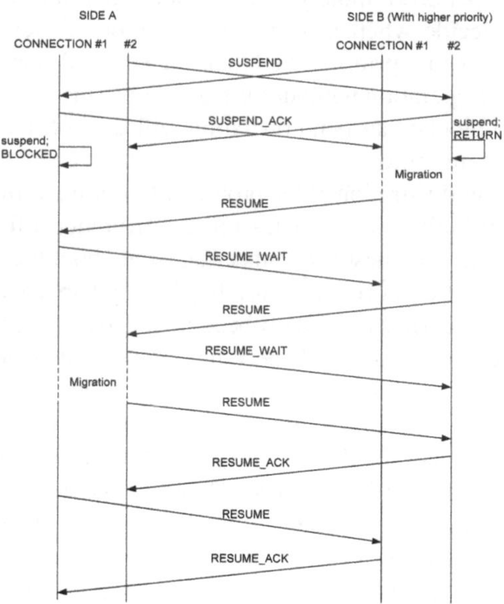

Figure 15.4: Time sequence of concurrent agent migration with multiple connections.

to a service proxy in NapletSocket controller. The proxy authenticates the agent and checks access permissions. After the security check passes, a NapletSocket or NapletServerSocket will be created by the proxy and returned to the agent.

Regarding the second issue, connection migration can be realized by the use of a socket ID. However, a plain socket ID couldn't prevent a third party from intercepting the information and exercising eavesdropping attacks. To this end, we applied Diffie-Hellman key exchange protocol to establish a secret session key between the pair of communicating agents at the setup stage of a connection. Any subsequent requests for suspend, resume, and close operations on the connection must be accompanied with the secret key. Such requests will be denied unless their keys are verified by remote peers. Since the key generated by Diffie-Hellman protocol is hard to break, NapletSocket connections are protected from eavesdropping attacks.

15.3.4 Socket Hand-Off

Recall that in NapletSocket, when a client connects to a server, it uses an agent ID to specify the destination. We need to find both host name and port number from the agent ID. To find host name, we can keep records of agents' traces and locate the host according to its ID. To find port number, we need to send a query message to the server. The server has to maintain a table indicating which agent uses which port and return the port to the client. Then the client can start a connection. In fact,

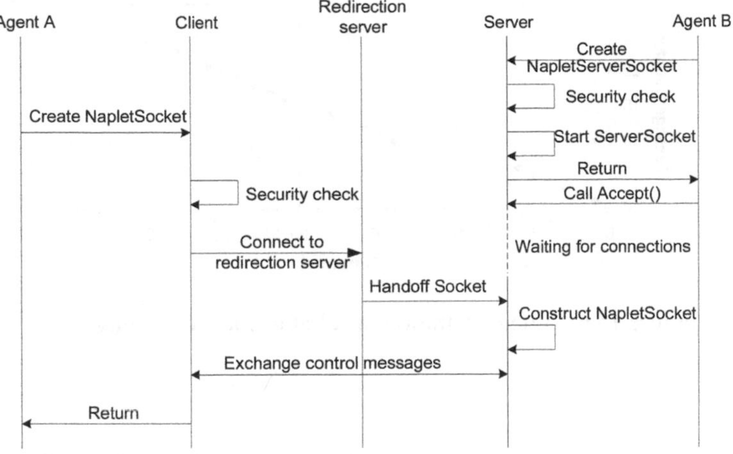

Figure 15.5: Socket hand-off in connection setup.

these operations can be saved using the socket hand-off technique: by connecting to the redirector at the server and indicating the desired agent. The server looks up which `NapletServerSocket` this request is for and redirects the current socket to it. Then the `NapletServerSocket` gets notified from blocking and constructs a `NapletSocket` according to the data socket it receives. This mechanism can save a round-trip time in querying host name and port number and there is no need for the server to maintain a table mapping ports to agents. Figure 15.5 shows the sequence diagram for socket hand-off in connection setup.

A similar mechanism also applies when resuming a connection. In this case, the client connects to a redirection server at the other side and sends a request to resume. The server then hands the socket connection to the suspended NapletSocket and wakes it up. Finally the notified NapletSocket updates its underlying data socket.

15.3.5 Control Channel

It is necessary to exchange control messages during state transitions of a NapletSocket connection. From a performance perspective, we used a separate channel for control messages and chose UDP as the transport layer protocol. Regarding the omission failures and ordering problems caused by UDP, we adopted a retransmission mechanism to provide reliable delivery on top of UDP. The basic idea is to use retransmission in case of failure. After sending a control message, the sender starts a retransmission timer and waits for an ACK from the receiver. If an ACK is received before time-out, the timer is canceled. If not, the message is retransmitted and a new timer for the message is set. Sequenced numbers are used to relate a reply to the corresponding request.

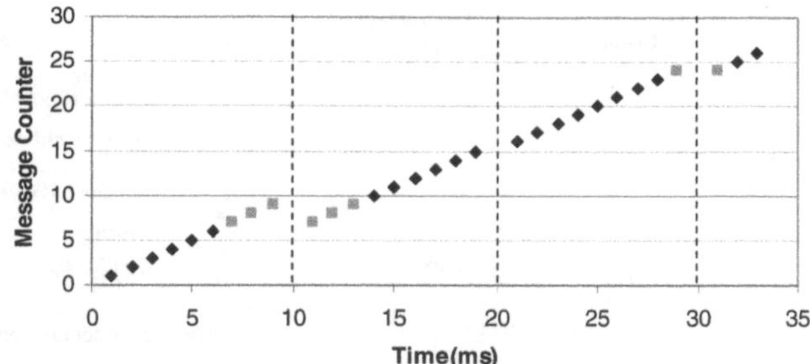

Figure 15.6: A demonstration of reliable connection migration.

15.4 Experimental Results of NapletSocket

In this section, we present an implementation of NapletSocket and its performance in comparison with Java Socket. All experiments were conducted in a group of Sun Blade 1000 workstations connected by a fast Ethernet. We first show the effectiveness and overhead of the implementation, focusing on the cost of its underlying operations such as open, suspend, and resume. We then evaluate the overall communication performance under various migration and communication patterns.

15.4.1 Effectiveness of Reliable Communication

The first experiment gives a demonstration of reliable connection migration with NapletSocket. A stationary agent A keeps sending messages at a rate of 1 millisecond (ms) to a mobile agent B. Each message contains a counter, indicating the message order. On the receipt of a message, agent B echoes the message counter back to A. Reliable communication requires that the mobile agent B receives the messages in the same order as they are sent.

Figure 15.6 shows the traces of the message counters received by the mobile agent B over the time. Agent B migrates at 10th, 20th, 30th ms. The dark dots show the messages read from the socket stream and the light dots are messages into or from message buffer in NapletSocket. In the first migration point, agent B migrates before it receives all of the messages in transmission. The undelivered three messages (7, 8, 9) are kept in the message buffer of NapletSocket. They are transferred together with agent B under the support of NapletSocket at the sender side and delivered to agent B by the NapletSocket at the receiver side after B lands. Similarly, the third agent migration involves transferring of one message at the same time.

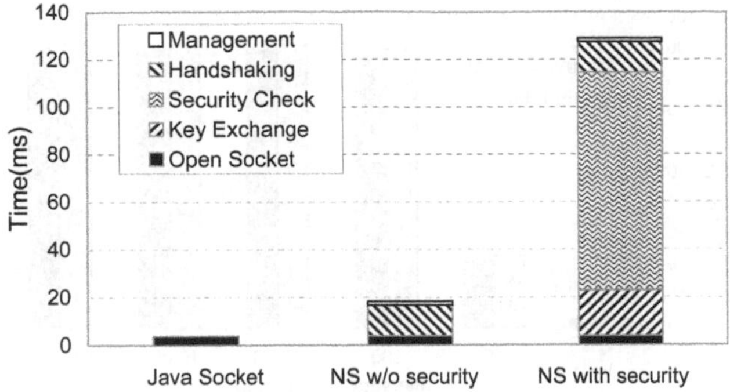

Figure 15.7: Breakdowns of the latency to open a connection.

Table 15.2: Latency to open/close a connection.

Connection Type	Open (ms)	Close (ms)
Java Socket	3.7	0.6
NapletSocket w/o security	18.2	12.5
NapletSocket with security	134.4	12.6

15.4.2 Cost of Primitive NapletSocket Operations

The second experiment focused on the cost of primitive operations defined in NapletSocket, including connection open, close, suspend, and resume. The cost of an operation refers to the interval between the time when it is issued and the time when it is completed.

We performed open and close operations with and without security checking for 100 times each. Table 15.2 shows the average time for each operation in different cases. For comparison, the costs for open and close operations in Java Socket are included. From this table, we can see that opening a secure mobile connection costs almost 40 times as much as that of a Java socket. Recall that establishment of a secure NapletSocket channel involves a number of steps: authentication, authorization, secret key exchange, handshaking and socket establishment. Figure 15.7 presents the breakdown of the cost. It shows that more than 80% of the time was spent on key establishment, authentication, and authorization. The cost would reduce to 18.2 ms without security support. It remains very high in comparison with the cost in Java Socket, but it is acceptable. It is because opening a connection is a one-time operation and the connection remains alive once established, which means that cost of opening a connection is amortized over agent migration.

Similarly, we recorded the cost of 27.8 ms and 16.9 ms for suspend and resume operations, respectively. The cost is mainly due to the exchange of control messages

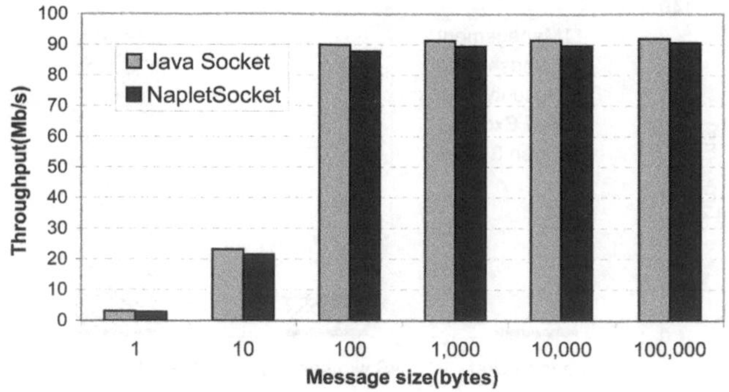

Figure 15.8: Throughput of NapletSocket versus Java Socket.

(handshaking), which makes up about 50% for suspending and 70% for resuming. Suspending a connection also requires checking whether there are any undelivered data in the input stream. Resuming a connection needs to set up a data socket and create I/O streams.

The benefit of provisioning a reliable connection can be seen by comparing the time required for reopening a connection with that of suspend/resume. If we close a NapletSocket before migration and reopen a new one after that, the total cost involved is about 147 ms. However, if we use suspend and resume instead, the cost is less than one third of the time for close and reopen operations. The total time saved by using suspending and resuming increases with the move of the agent.

15.4.3 NapletSocket Throughput

In the third experiment, we tested the NapletSocket throughput by the use of TTCP [61] measurement tool, in which a pair of TTCP test programs call Java Socket methods to communicate messages of different sizes as fast as possible. Because NapletSocket bears much resemblance to Java Socket in their APIs, we developed a simple adaptor to convert TTCP programs into NapletSocket compliant codes.

Figure 15.8 presents the NapletSocket throughput between two stationary agents. The throughput of Java Socket is included for comparison. Each data point is an average of seven independent runs. This figure reveals that the NapletSocket throughput degrades slightly (less than 5%). This degradation is mainly due to synchronized access to I/O streams. With the increase of message size, the performance gap becomes almost negligible.

To test the impact of agent mobility on NapletSocket throughput, we designed two migration patterns, in which a pair of agents keep communicating to each other, while they are traveling around. One is *single migration* where one agent remains stationary and the other keeps moving at a certain rate. The other is *concurrent migration* in which both agents travel simultaneously along their own paths and com-

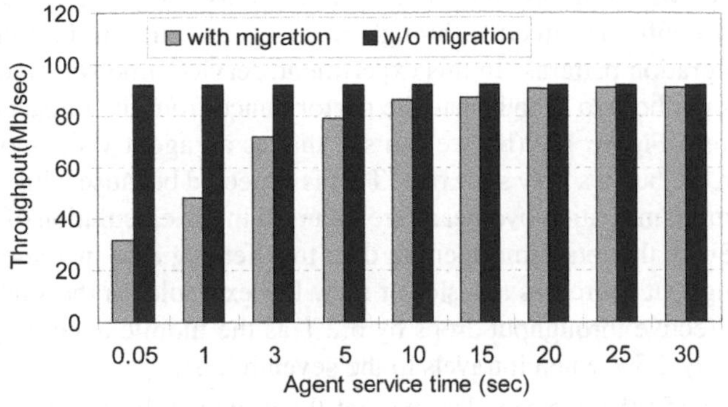

(a) Impact of migration frequency

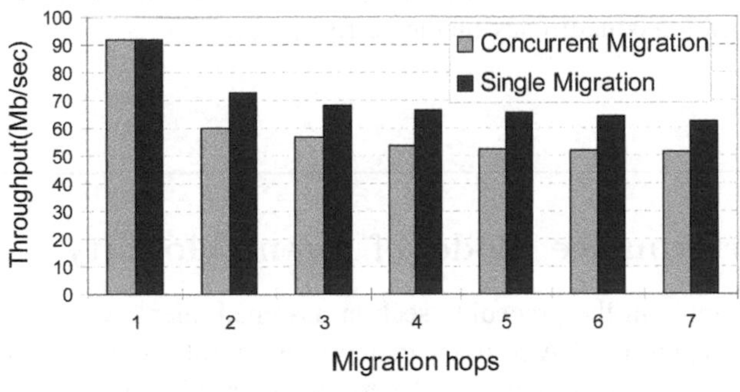

(b) Impact of migration hops

Figure 15.9: Effective throughput of NapletSocket in connection migration.

municate to each other at each hop.

Expectedly, the NapletSocket throughput tested by TTCP is dependent on agent migration frequency (i.e., service time at each hop). We refer to the total traffic communicated over a period of communication and migration time as *effective throughput*. Figure 15.9(a) shows the effective throughput with different migration frequencies in a single migration pattern. In this experiment, we assumed a constant message size of 2 KB.

From Figure 15.9, we can see that the measured effective throughput between stationary agents goes up to 92 Mbps. The throughput increases as the agents spent more time in communicating in each host, starting from 32 Mbps when the service time is 1 second to the maximum value when the agent stays in a host for more than 10 seconds. If an agent stays in a host for long enough time, the effective throughput gets very close to the one without migration. That implies that the effect of agent and connection migrations on throughput becomes negligible when an agent migrates at a low frequency.

Furthermore, we examined the impact of migration hops on throughput. Figure 15.9(b) presents the effective throughput as agents migrate in both single and concurrent migration patterns. In this experiment, service time was fixed to as large as 20 seconds per host so as to isolate the performance from the impact of migration frequency. From Figure 15.9(b), we can see that as an agent visits more hosts, the throughput drops, but at a very slow rate. This is expected because with the migration of the agent, more migration overheads are incurred in the calculation of the effective throughput. Since the total time spent in data transferring also increases at a higher rate, the throughput decreases at a slower rate. For example, in the single migration pattern, the effective throughput drops by 6.2% as the mobile agent migrates to the third host and by 2.5% when it travels to the seventh host.

From Figure 15.9(b), we can also see that the effective throughput in concurrent migration is smaller than that of single migration. It is because concurrent migration incurs more overheads. In the next section, we will have more discussions about the impact of migration concurrency on the performance.

15.5 Performance Model of Agent Mobility

The experiments in the preceding section assumed agents communicate to each other "as fast as possible." Also in the concurrent migration pattern, both agents actually migrated at the same fixed rate due to the assumption of constant service time at each host. The objective of this section is to investigate the impact of the communication rate as well as the agent migration concurrency on effective throughput of NapletSocket.

15.5.1 Performance Model

Consider two mobile agents, say A and B, which are connected via a NapletSocket connection. Both of them are traveling around a network at various rates. Without loss of generality, we assume agent B has a higher priority than A. At each host, the agents process their tasks for a certain time and communicate with each other for synchronization, as shown in Figure 15.10. Associated with each agent migration is a connection migration.

It is known the effective throughput of NapletSocket is determined by the cost for connection migration as well as the overhead for agent migration. Since the cost for agent migration is application dependent (being dependent upon the code and state sizes), we develop a model for connection migration, denoted by $T_{c-migrate}$, instead.

Recall that a connection migration starts with a suspend request and ends with a resume operation. A suspend operation changes the state of a connection from ESTABLISHED to SUSPEND. A resume operation changes the states of a connection

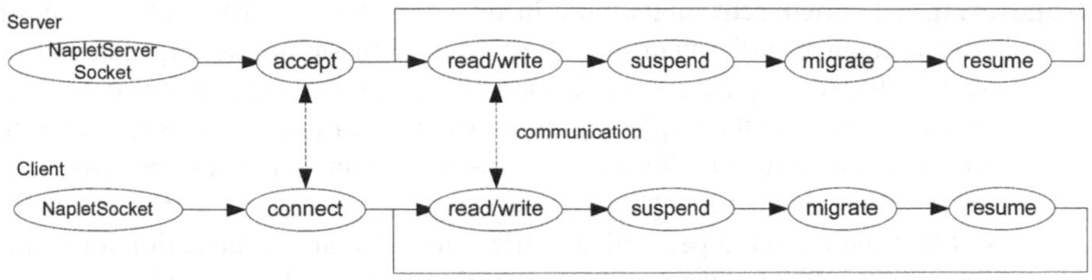

Figure 15.10: Migration/communication pattern using NapletSocket.

from SUSPEND back to ESTABLISHED. Denote $T_{suspend}$ and T_{resume} as the costs for suspend and resume operations, respectively. It follows that

$$T_{c-migrate} = T_{suspend} + T_{resume}. \tag{15.1}$$

Notice that the costs for suspend and resume operations are related to agent migration concurrency. In the case that only one endpoint of a connection is mobile, both $T_{suspend}$ and T_{resume} are constant. In the case of concurrent migration where both endpoints of a connection are mobile, $T_{suspend}$ and is dependent upon the ordering of their requests. Let t^a_{begin} and t^b_{begin} denote the request time of the two agents A and B and their interval $\tau = |t^a_{begin} - t^b_{begin}|$. If the interval τ is large enough for the first suspend to complete before the second suspend is issued, it becomes the case of single agent migration. Otherwise, the two suspend operations are performed concurrently.

As we discussed in Section 15.3.1, concurrent migration can be further distinguished between two cases: overlapped and nonoverlapped. In the following, we analyze the cost for suspend operation in these two cases.

Overlapped concurrent migration. For agent B with a higher priority, its suspend cost $T^b_{suspend}$ is the same as in the single migration pattern. For agent A with a lower priority, its suspend operation couldn't finish until it receives a SUS_RES message from B, indicating the completion of B's migration, as shown in Figure 15.3(a). Hence, the arrival time of SUS_RES at agent A

$$t_{sus\_res} = t^b_{begin} + T^b_{suspend} + T^b_{a-migrate} + T_{control}, \tag{15.2}$$

where $T^b_{a-migrate}$ refers to the migration time of agent B and $T_{control}$ is the latency for delivery of a control message between agent B and A. Consequently, the suspend operation of agent A can be finished within the time of $t_{sus\_res} - t^a_{begin}$. Since B's migration is overlapped with A, the cost for suspend of agent A can be approximated as

$$T^a_{suspend} = T_{control} + T^b_{suspend} + \tau. \tag{15.3}$$

Nonoverlapped concurrent migration. In this case, we assume agent A issues a suspend request earlier than agent B's. According the timing sequence in Figure 15.3(b), A's request gets confirmed and hence its suspend operation takes the same time as in the single migration pattern. For agent B, its suspend operation won't be issued until a RESUME message from agent A is received. The waiting time is equal to $T^a_{suspend} + T^a_{a-migrate} + T_{control} + \tau$. In fact, B's waiting time is overlapped with A's migration. As far as connection migration is concerned, B saves the cost for suspend operation. Hence, we have

$$T^b_{c-migrate} = T_{resume} + T_{control} + \tau. \tag{15.4}$$

15.5.2 Simulation Results

The performance model in the preceding section reveals that the cost for connection migration $T_{c-migrate}$ depends on the suspend starting time t^a_{begin} and t^b_{begin} and their interval τ, in addition to the cost for delivery of a control message ($T_{control}$) and the cost for suspend and resume operations ($T_{suspend}$) and (T_{resume}). The starting time T_{begin} is in turn determined by the agent migration pattern, characterized by migration frequency.

In this simulation, we set $T_{control}$, $T_{suspend}$, and T_{resume} to 10 ms, 27.8 ms, and 16.9 ms, respectively, as we measured in our experiments in Section 15.4.2. In addition, we set the cost for agent migration $T_{a-migrate}$ to be 220 ms. We evaluated the impact of migration frequency by modeling it as a random variable. We assumed the random variable follows an exponential distribution with an expectation of μ.

Figure 15.11 shows the cost for connection migration of NapletSocket with the change of mean service time for agent A (i.e., $1/\mu^a$). Plots for different service time for agent B $1/\mu^b$, relative to $1/\mu^a$, are also presented. When two agents migrate with a very high speed (i.e., small service time), there are more chances for concurrent connection migrations. When they migrate with a low speed (i.e., large service time), a single connection migration most likely occurs. In both cases, the cost for connection migration, $T_{c-migrate}$, remains unchanged for the high-priority agent. In contrast, the low-priority agent experiences a little more delay when both of the agents migrate at a high speed.

From Figure 15.11, we can also see that the lowest latency for both high- and low-priority agents happens around the point where their starting time interval τ is larger than $T_{control}$, but not large enough to become single migration patterns. The plots with different settings of μ^b/μ^a also show that given an A's migration rate, an increase of the ratio means agent B migrates faster so that when agent A suspends a connection, it has more chances of meeting an ongoing suspend request from B. This leads to a block of agent A's suspend requests and the overall cost for agent A's connection migration gets decreased.

Finally, we examine the impact of message exchange rate on the cost for connection migration. Instead of measuring the cost by time, this simulation uses the metric of the number of control messages involved in each connection migration, relative to the number of data messages communicated through the established connection.

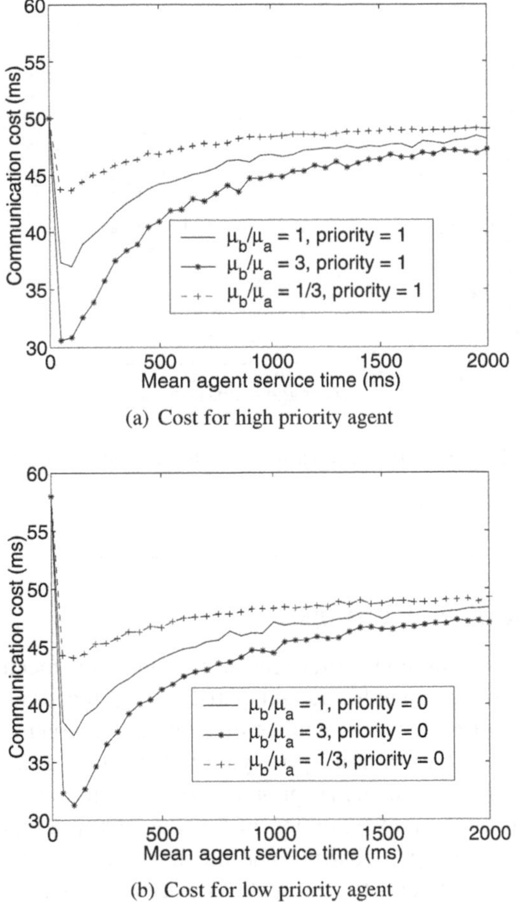

(a) Cost for high priority agent

(b) Cost for low priority agent

Figure 15.11: Connection migration cost of NapletSocket.

The message exchange rate, denoted by λ, refers to the number of data messages transmitted in a time unit. We define $r = \lambda/\mu$ as a relative message exchange rate, with respect to migration frequency μ. Figure 15.12 shows the connection migration overhead with different combinations of message exchange rates and migration frequencies.

It is clear that for a fixed ratio r, when the message exchange rate is small, the agent issues relatively more control messages to maintain a persistent connection and hence more overhead incurs. As the message exchange rate increases, the overhead is amortized over each communication. When the ratio r decreases to as low as one, which means the agent communicates once in each host, the overhead for persistent connection is always above 80% no matter how large the message exchange rate is.

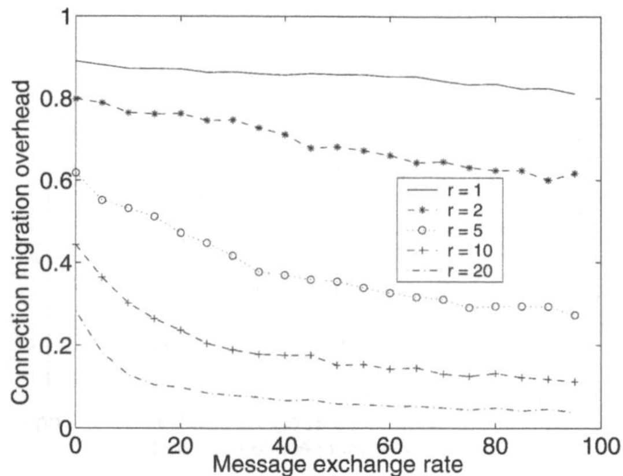

Figure 15.12: Connection migration overhead in agent communication.

15.6 Concluding Remarks

Mailbox-based asynchronous persistent communication mechanisms in mobile agent systems are not sufficient for certain distributed applications like parallel computing. Synchronous transient communication provides complementary services that make cooperative agents work more closely and efficiently. This chapter presents a connection migration mechanism in support of synchronous communication between agents. It supports concurrent migration of both agents in a connection and guarantees exactly once message delivery. The mechanism uses agent-oriented access control and secret session keys to deal with security concerns arising in connection migration. A prototype of the mechanism, `NapletSocket`, has been developed in the Naplet mobile agent system. Experimental results show that the NapletSocket incurs a moderate cost in connection setup, mainly due to security checking, and marginal overheads for communication over established connections. Furthermore, we investigate the impact of agent mobility on communication performance via simulation. Simulation results show that NapletSocket is effective and efficient for a wide range of migration and communication patterns.

Current work focuses on reliable problems caused by the presence of agent mobility. It has no support for detection and recovery from link or host failures. As part of ongoing work, we are going to extend the NapletSocket for fault tolerance.

Chapter 16

Mobility Support for Adaptive Grid Computing

Existing computational grid services are often organized in a remote or volunteer computing model. In the remote computing model, clients request services by sending data to servers; in the volunteer computing model, clients distribute their tasks to volunteer servers on demand. The former is limited to services that are predefined by the servers and the latter is unable to provide guarantees of service qualities from the perspective of applications. This chapter is devoted to a novel mobile agent-based push methodology to reverse the logic of pull-based task distribution in the volunteer computing model. It allows clients to dispatch their compute-intensive jobs as agents onto the servers so as to provide clients with customized computing services on the Internet.

16.1 Introduction

The idea of harnessing computational power of networked computers has long been an active area of research. High throughput clusters rely on job queueing and scheduling services to harness computing power of lightly loaded computers. Job queueing or scheduling systems like Condor [199] and LSF [355] has covered a wide range of needs, from traditional batch queueing to load sharing and cycle stealing. There are also parallel programming environments that provide task scheduling, load balancing, and even fault tolerant services in support of parallel applications on clusters [33].

There are two primary research objectives for high-performance computing on the Internet. One is to seamlessly integrate networked computers together to form a computational grid [116]. It mostly focuses on constructing a superserver out of networked computers and on harnessing computing powers of the Internet. For example, Globus [115] provides a big of mechanisms, including multiparty security, resource location, GRAM resource management, and Nexus multiprotocol communication library, for integration of networked resources; Legion [142] aims at a single image superserver by addressing issues like security, scalability, programmability, fault tolerance, and site autonomy within a reflective object-based metasystem. The other objective is to provide an easy-to-access Web-based computing framework for clients and servers to match their demands with supplies. The server can be a single machine, distributed servers, or even a computation grid on the Internet.

This chapter is mostly targeted at the second objective. Casanova and Dongarra categorized the Web-based computing architecture into three classes: remote computing, code shipping, and proxy computing [49]. Following the remote computing model, PUNCH [167], NetSolve [49], and Ninf [279, 1] assume that programs persistently reside on servers. Users request services by sending data to the server, which executes the code correspondingly and sends results back to the users. Such network-enabled solvers allowed users to invoke advanced numerical solution methods without having to install sophisticated software. Another good example is Biology WorkBench [227]. It integrates many standard protein and nucleic acid sequence databases, and a wide variety of sequence analysis programs into a single Web interface. Like the remote-procedure-call, the remote computing model is limited to fixed grid services, and clients cannot have customized services as required by special characteristics of their input data.

The code shipping model, also referred to as volunteer computing in Bayanihan [273], was first demonstrated by applications in cracking of 56-bit DES encoded messages [84] and in the search of the first million digit prime number [281]. The applications were relying on a coordinator that maintains a pool of parallel tasks and dispatches the tasks to participants on demand. There are recent experimental software systems in support of the code shipping model. In Charlotte [23], when a program reaches a parallel step, it registers itself with a specific daemon process, which creates an applet for each parallel task and maintains a list of the applets in a Web page. Any machine can visit this homepage using browsers and download applets for execution if it wishes to donate its cycles. Similarly, in Javelin [63] tasks are declared as Java applets and maintained in a broker accessible to registered computing servers. The broker makes a trade between resource demands and supplies between clients and servers.

A defining characteristic of code shipping model is passive. Tasks declared as Java applets are just waiting for visits from voluntary servers. It is similar to the idea of global work stealing in Atlas [19]. Atlas provided a framework for idle servers to steal threads from those that are busy. Unlike other Web-based computing systems, any Atlas machine can be either a server or a client and clients are closely coupled with servers. The passive pull or volunteer execution model has been proved to be effective for applications that are of common interest to a group of users. However, it cannot provide any guarantee of the service quality from the perspective of applications.

Proxy computing is built upon the concept of mobile objects and process migration [225]. Process or fine-grained thread migration concerns the transfer of user-level processes or threads across different address spaces in "closed" environments. Because the Web is an open and stateless environment, in which clients and servers may join or leave anytime without critically affecting other participants, the traditional process/thread migration technology cannot be applied for computing services on the Internet. The mobile agent approach goes beyond the way of proxy computing, because the running state is to be migrated together with the code and data.

In [340], we developed an agent-oriented grid computing framework, TRAVELER, in support of a task "push" methodology. The push model allows users to dispatch their jobs as agents and enable the agents to migrate between the available servers for load balancing, fault tolerance, and improving data locality. It reverses the logic of pull-based task distribution in volunteer computing and supports customized computing services from the perspective of applications. In the following, we give the design and implementation of the framework.

16.2 An Agent-Oriented Programming Framework

16.2.1 The Architecture

Figure 16.1 shows the TRAVELER agent-oriented programming framework. Under the framework, clients declare their compute-intensive jobs as mobile agents. They request services by dispatching the agents to a broker via a Virtual Machine Interface (VMI). The VMI is a stationary agent that provides the client with a perception of running on a grid with unlimited computational resources. It can be preinstalled on client machines or created on demand. The broker executes trades between clients and registered servers and forms a virtual machine out of the available servers to perform each computational agent. The virtual machine can be floating over the space of available servers with migration of its agent. The machine can also be reconfigured to adapt to the change of the server capacities. For scalability of the services, the system can be configured with more than one broker. Each broker serves regional clients and servers or nationwide domain-specific clients and servers. Brokers are organized in a hierarchical way to form a computational grid.

Success of the agent-oriented framework relies on a simple well-defined programming interface. The programming paradigm is based on a `TaskAgent` base class and a server-side agent docking class `AgentServer`. The `TaskAgent` class encapsulates basic agent properties like agent identifier (ID), credential of the originator, itinerary, and footprint. It implements Java `Runnable` and `Cloneable` interfaces to support multithreaded agents and their mobility for high-performance computing on symmetric multiprocessing (SMP) servers. Users transform their jobs into application-specific agents by extending from the `TaskAgent` base. The compute agents are dispatched to a broker via a stationary VMI agent. Following is a simple application example `MyExample` that creates a VMI on demand (line 4) and an `LUAgent` for matrix factorization (line 5). The `LUAgent` is dispatched by the VMI object (line 6).

```
1) public class MyExample implements Runnable{
2)       public MyExample() { ... };
3)       public void run() {
4)             VMI luVMI = new VMI( broker );      Create a VMI on demand
5)             LUAgent ft = new LUAgent( int size, double[][] matrix );
```

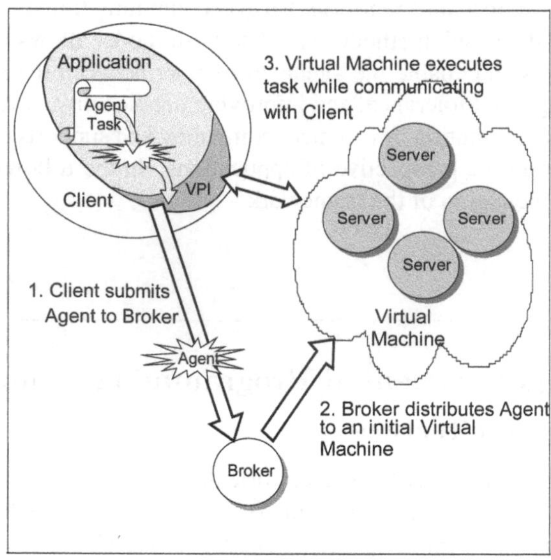

Figure 16.1: Architecture of an agent-oriented gid computing framework.

```
            luVMI.dispatch(ft);                    // Dispatch the agent to broker
        }
    }
```

The VMI agent provides *stateless, asynchronous, and location-independent* communication channels between the users (originators) and their computing agents so that users can go off-line, if needed, after they dispatch task agents and agent migrations become transparent to the users. Through the channels, the users (originators) can retreat or terminate their agents or inquire the agent states. Location-independent communication is realized by an *agent naming directory* service provided by the broker.

The `AgentServer` class runs a customized `AgentSecurityManager` to protect servers from possible attacks by alien computing agents. The activities that an agent can perform could be limited in much the same way as a Java applet's activity is limited. By limiting the scheduling priority of alien agent threads and their spawned thread groups, the `AgentSecurityManager` monitors the agent executions and prevents them from any "denial-of-service" attacks.

The `AgentServer` class contains references to an array of `AgentContext` objects, each of which defines an agent execution context. The context includes interfaces to accessible local system resources (e.g., CPU, memory, and temporary storage) and software services in a server and FIFO queues for synchronous and asynchronous communications between agents. For security reasons, alien agents are not allowed to communicate with their peers or VMI agents directly. Any inter-agent communication must be handled by an `AgentContext` object. Agent communication languages like KQML [110] and FIPA's ACL [111] were designed for interactions between autonomous agents that originate in different places. In high-performance mobile computing, communications are restricted to be between cooperative agents created by the same user and for the same job. It allows the agents to communicate with each other directly after they establish communication channels via `AgentContext` objects.

We note that the `TaskAgent` and `AgentServer` bears much resemblance to `Naplet` and `NapletServer` of the Naplet system in Chapter 12. However, the Naplet system was designed for general purpose network-centric distributed applications, while TRAVELER was in support of push methodology in grid computing.

Due to the autonomy and mobility of the agents, the "push" methodology has the following characteristics:

- Survival of intermittent or unreliable Internet connections. Traditional distributed applications rely on reliable network connections throughout their lifetime. If the connection goes down, the client often has to restart the application from the beginning. Since an agent can be programmed to satisfy one or more goals, even if the object moves and loses contact with its creator, the infrastructure will allow clients to dispatch computational agents into the Internet and then go off-line. The agent will reestablish the connection to its originator and present results back when it finishes its assigned task. Survival of intermittent connections is especially desirable for long-lasting compute-bound agents.

- Adaptive and fault tolerant execution model. Traditional client/server applications need to specify the roles of the client and the server very precisely, based on some predicted network traffic information, at their design time. Due to the instability of the Internet, performance of the applications often fluctuates unpredictably. The proposed infrastructure will allow computational agents to move themselves autonomously from one machine to another to harness the idle CPU cycles, balance the workload of machines, and enhance data locality. Persistent state associated with a mobile agent will also help recover computational tasks from their failures.

16.2.2 Strong Mobility of Multithreaded Agents

A mobile agent by its nature shall be able to suspend its execution threads before migration and resume the executions at suspended points when the agent reaches a destination. We refer to this type of mobility as strong mobility. To support strong mobility, the agent must be able to carry its execution state in both heap and stack, together with its code, while traveling. Since Java Virtual Machine (JVM) does not allow programs to directly access and manipulate execution stacks for security reasons, the execution state and program counters of the threads are not serializable for migration. As a result, each thread of a Java-based agent has to start over at the beginning of its run() [1] method upon arrival at a new site. We distinguish this mobility as weak mobility, in contrast to strong mobility. It is because of this reason that few of Java-based agent systems have support for strong mobility.

Weak mobility has proven to be sufficient for most distributed, information oriented applications. For example, shopping agents, which are to check the availability and prices of some goods, obviously have no requirements for strong mobility because they are expected to perform atomic operations at each site; agents for network administrations normally perform different operations at different sites and they have no demands for strong mobility either. In contrast, strong mobility is highly demanded for long-running compute agents.

Strong mobility for arbitrary Java-based agents is hardly viable without changing the JVM. For interoperability, we present an application-level code transformation mechanism to avoid the use of stack to keep the running state at migration points. The basic idea is to transform the agents into strongly mobile agents by inserting `onMove` checkpoints at potential migration places. `onMove` is an application-specific hook provided by `TaskAgent` base class. It records the stack state of each thread into a group of instance variables in heap when migration is needed. Upon arrival of a new server, each thread restores its stack state from the set of instance variables and resumes executions from where they are suspended.

This code transformation mechanism with checkpointing restricts agent migration to occur in presumed places. A question is where the checkpoints should be placed. In principle, they should be in places where there is minimal stack state. A good checkpoint will not only save heap space, but also simplify the code transformation procedure. The following `LUAgent` class defines a mobile agent for matrix LU factorization. The algorithm is an iterative process, running over the diagonal elements of matrix A (k–iteration). We assume migration to be between k-iterations. As a result, the `onMove()` checkpoint only needs a single heap variable `curPivotRow` (declared as a nonprimitive type `Integer`) to save the value of k-index. Such a code transformation can be carried out either by programmers or by a preprocessor.

```
public class LUAgent extends TaskAgent {
    double[][] A; Integer n;
    Integer curPivotRow;
    public LUAgent(int s, double[][] m) {
        n = new Integer(s); A = m;
    }
    public void run() {
        int in = Integer.intValue(n);
        int kBegin = Integer.intValue(curPivotRow);
        for (int k=kBegin; k<in; k++) {
            for (int j=k+1; j<n; j++)
                A[k][j] = A[k][j]/A[k][k];
```

[1]Note that run() method represents the only entry point for each Java thread.

```
            for (int i=k+1; i<in; i++)
                for (int j=k+1; j<in; j++)
                        A[i][j] = A[i][j] - A[i][k]*A[k][j];
            onMove();
        }
    }
}
```

It is widely perceived that Java is platform neutral. It is true for single threaded codes. For multithreaded programs, it may not be the case. It is because Java language leaves thread scheduling unspecified. That is, different JVMs may have different thread scheduling implementations. For example, JVMs on Microsoft Windows and Sun Solaris support time-slicing and nonpreemptive scheduling policies, respectively. Consequently, a group of suspended threads on their `onMove()` checkpoints may resume in a different execution order on a remote machine. The question of where to place checkpoints in each thread becomes crucial.

In this study we gear mobile agents toward a popular multiphase bulk synchronous computation model. A computation in this model proceeds in phases. During each phase, threads perform calculations independently and then communicate with others through instant access to shared variables. The phases are separated by global barrier synchronization. Bulk synchronous computations arise frequently in scientific and engineering applications. Like code transformation for `LUAgent`, we insert *collective* `onMove()` checkpoints between phases. Due to the presence of a global barrier between phases, the threads of a compute agent are guaranteed to be suspended and resumed in consistent states, independent of their execution orders.

Similar ideas were extended to compute agents on distributed servers. Cooperative task agents are run in a single program multiple data (SPMD) paradigm on distributed servers. In bulk synchronous computations, the agents need to communicate with their data-dependent peers in each phase. The global barrier between phases will evacuate all pending communications in a phase. Message evacuation and agent migration are realized by `AgentServers`.

16.3 Distributed Shared Arrays for Virtual Machines

To support multithreaded agents on distributed virtual machines, the TRAVELER implements a distributed shared array (DSA) run-time support system, as an integral part. The DSA system provides a Java-compliant interface to extend multithreaded programming to clusters of servers. It shares a common objective with Global Array [230] to combine the better features of message passing and shared memory for a fairly large class of regular applications. It allows programmers to explicitly specify globally shared arrays and their distributions. Shared arrays are distributed between threads, as either regular or as the Cartesian product of irregular distributions on

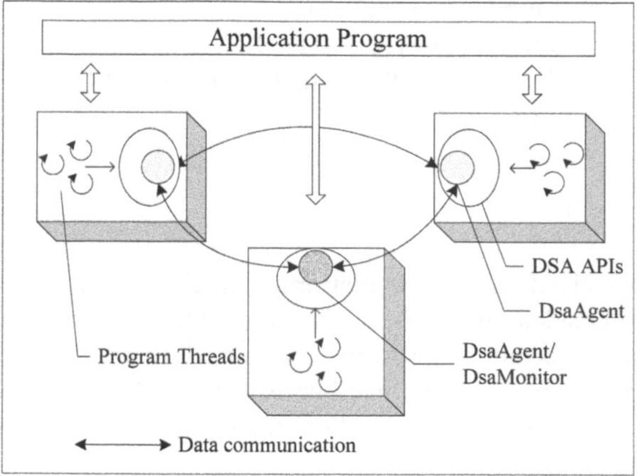

Figure 16.2: The distributed shared arrays architecture.

each axis. It makes a better trade-off between ease of programming and execution efficiency.

16.3.1 DSA Architecture

Since the array data structure is the most common data structure used by scientific and engineering application programs, it becomes a primary building block of the DSA system. The array is partitioned into smaller data blocks and distributed across cluster nodes. The DSA run-time system provides a single system image across the cluster for the DSA. It relieves programmers from run-time scheduling of internode communications and the orchestration of the data distribution. Remote data access is supported by the DSA run-time system and hidden to application programs.

As shown in Figure 16.2, the DSA run-time system comprise a main API component known as the `DsaAgent`. A distributed virtual machine consists of a group of `DsaAgents`, each running at a participant node. Multiple virtual DSA systems could coexist spanning and sharing multiple available nodes, but each system has its own distinct group of `DsaAgents`. The `DsaAgents` are responsible for local and remote access to shared objects. A distinguished `DsaAgent` acts as `DsaMonitor` that holds the responsibility of managing the interconnected nodes.

We summarize the characteristics of the DSA programming model as follows:

- The DSA model supports multithreaded programming in a SPMD paradigm. The concurrent threads running on different cluster nodes share arrays distributed among the nodes.
- Threads in the same node can communicate through local shared variables, while communicating with other threads on different nodes through DSA synchronous remote operations. On the other hand, accessing shared array fol-

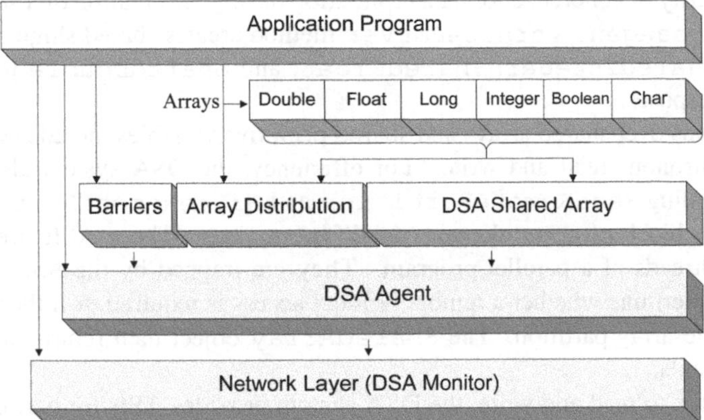

Figure 16.3: DSA API components.

lows the shared address space programming model. The threads can access any parts of the shared array using loads (read) and stores (write) operations with the same array indices as in the sequential program.

- Access to any array element is transparent to parallel programs. The DSA run-time system supports remote array access. The DSA system also supports replication of user-defined array partitions to reduce remote data access cost.

- The application program specifies how a matrix is physically distributed among the nodes and how the array layout is assigned to running threads. It also controls the granularity of cache coherence.

16.3.2 DSA APIs

The DSA run-time support system provides a Java-compliant programming interface to extend multithreaded programming to clusters. The DSA programming model is based on `SharedArray` and `SharedPmtVar` classes, which implement a set of primitive operations for access to user-defined shared arrays and shared objects of primitive types and management of the shared data structure. Figure 16.3 shows the DSA API organization.

`SharedArray` objects are created by a `DsaAgent`'s `createSharedArray` method. Since Java doesn't support operator overloading, the `DsaAgent` actually provides extensions of `SharedArray` for different element types. For examples, `createIntSharedArray` and `createFloatSharedArray` methods return references to distributed shared integer and float arrays, respectively. `SharedArray` objects can be distributed between threads in different ways. Currently, one and two dimensional block decompositions are supported. In addition to the way of block decomposition, programmers can also specify array partition size as the granularity of coherence for data replication during the creation of a shared array. Similarly, `DsaAgent`'s `createPmtVar` method creates shared singular variables of types `SharedInteger`, `SharedFloat`, and `SharedDouble` for synchronization purposes.

Operations over shared arrays and shared primitive variables include synchronous and asynchronous read and write. For efficiency, the DSA system also provides APIs, including `ReadRemotePartition` and `WriteRemotePartition`, for access to a block of array elements of different sizes. These APIs are called by spawned threads of a parallel program. They are trapped by the `SharedArray` object to determine whether a remote or local access is required or if there is a valid cache of the array partition. The `SharedArray` object then returns the required array elements.

In addition to read and write, the DSA system provides APIs for barrier synchronization between application threads. `LocalBarrier` and `GlobalBarrier` are two built-in objects for this purpose. The local barrier is used to synchronize threads residing on the same node and the global barrier is for threads in the entire virtual system. Programmers can also create their own barrier objects, via the method createBarrier, for synchronization between any group of threads. The application program has direct access to the distribution component of the DSA system. This allows more optional control over the distribution/caching mechanism performed by the DSA. In the following, we illustrate the DSA programming model through an example of parallel inner product, as shown in Figure 16.4.

The `InnerProduct` class instantiates `DsaAgents` in the constructor. The `DsaAgents` object register their services to a `DsaMonitor` and together form a DSA virtual. DSA access occurs in worker threads of the parallel application. A global barrier ensures the synchronization of this step. The worker threads proceed and generate different access patterns of the shared array. The `DsaAgent` objects trap the array accesses and determine, with the help of a cache coherence mechanism, whether remote access is needed or the data partition is available locally. In case of a remote access, the `DsaAgent` object contacts the remote counterparts and requests the data partition.

16.3.3 DSA Run-Time Support

Recall that the `DsaAgent` is responsible for local and remote access to distributed arrays and for handling coherence protocols between replicated array partitions. The DSA agent is run as a daemon thread to take advantage of the lightly loaded processors within a participant node. The `DsaAgent` features a directory-based cache coherence in support of replication of user-defined array partitions and a communication and scheduling proxy mechanism for reducing network contention. The DsaAgent is developed as a mobile agent so that the DSA-based virtual machine can be reconfigured to adapt to the change of resource supplies and demands.

```
import DSA.Agent.*;
import DSA.Arrays.*;
import DSA.Sync.*;
import DSA.Distribution.*;
import DSA.Communication.*;

public class InnerProduct extends TaskAgent {
  private static int numThreads = 0, arrayDimension = 0, partitionSize = 0;
  private static DsaAgent dsa = null;
  private static FloatSharedArray matrixA = null;
  private static FloatSharedArray matrixB = null;
  private static FloatSharedArray resultMatrix = null;
  protected static DsaBarrier stageBarrier = null;

  public InnerProduct(int dim, int size, int nthreads) {
     arrayDimension = dim;
     partitionSize = size;
     numThreads = nthreads;
                                                    // Creat the DSA and create the shared arrays.
     dsa = new DsaAgent(numThreads); dsa.initialize();    // 2D arrays are created.
     matrixA = dsa.createSharedFloatArray("matrixA", arrayDimension, arrayDimension, partitionSize);
     matrixB = dsa.createSharedFloatArray("matrixB", arrayDimension, arrayDimension, partitionSize);
     resultMatrix = dsa.createSharedFloatArray("resultMatrix", arrayDimension, arrayDimension, partitionSize);
     stageBarrier = dsa.CreateGlobalBarrier("globalBarrier");
  }

  public void run() {
     InitializeMartixAandMatrixB();              // initialize the data set for both matrices.
     Thread threads[] = new Thread[numThreads];
     for (int i=0;i<numThreads;i++) {
        threads[i] = new Thread(new Worker());
        threads[i].start();
     }
     for (int i=0; i < numThreads; i++)
        threads[i].join();
     dsa.finalize();
  }

  private static class Worker implements Runnable   {
     public void run() {
        float sum = 0;
        arrayDistribution matrixALayout = matrixA.getArrayLayout();
        arrayDistribution matrixBLayout = matrixB.getArrayLayout();
        stageBarrier.barrier();                  // All threads across the nodes will start at the same time.
        try {
           for (int i = matrixALayout.startingIndexAtThread(); i < matrixALayout.maxIndexAtThread(); i= i++)
              for (int j = matrixBLayout.startingIndexAtThread(); j < matrixBLayout.maxIndexAtThread(); j=j++ ) {
                 sum = 0;
                 for (int k = 0 ; k < arrayDimension ; k++)
                    sum += matrixA.read(i, k) * matrixB.read(k,j) ;
                 resultMatrix.write(i,j,sum);
              }
        } catch(Exception e){}
     }
  }
} // end of InnerProduct class
```

Figure 16.4: An example of DSA parallel programs.

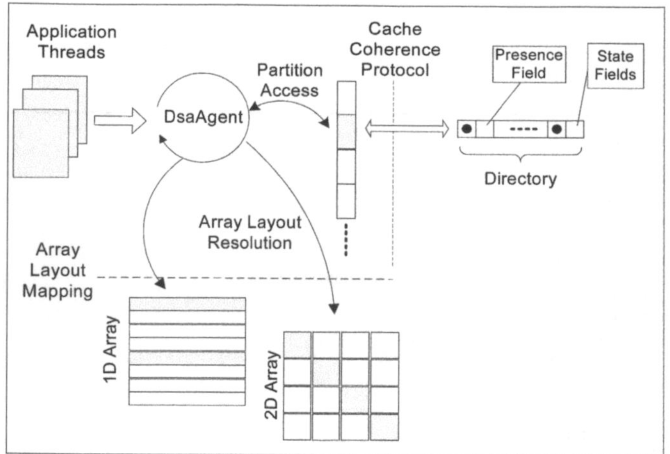

Figure 16.5: DSA directory-based cache coherence.

16.3.3.1 The DSA Cache Coherence Protocol

The server cluster architecture is characteristic of deep memory hierarchy and nonuniform memory access cost at different levels. In particular, remote memory access across nodes often costs up to three orders of magnitude more time than local memory access within a node. An important technique to reduce the remote access time is data replication. In literature, there are many studies on data replication and its related cache coherence protocols. Most of the coherence technologies assumed a fixed coherence unit. In contrast, the DSA system support replication of user-defined array partitions and the partition size can be specified during the creation of a shared array according to the virtual machine configuration. For the purpose of load balancing, the partition size is often set to the array size divided by the total number of participating threads.

The DSA system follows a directory-based cache coherence protocol to manage cluster wide data replications. In this protocol, each partition is designated a home node. A home node tracks the sharers of its partition in a directory structure. Due to the tightly coupled environment of interconnected cluster nodes, the home node will update that state of the partition according to the remote operations that it receives. The state of a cached partition is determined explicitly in the cache directory, as shown in Figure 16.5 . A unique ID, known as the partition index, is assigned to each SharedArray partition. The cache directory keeps track of each partition status using 2 bits on each partition entry. Partition status can be Shared, Exclusive, and Invalid. Shared status means the current node has an up-to-date copy of the partition but other nodes share it. Exclusive status means the current node is the only node that has an up-to-date copy of the partition. Invalid status means some other node is changing the partition and the current copy is out of date. Each home node keeps track of who

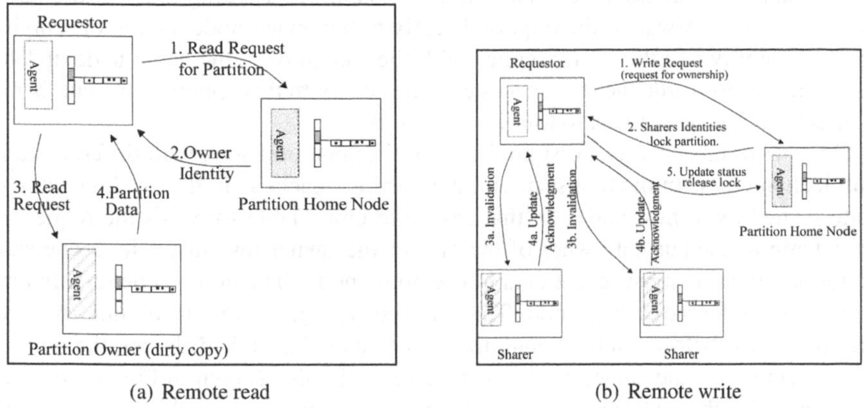

(a) Remote read

(b) Remote write

Figure 16.6: Cache coherence operations.

has an exclusive copy and who are the sharers for its assigned partitions by using a set of presence bits. Each bit corresponds to a unique DSA node. On receiving a request for a SharedArray element, the local DsaAgent will determine the availability of the requested element. It will check through the cache directory whether a valid copy of the requested partition exists in the local memory. The DsaAgent delegates the call to the cache coherence algorithm to determine whether the read or write access is a cache hit or miss. Upon a cache miss, a partition request call propagates to the home node. The home node depending on the status of the partition in request, makes share transaction on request for exclusive copy, update transaction on shared reads, or invalidation transaction on writes. Ownership of partitions moves to exclusive node on write request. The home node directory maintains the sharers nodes, and the exclusive node of a partition. Figure 16.6 shows basic operations of the coherence protocol.

The cache coherence protocol introduced over the set of participant nodes generates network communication transactions. The transactions can be requests, acknowledgments, or cache partitions along with some control attributes. Although these transactions are part of the communication activities that the DSA system tends to hide and their costs are main concerns for the DSA efficiency. The coherence protocol has no control over the cost of individual network transactions. But the protocol design and algorithm have more control over the number of network transactions required to handle read or write cache misses by reducing the number of network transactions generated per cache miss operation, the demand for the network bandwidth decreases, and the contentions on the network resources.

On a read miss, instead of following the strict request–response approach introduced earlier, more efficient intervention forwarding is implemented. In the first scenario the home node replies to the requester with the identity of the owner, the requester then contacts the owner to share the data, and the owner also sends a revision of the data to the home node. With the intervention-forwarding scheme, the home node delegates (forwards) the request directly to the owner node instead of replying with its identity. As the owner replies with the shared data, the home node updates its cache and forwards the replies to the requester. With this approach the number of transactions has be reduced to four instead of five.

On a write miss, we can distinguish two different situations with different communication costs. In the first scenario, the partition cache is in the invalid exclusive state owned by a third node. In that case, the home node forwards the request of ownership to the current owner of the cache; the owner invalidates its cache state and reply with the most recent data to the home node. The home node then returns that cache to the requesting node. This case behaves just like intervention forwarding mentioned before; and although the cache state change is different, the number of transactions is the same. In the second scenario also shown in Figure 16.6, the cache is in the shared state, and shared by more than one node; in this case, the home node takes the necessary steps to invalidate all the sharers of that cache, and upon completion, the cache ownership is handed out to the requesting node.

Besides reducing the number of network transactions per read/write misses, there are two main advantages by making the home node the center of these coherence protocol transactions. When a request arrives at the home node, the directory coherence lookup determines whether the home node has a valid cache of the requested partition in memory; in most read miss cases hopefully that will be the true, since a cache partition on shared state always guarantees a valid data on the home node, thus reducing the number of transactions even further. So taking the case of two consecutive read miss requests on the same cache partition from two nodes, the first request will generate four transactions; while the second will only introduce two, since the home node will reply with a valid cache directly and does not need to forward the request to any other node since the cache is already in valid shared state.

The second advantage is ensuring consistency by the locking mechanisms associated with a cache invalidation/update on the home node, thus behaving like an atomic transaction that might consist of multiple transactions. So, if another requester from another node inquires the same cache partition, the lock on the home node ensures the correctness of the data and eliminates the chances of race conditions.

Finally, we note that one of the objectives of the DSA system is to give programmers more control of data distribution while hiding the details of orchestrations for thread safe data accesses. However, a strict consistency for a global total order between read/write operations would incur extremely high more read/write overhead. Relaxed consistency is to make a good trade-off between efficiency and programmability. In a relaxed consistency, it becomes the programmer's responsibility to make sure that the application threads have safe access into any shared objects, as they will normally do in a multithreaded programming model. This is a simple and intuitive rule because any thread safe application will follow to make sure no race condition exists for correct data access. The DSA run-time support system assumes a relaxed consistency in the access of shared array partitions. The cache coherence protocol ensures thread safe access to shared array partitions by exploiting concurrency between reads and writes.

16.3.3.2 Mobility of the DsaAgent

Throughout the lifetime of a parallel computation, it may exhibit a varying degree of parallelism and imposes varying demands on the resources. Availability of the computational resources of its servers may also change with time, in particular, in multiprogrammed settings. In both scenarios, a virtual machine must be reconfigured to adapt to the change of resource demands and supplies. `DsaAgent` class defines a distributed virtual machine service. Its mobility opens a new dimension for service migration. Chapter 17 presents the detail of service migration for reconfigurable virtual machines.

16.4 Experimental Results

The evaluation of the TRAVELER was done in three major aspects. First was the time for establishing a parallel virtual machine, including the cost of RMI. Second was the cost of local (within a server) and remote (across servers) data access over the DSA. Last was about TRAVELER's overall performance in the solution of two applications: LU factorization and Fast Fourier Transformation (FFT).

All the experiments were conducted on a cluster of four SUN Enterprise Servers. One machine was 6-way E4000 with 1.5 GB of memory and the other three were 4-way E3000 with 512 MB of memory. Each processor module had one 250 MHz UltraSPARCII and 4 MB of cache. The machines were connected by a 155 Mbps ATM switch. We designated one 4-way machine as the broker and others for parallel virtual machines. Clients were run either in the same machine as the broker or in workstations of a remote local area network. All codes were written in Java and compiled in JDK 1.1.6.

16.4.1 Cost for Creating a Virtual Machine

Establishing a parallel virtual machine on a cluster of servers involves three major steps: a client submits task agents to a broker, the broker executes trades, and the broker dispatches the agents to selected servers. The client and broker can be run either on the same machine or on different machines.

In this experiment, we measure the cost for the DSA construction as a stand-alone run-time support system. As the DSA system starts, the first thing is to create DsaAgents on participating nodes and form a DSA network (virtual machine) across the machines. When a shared array is created, the DsaAgent provides detailed information about the array (e.g., array size and partition size) to its cache coherence protocol for efficient management of remote data access. Table 16.1 shows the DSA start-up overhead on various virtual machine sizes and the cost for creation of shared double arrays. Assume the arrays are partitioned in two-dimensional block decomposition and the partition size is 32 x 32.

Table 16.1: DSA start-up overhead and cost for creation of shared arrays.

VM Size (nodes)	One	Two	Three	Four
DSA network startup (ms)	93	430	583	624
Creation (512 x 512) (ms)	104	92	78	67
Creation (1024 x 1024) (ms)	148	127	108	89

Table 16.1 shows that as the DSA virtual machine size increases, its construction time increases. This start-up overhead includes the transmission time of the DSA network configuration information from DsaMonitor to DsaAgents and the registration time from the DsaAgents to the DsaMonitor. That is why construction of a virtual machine of two nodes costs much more than a single node machine. When the virtual machine size increases from two to four nodes, the start-up overhead increases at a sublinear rate because of overlaps of registrations from different DsaAgents. From Table 16.1, we can also see that the cost for creation of shared arrays decreases as the virtual machine expands. It is because all the partitions are equally distributed among the cluster nodes and the number of array partitions per node decreases as the virtual machine size increases.

16.4.2 Cost for DSA Read/Write Operations

The DSA run-time support system provides a high-level programming abstract, built over the RMI communication layer. It is essential to know read/write overhead due to the DSA support, in comparison with RMI communication cost. Moreover, we developed a RPC wrapper over Java sockets to perform the same DSA read/write operations.

In the case of RMI communication, we tested the cost for remote read/write for a number of times, each with access to different block sizes, and took an average of the costs. In order to perform equivalent remote access in the DSA system, all read and write operations over a shared array have to be cache miss. To the end, we accessed the first element of each partition and set the stride of consecutive data accesses to be the shared array partition size. At the beginning of the test code, one DsaAgent walked through all the partition and requested exclusive ownerships for the partitions. Then the second DsaAgent started the actual test and all its references were guaranteed to be read/write miss. The RPC wrapper was implemented through a two-way Java socket communication, which performs solely read/write operations of various block sizes.

Figure 16.7 shows remote read/write cost via the DSA system, in comparison with the approaches of RMI and RPC wrapper. It can be seen that the DSA read/write miss incurs small percentages of overhead over the RMI access of partitions of medium and large sizes. For large partitions of 128 x 128 size, the DSA read/write miss incurs a marginal overhead. But for small partitions of 8 x 8 and 16 x 16 sizes, the

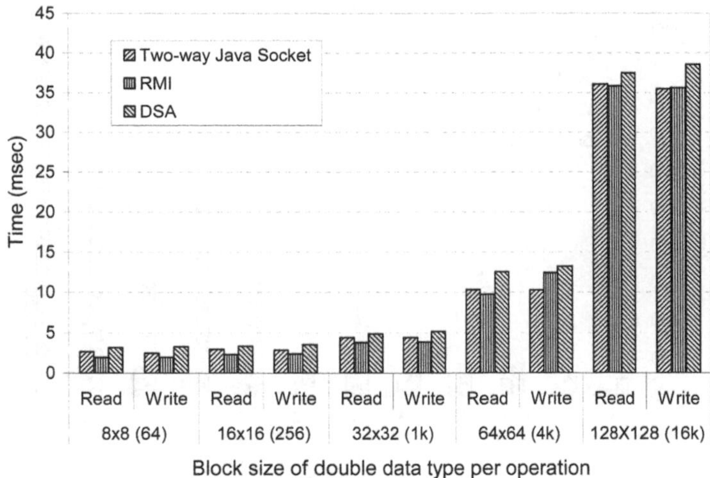

Figure 16.7: Remote Read/Write overhead in DSA, RMI, and Socket wrapper.

DSA remote access incurs overhead as high as 60%. Figure 16.7 also shows that the RPC wrapper over two-way Java socket communication is more expensive than RMI. The RPC wrapper is based on two object transfers between client and server over a Java socket and each requires marshaling and unmarshaling of serialized objects over the network. The cost for marshaling/unmarshaling is the dominant factor for both RMI and Java socket, as they both rely on object serialization. But in the RMI case, the remote read/write needs to send array indices of primitive types. Since the transferring of primitive data needs no serialization, its cost is much less than the object marshaling/unmarshaling as required by the RPC wrapper.

In the third experiment, we tested the average cost for data access due to the impact of partition replication. Since the write operation involves cache invalidation protocol, we considered writing of single array elements of a shared double array. We assumed a requester node to be performing all the writes and waiting until all other participating nodes read shared array elements and become active sharers. The requester node then wrote elements into the shared array consecutively by a stride of various sizes so as to ensure that not all writes would be hit once an array partition becomes exclusive. The requester kept performing writes for 1 second. The test was repeated under different configurations with different partition size, strides, and number of participating nodes:

- The number of participating nodes started from two up to four.
- The partition size started from 8 x 8 and increases up to 32 x 32.
- The stride of consecutive write indices started from one up to partition size in a dimension. For example, for the partition of size 8 x 8, the maximum stride is set to eight.

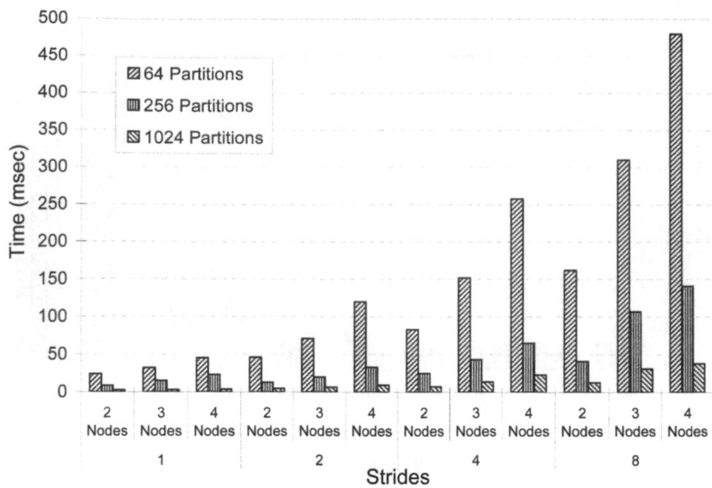

Figure 16.8: Average cost for a write to distributed shared arrays.

Figure 16.8 shows the average cost for a write in different configurations. It reveals that the average cost for a write increases as the number of participating nodes. For example, in the case of stride of 2 and block size of 1024, as the number of participating nodes increases from 2, 3, to 4 nodes, the average write cost increases from 4.7 ms, to 6.5 ms and 9.4 ms, respectively. This is because the partition was set to be shared by all the nodes, except the requester, before its write. The more the sharers are, the more nodes are involved in cache coherence protocol for invalidation.

Figure 16.8 also shows the advantage of a large partition. As the partition size increases, there would be more cache hits than misses. Consequently, there would be fewer demands for cache invalidation and less average write cost. For example, in a 4-node virtual machine, the average time for writes in a stride of 2 is reduced from 120 to 9.4 microseconds as the partition size increases from 64 to 1024.

16.4.3 Performance of LU Factorization and FFT

In addition to micro-benchmark, we evaluated the overall performance of the DSA system in two applications: LU factorization of matrices and FFT.

16.4.3.1 LU Factorization

LU factorization is a kernel for the solution of systems of linear equations. It loops over the matrix diagonal elements and performs pivotal row scaling and submatrix calculations at each iteration. We considered a double array of size $1024x1024$. We exploited parallelism within the iteration of the outmost diagonal loop and experimented with a two-dimensional block cyclic decomposition approach with various

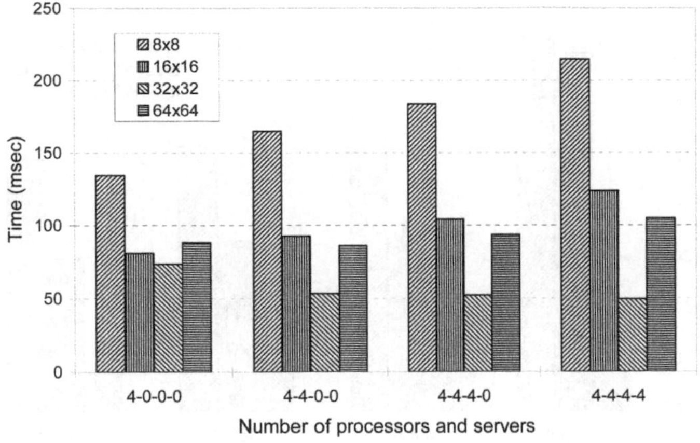

Figure 16.9: Parallel LU factorization of matrices using different partition sizes.

block sizes for balanced task assignment.

Figure 16.9 presents the results in different configurations of the DSA virtual machine. It is known that the partition (or block) size determines the ratio of computation to communication as well as the overhead of cache coherence protocol. Decomposition with small size partitions implies less calculation before a thread proceeds into a communication phase, possibly requesting data from threads in remote nodes. This explains the poor scalability of the algorithm for small partitions of 8 x 8 and 16 x 16. As partition size gets larger, the ratio of computation to communication increases and the average remote access time decreases, as shown in Figure 16.8. However, large partitions lead to a severe load imbalance problem. For example, even in a single node machine (4-way SMP node), partition size 64 x 64 yields worse performance than partition size 32 x 32, both assuming 4 threads running concurrently. With the increase of the virtual machine size and the number of participating threads, the load-balancing problem becomes even more severe. Meanwhile, each thread gets fewer tasks to do. Figure 16.9 shows the partition size of $32x32$ gives a relatively good trade-off, although the algorithm gains marginal improvements due to cluster computing. The algorithm showed a performance gain of 27% when the virtual machine expanded from a 4-way SMP node to two SMP nodes.

There are many reasons for the poor overall performance. To isolate the factors of the DSA overhead, cluster architecture, and Java programming environment, we compared the LU factorization performance of the DSA version with a Java socket implementation and an MPI/C version and presented the results in Figure 16.10. The DSA overhead over the pure Java socket implementation is due to: (1) RMI communication, and (2) cache coherence protocol. As the virtual machine expands, there are more DSA nodes sharing cached partitions, which leads to a higher access miss rate. Figure 16.10 also demonstrates the performance gap between Java and MPI/C languages.

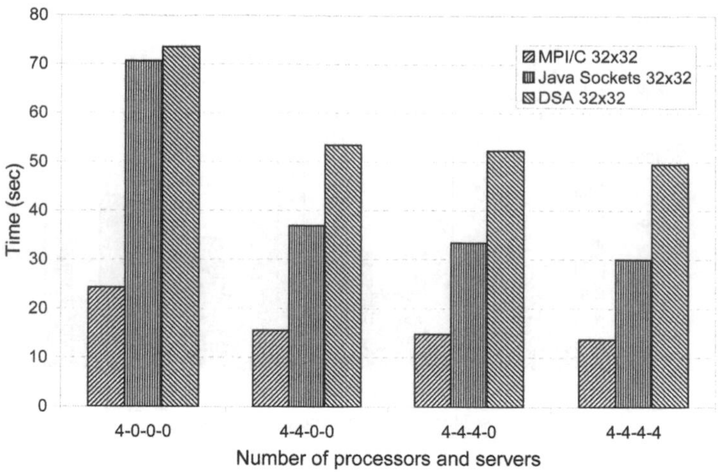

Figure 16.10: Parallel LU factorization of 1024 x 1024 double matrices in DSA, Java socket, and MPI/C.

16.4.3.2 FFT Application

FFT is one of the fundamental problems in digital signal processing. It is often used in many scientific applications such as time series and wave analysis, convolution, and image filtering. FFT algorithms compute the discrete Fourier transform on $N = (2n)$ points in $O(nN)$ time. An n-point FFT algorithm involves n stages of computation. In each stage, every point would be updated once, involving one complex multiplication and one complex addition. In a multithreaded FFT, each thread is assigned a portion of the data set and performs independent updates on its own data points. Between stages, the threads access to other portions of the data set. We tested the FFT algorithm with partition size varying from 8 x 8 up to 128 x 128.

FFT has a different access pattern from LU factorization in that an assigned home node is not the only one that would modify its partition data. This leads to more costly write misses and requires more invalidation operations in order to maintain coherence of cached partitions.

Figure 16.11 presents the performance of a 1024-point FFT with different partition sizes under the DSA support. With the expansion of the DSA virtual machine, the system incurs more overhead for ensuring cache coherence and results in more costly read and write misses. The ratio of computation to communication increases as the partition size increases. When the ratio goes up to a point where the partition size is of 64 x 64, the FFT application starts to observe slight performance gains when the virtual machine is expanded from a single node to a two-node system. But as the partition size goes up to 128 x 128, the communication cost exceeds the advantage of having large computation segments and consequently leads to a loss of overall performance. Due to the limitation of the problem size, the maximum number of

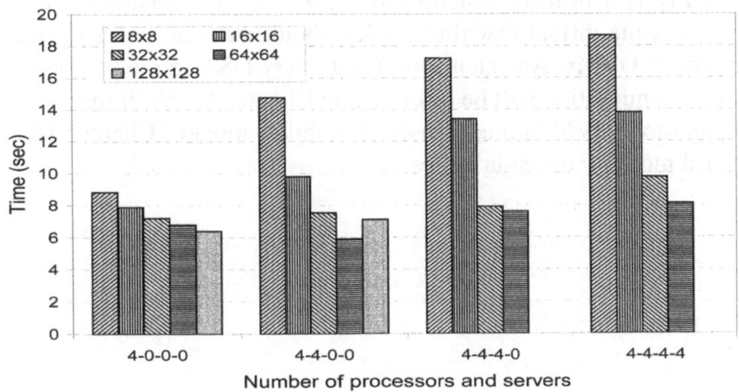

Figure 16.11: Parallel FFT for 1024 x 1024 double arrays on different virtual machine configurations.

worker threads that the system can be configured in the case of partition size of 128 x 128 is limited to eight.

Finally, we note that as in the case of LU factorization application, performance hardly gains with the FFT algorithm. It is mainly due to the high cost for remote data access in the cluster environment. However, Figure 16.11 shows significant impacts of the coherence granularity on the overall performance. The DSA system benefits user-defined coherence granularity in the form of partition sizes.

16.5 Concluding Remarks

In this chapter, we have presented a mobile agent based "push" methodology and a resource brokerage infrastructure, TRAVELER, in support of grid computing. The TRAVELER system provides an agent wrapper to simplify agent-oriented programming. Agents are dispatched to a resource broker for services. Upon receiving an agent, the broker executes trades and forms a parallel virtual machine over available servers to execute the computational agent. Due to the commonplaceness of multiprocessor servers and the increasing popularity of server clusters, the TRAVELER supports multithreaded agents for high-performance computing on clusters of servers. Agents are cloned on each server and run in the SPMD paradigm. Agent-oriented cluster computing is supported by an integral distributed shared array (DSA) run-time system, which forms a distributed virtual machine (DVM) for a hybrid message-passing and shared-address-spacing programming model.

Due to the dynamic nature of the Internet, grid computing services on the Internet ought to rely on run-time support for being adaptive to the change of the environment. Mobility is a primary function to support such adaptation. In addition to agent migration, the virtual machine service should be mobile for the construction of reconfigurable DVMs. An extension of stationary DSA to mobile DSA for virtual machine service migration will be discussed in Chapter 17. Both forms of migration in a reconfigurable DVM defines a hybrid mobility model. Chapter 18 deals with related hybrid mobility decision issues.

Chapter 17

Service Migration in Reconfigurable Distributed Virtual Machines

Virtual machine (VM) technology provides a powerful layer of abstraction and allows multiple applications to multiplex the resources of a virtual organization in grid computing. The grid dynamics requires the VM system be distributed and reconfigurable. This chapter presents a service migration mechanism, which moves the computational services of a distributed shared array (DSA) VM to available servers for adaptive grid computing. In this way, parallel jobs can resume computation on a remote server without requiring any service pre-installation. The DSA VM with mobility support is referred to as a Mobile DSA.

17.1 Introduction

Computational grids can integrate geographically distributed resources into a seamless environment [116]. In one important grid scenario, performance-hungry applications use the computational power of dynamically available nodes. Compute nodes may join and leave ongoing computations, and they may have different configurations, such as processor speed, memory space, and network bandwidth. Many intricate problems will emerge when high-performance applications are executed in such dynamic environments. One of the major challenges is to provide a run-time system that supports efficient execution and communication while achieving the adaptivity to the grid dynamics.

VM technology provides a powerful layer of abstraction for resource management in grids [109]. It enables user applications to become strongly decoupled from the system software of the underlying resourccs and other applications. This property facilitates the development of grid applications by allowing programmers to use well-defined service interfaces. However, the conventional VMs are not easily manageable and scaled in wide area environments, especially across administrative domains. Moreover, since the system services are monolithically placed, they may be far away from their requesters. This leads to high communication overhead and delay, which violate the performance requirements of parallel applications. A remedy for these problems is to construct a distributed VM, by factoring VM services into logical components, replicating and distributing them throughout a network [290].

A distributed VM can be viewed as a set of virtual servers running on top of multiple physical servers. It is certainly possible to deploy its components as static computing units. However, when a server running a parallel job is reclaimed by its owner, the remaining processes of the same job have to be stopped. To make progress, a parallel application requires that a group of servers be continuously available for a sufficiently long period of time. If the state of a large number of servers rapidly oscillates between available and busy, a parallel computation will make little progress even if each server is available for a large fraction of time. Moreover, the application has to terminate when a server failure occurs. Research in [177] shows the CPU availability and host availability are volatile in grids. Adaptivity is fundamental to achieving application performance in such dynamic environments [101]. Therefore, the abstraction layer provided by a distributed VM would not be fully exploited unless it can be instantiated and reconfigured at run-time to tackle the grid dynamics.

The existing approaches to constructing reconfigurable systems, such as process (thread) migration [225] and agent migration [126], have significant limitations. They only move the execution entities among servers and leave the supporting services behind. Their effectiveness is based on two assumptions: (1) there are servers with idle resources; (2) the underlying run-time support system, as a form of grid services, has already been running on the destinations. On the other hand, parallel applications strive to distribute computational jobs evenly among servers. As a result, it is difficult for an overloaded server to steal idle CPU cycles from another one with similar processing capacity. Things become even worse when a server is going to fail. In that case, all of the computational jobs on the failing node will swarm into other existing servers. It will lead to cascaded server overloads and the entire parallel application has to be stopped, if no new servers are added to the system. As to the second assumption, it is not proper to preinstall and execute all the possible services on each grid computer, across different administrative domains. This is particularly the case for volunteer-based grids like SETI@home project (`setiathome.ssl.berkeley.edu`).

In [124, 125], we proposed a scalable and effective approach, *service migration*, to achieve adaptive grid computing. The semantics of service migration is to suspend the residing execution entities, stop the service programs of a virtual server on the source node, migrate the run-time service data and states along with the execution entity states to a destination node, initiate a new virtual server with restored services on the destination, and resume application execution. In this way, parallel jobs can continue computation on an available server without requiring any service preinstallation, in face with server overload and/or failures. Service migration provides a general approach to reconfigurable distributed machines. It can be applied to many distributed VM systems, such as the Java/DSM [347] and the Global Object Space [362]. However, since they use modified Java Virtual Machines (JVMs), the portability and interoperability problems complicate service migration. In contrast, the DSA system in Chapter 16 is a run-time support system on top of intact JVMs. It provides virtually shared arrays for distributed compute nodes. Each DSA virtual server is constructed as a Java object. The service encapsulation and system simplicity facilitate the service movement. As an illustration of the service migration mechanism, we incorporated it into the DSA and designed a mobile DSA (M-DSA). Like the DSA system, the M-DSA system supports a Java-compliant distributed VM to accommodate parallel computation in heterogenous grids. Service migration allows reconfiguring the VM and makes it adaptive to the dynamically changing environments.

Service migration complements other migration techniques moving execution entities. A hybrid migration infrastructure will make a system become more adaptable to dynamic environments. In the following, we present the details of the design and implementation of the M-DSA system.

17.2 M-DSA: DSA with Service Mobility Support

Service migration achieves adaptive grid computing by moving run-time support services from an overloaded or failing server to an available one. As one of its applications, we developed a M-DSA, which extended a DSA run-time system in Chapter 16 for virtually shared arrays with mobility support.

17.2.1 M-DSA Architecture

Recall that the DSA services provide a single system image across a set of compute nodes for DSAs. It relieves programmers from run-time scheduling of internode communications and the orchestration of the data distribution. Each DSA virtual server stores a portion of a shared array. Remote data access is supported by the DSA run-time system and hidden to application programs. The DSA system comprise a main API component known as the `DsaAgent.` Each virtual server involved in a parallel application has a DsaAgent. The DsaAgents are responsible for local and remote access to shared array elements. Parallel jobs are constructed as computational agents, distributed among the virtual servers. The original `DsaAgents` are stationary and cannot move between servers. As a result, a parallel application will make little progress or even be terminated, when servers become overloaded or failed.

Service migration enables the construction of a reconfigurable distributed VM in grid computing. It allows moving a virtual server to a remote computer and continuing computational services for load balancing and fault resilience.

To support service migration, we redesigned the `DsaAgent` as a mobile object and meanwhile retained its original features. The extended DsaAgents constitute a set of virtual servers in a distributed VM, as depicted in Figure 17.1. Migration-enabled methods allow a DsaAgent to perform service migration to transfer DSA services among physical servers. We call the DsaAgent on a server that starts a parallel application as the *coordinator*. It is responsible for receiving migration requests from the LoadMonitor, and triggering a service migration at some potential migra-

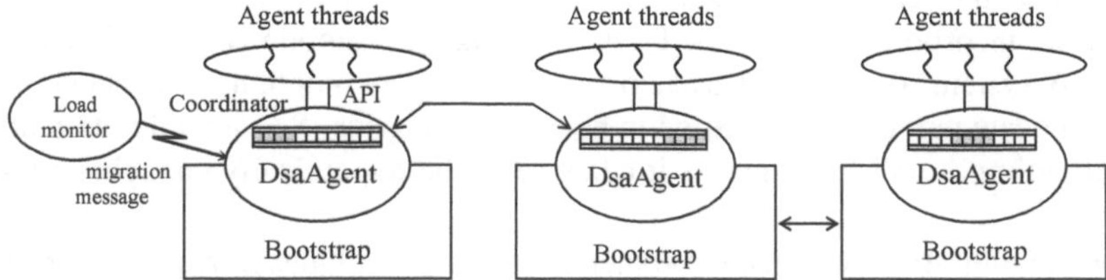

Figure 17.1: The M-DSA architecture.

tion point. The LoadMonitor makes migration decisions based on the performance information collected by sensors on each virtual server, during the application execution. To accommodate service images transferred from remote nodes, a *Bootstrap daemon* is prerun on the destination node. It receives the incoming service image and initiates a new DSA virtual server with the restored services.

In order to enable different applications multiplexing a grid computer, we allow multiple virtual servers to reside on a physical server. They provide different run-time services and accommodate the corresponding computational agents. The M-DSA runs on top of the JVMs, which facilitate the service migration in heterogeneous environments.

Like DSA, the M-DSA system can be used stand-alone or be integrated with other run-time systems like TRAVELER in Chapter 16.

17.2.2 An Example of M-DSA Programs

Like DSA programming, execution of an M-DSA program involves three main steps: (1) to create a DsaAgent on each server to manage the shared arrays; (2) to create the required shared arrays; (3) to access the shared array through read/write operations. Figure 17.2 shows an M-DSA program skeleton. DSA access occurs in working threads of the parallel application, which create shared arrays. The working threads proceed and generate different access patterns of the shared array. The DsaAgents trap the array accesses and determine, with the help of a coherence mechanism, if remote access is needed or the data partition is available locally. In case of a remote access, the DsaAgent contacts the remote counterparts and requests the data partition. The bold statements in this example are inserted for the possible service migrations. A barrier operation ensures the synchronization of a computation step. It also represents a potential migration point. The variable `phaseNum` records the phase index of the next computation step. Its value, along with the set of `if` statements, ensures the working thread will correctly resume its computation after service migration. A preprocessor can insert these migration-oriented statements when scanning the application program, as discussed a lot in the compilation technology. We will present the details of service migration in the next section.

```
class AgentThread implements java.io.Serializable, Runnable {
    private DsaAgentImpl dsaAgent;
    private int arraySize, blockSize;
    private Barrier phaseBarrier;
    private SharedArray array;
    private int lowerIndex, upperIndex, step;
    private int phaseNum=0;

    public AgentThread (DsaAgentImpl dsaAgent, int aSize, int bSize) {
        this.dsaAgent = dsaAgent;
        this.arraySize = aSize;
        this.blockSize = bSize;
        array = dsaAgent.createSharedDoubleArray("arrayName", aSize, bSize);
        phaseBarrier = dsaAgent.createBarrier("barrierName");
    }

    public void run() {
        int index;
        if (phaseNum < 1) {
            lowerIndex = ...;   upperIndex = ... ;   step = ...;
            for (index=lowerIndex; index < upperIndex; index+=step)
                array.write(index, index*1.0);
            phaseNum ++;
            phaseBarrier.barrier();
        }
        if (phaseNum < 2) {
            lowerIndex = ...;   upperIndex = ...;   step = ...;
            for (index=lowerIndex; index < upperIndex; index+=step)
                double val = (double)array.read(index);
            phaseNum++;
            phaseBarrier.barrier();
        }
    }
}
```

Figure 17.2: An example of parallel program in M-DSA

17.3 Service Migration in M-DSA

Parallel applications assume different workload characteristics on different computation phases. At the same time, the availability of grid resources also changes with time. The combination of these two factors requires to dynamically move the services and computation among computers to tackle server overload and failures.

A service migration involves collecting and packing the states of run-time services and agents, transferring them to a destination node, reinitiating the services, and resuming agent execution at the statement prior to migration on the new virtual server. Compared with process/thread migration [225], service migration deals with not only moving the execution context of each process/thread, but also requiring the reconstruction of the execution environment on a remote server. It introduces more technical challenges as a consequence.

17.3.1 Performance Monitoring for Service Migration

The overheads associated with packing, transferring, and rebuilding run-time services of a virtual server are relatively high. Parallel applications usually have strict performance requirements. Therefore, agents on a virtual server should have a sufficient amount of computation and two consequent service migrations should keep distant spans in order to mitigate the impact of migration overhead.

If the memory model of the distributed VM system is the traditional sequential consistency, then the system appears like a multiprogrammed uniprocessor and DSA services can be migrated randomly with a guarantee of correctness of resumed execution. However, for better performance, the M-DSA system adopts a relaxed memory model to reduce both the number of messages and amount of data transfers between VMs. When the source server performs a service migration to a destination, the shared array partitions managed by the corresponding DsaAgent cannot be accessed during the migration. As a result, the memory consistency is complicated and it is similar to the consistency problem in "partitioned networks" [79]. In such a model, some virtually shared data between two consecutive synchronization points could be in inconsistent states. If agents access such data managed by the migrated DsaAgent, their computation may be incorrect, especially when they are used as input. To ensure correctness, service migration can only be allowed at synchronization points or barriers, as illustrated by the example in Section 17.2.2.

To exploit the benefits of service migration and minimize the impact of migration overhead, we need to monitor the dynamic performance of running applications and make migration decisions based on these accurate measurements. However, most of the existing process/thread migration approaches circumvent the execution monitoring issue, and their feasibility is questionable as a consequence.

In M-DSA, we utilize the Autopilot [260] toolkit to tune the migration timing according to the actual application progress. Each virtual server incorporates an *Autopilot sensor*, which extracts quantitative performance data of the corresponding application. The application program is instrumented with synchronization method calls and the DsaAgents record the execution time of each computational phase at those points. To avoid introducing a heavy calculation burden for the LoadMonitor and much global information of the system, we specify an attached function for each sensor to normalize the execution time by the processing speed of the physical server. When the LoadMonitor has collected the performance data from all the sensors, it compares the data with an overload bound to make migration decisions. This threshold is a positive real number. Its value can be predefined or dynamically adjusted to tune to heterogeneous systems. When a service migration is decided, we need to find an appropriate destination server to accommodate the migrated virtual server. This involves the resource management in M-DSA.

During service migration, the run-time data and states of a virtual server are transferred to a remote physical server. As a consequence, the monitoring infrastructure should be adjusted. This adjustment is small, because there are no direct connections among sensors. What we need to do is create a sensor for the new virtual server on the destination. Then the Autopilot toolkit will register properties of this sensor to a designated *Autopilot manager*, which resides on the LoadMonitor and acts as a name server for clients to look up sensors.

17.3.2 Service Packing and Restoration

Service migration involves the movement of both run-time services and computational agents. Although they can be carried out separately, certain data and state information should be handled properly. For instance, the shared arrays are created and referenced by agents, but they are allocated and managed by the virtual server. In this section, we discuss the policies for wrapping up and restoring various JVM run-time data structures for service migration. The objective is to minimize the amount of data and states to be captured, while ensuring the correct resumption of services after migration.

The JVM specification [198] defines several data areas that are used during program execution, including the heap, method area, and JVM stack. They together form the execution context of a run-time service on a virtual server. To allow different applications to multiplex resources of a computer, we let the Bootstrap daemon on a node capture the service data and states, and wrap them into an application-independent form, *service pack*, which contains components from the following areas, as illustrated in Figure 17.3.

The heap area for a virtual server stores the dynamically created shared arrays and associated control structures, such as the status and owner address for each array block. Since we apply the SCI-like directory-based coherence algorithm, each virtual server acts as the home node for a portion of a shared array and the owner information changes when a remote agent exercises a write operation to a managed block. Besides, the status of an array block may also be altered, as it elements may be read and written at run-time. As a result, most data structures maintained in the heap area have to be contained in the service pack. An exception is the locks for block accesses, because they can be created and initiated on the new virtual server.

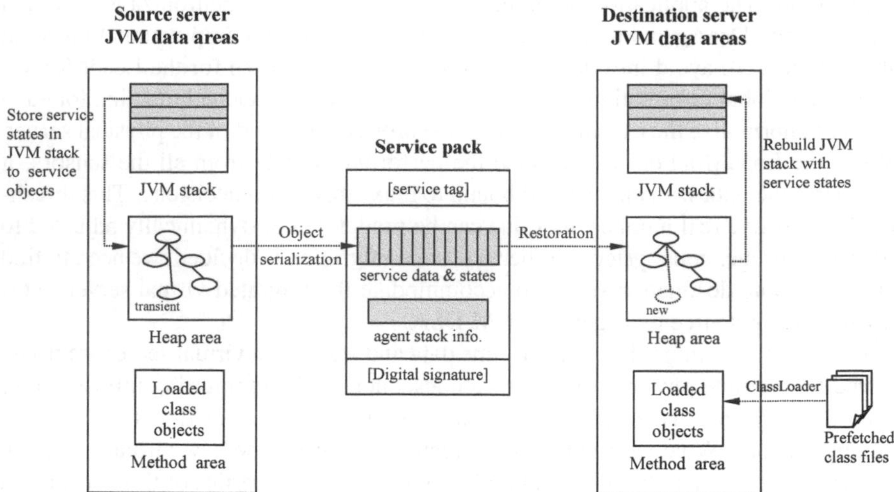

Figure 17.3: Service wrapping up and restoration in M-DSA service migration.

In M-DSA, we include the shared arrays in service pack instead of the agent image, because they are allocated in the virtual server address space and accessed by computational agents only via well-defined APIs. After the virtual server is restored on the new destination, the agent references to a shared array will be updated by using the array name to locate the actual array object.

The method area stores the per-class structure for each loaded class. It can be rebuilt at the destination when all the referenced classes are loaded. So, we do not include information of the method area in the service pack. To retrieve a remote class file, the traditional approach is to invoke the Java *ClassLoader* to load the class at run-time. Thus, the class transmission time will contribute to the service migration overhead. Instead, M-DSA applies a *prefetch* technique to let the new destination server retrieve the virtual server and application program codes in parallel with the transmission of a service pack, from a nearby server. We realize this technique with the aid of LoadMonitor. When the LoadMonitor determines a service migration should be performed at a synchronization point, it selects an existing physical server that is in the same local area network as the new destination, i.e., their subnet addresses are identical. If there are multiple candidates, we choose the one with the least workload, according to the application performance data collected by Autopilot sensors. Then, a migration message containing the network address of this nearby server, along with the migration source address, is sent to the Bootstrap daemon of the new destination. The daemon contacts the nearby server for program codes, while waiting for the arrival of service pack from the source. To reduce the number of control messages exchanged for code transfer, we group the compiled class files used in virtual server initiation and agent restoration into a JAR package, and transmit it to the new destination in one piece during a service migration.

A JVM stack maintains the frames for virtual server method calls. Each virtual server provides the shared array access services to computational agents by defining a set of access methods, and these services will be reinitiated on a new physical server after migration. Therefore, the JVM stack can be rebuilt on the destination without transferring the stack frames. However, there is some control information, such as the shared arrays' names, the number of agents, and their references, that should be kept in a service migration. Since the potential service migration points are at the computation phase boundaries, this control information is well stored in the local variables of the virtual server object when a service migration occurs. This simplifies the stack capture, because we only need to transfer the virtual server object containing the necessary control information to the destination, and restore the object there. The Java *object serialization* mechanism can achieve this end. It packs the virtual server object along with other referenced objects, allocated in the heap area, and at the same time discards those unnecessary components specified by the *transient* quantifier. The restoration procedure is performed reversely, and the serialized objects are interpreted according to the prefetched class files loaded by a Java ClassLoader. Besides, we avoid reinitiating the object members containing the above control information by using a migration indicator, and the transient components need to be created and initiated explicitly.

We integrate the service wrapping function in a *checkpointing* method to construct the service pack. It also allows us to write the packs to hard disks, and perform encryption and/or compression before transferring them to remote nodes. If there are multiple virtual servers on a physical server, the Bootstrap daemon wraps them into separate service packs and migrates them one by one. Correspondingly, they will be restored sequentially on the destination server after migration.

17.3.3 Computational Agent State Capture

A computational agent contains one or a set of threads carrying out parallel jobs. Its migration can be realized by some existing thread migration techniques, such as [362]. But this requires that the supporting services have been running before thread migration occurs. One way to fulfill this requirement is to perform service movement first, and then migrate the agent threads after the service has been restored on the new server. It is clear two sequential migrations introduce significant delay to the parallel computation. In M-DSA, we transfer the states of computational agents along with the service pack to reduce migration overhead. The details are as follows.

Each virtual server contains a data structure recording all its residing agents and having a reference to each agent object. During service packing, the virtual server object is serialized. As a result, all its referenced nontransient members, including the agent objects, will be included in the service pack. When the service pack is transferred to a new server, the agent thread objects will also be moved there. A challenge in agent movement is how to ensure its threads will resume execution just at the points where they are stopped.

The program counter (pc) register, in the per-thread run-time data area, contains the address of the JVM instruction currently being executed by a thread. It indicates

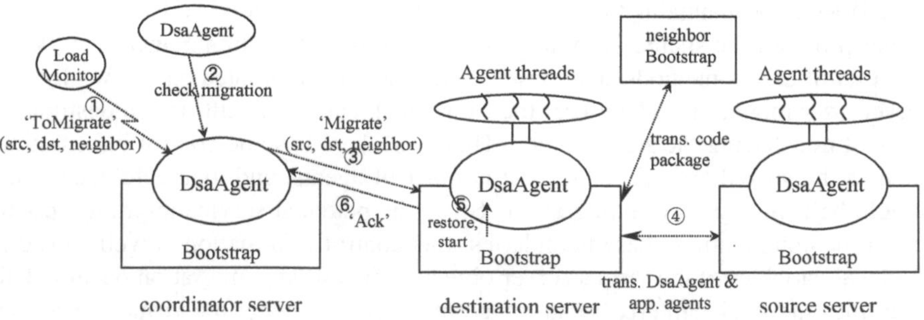

Figure 17.4: Service migration in M-DSA system.

the start point of computation after service migration. Since the potential migration points are at the computation phase boundaries, we apply a portable approach to preserve the pc register information. We represent a thread's pc register by a set of integer numbers stored in `phaseNum` (refer to the example in Section 17.2.2). After an agent thread has performed a computation phase, the `phaseNum` is increased by one to indicate the index of next phase. Then, a `barrier` method carries out a synchronization operation. It also makes the coordinator DsaAgent check whether there is any service migration request message received from the LoadMonitor. If such messages are found, the run-time support system will help perform the service migration and the computation will be resumed on the destination server. To let the agent thread locate the right computation phase after migration, an `if` statement is inserted at the beginning of each phase to compare the value of `phaseNum` with the phase index of the following computation. Positive values of `phaseNum` indicate the agent was migrated and has executed on another server.

Unlike service packing, we extract the agent thread execution state stored in the JVM stack frames. We adopt the state capture and restoration approaches proposed in [309] to rebuild thread stack on the a server after migration. At the same time, the thread pc register is set according to the value of `phaseNum` and the set of `if` statements. Thus, an agent can resume execution at the correct computation phase after a service migration.

17.3.4 Service Migration: A Summary

In summary, we present the complete service migration algorithm in Figure 17.4. Specifically,

1. When the LoadMonitor detects some server in the system is overloaded or is unavailable, based on the application performance data collected from Autopilot sensors, it issues a "ToMigrate" message, along with the addresses of the source and destination servers, and a neighbor server to the coordinator, which will record this message in a message queue.

2. When the parallel computation reaches a synchronization point, the coordinator is requested to perform migration checking, which finds out whether there is a migration request to be served at this synchronization point.

3. If a "ToMigrate" request is found in its queue, the coordinator will send a "Migrate" message to the Bootstrap daemon of the destination server, denoted as $bootstrap_{dst}$, and wait for the acknowledgment of migration completion.

4. Then bootstrap$_{dst}$ contacts the Bootstrap daemon on the source server, from which the DSA service pack and agent images will be transmitted to the destination. The related program codes are prefetched from the neighbor server.

5. The bootstrap$_{dst}$ loads the retrieved class files and builds the method area in the main memory. The DSA service and agents are reestablished by the restoration procedure. It also starts the DSA service by running the newly initiated DsaAgent. Computational agents are triggered to resume their parallel jobs at the proper computation phase.

6. As soon as the service migration is successfully completed, the bootstrap$_{dst}$ sends an acknowledgment message back to the coordinator.

17.4 Interface to Globus Service

The Globus Toolkit [115] is the de facto standard for grid computing. It provides an infrastructure and tools to manage grid resources. To construct a reconfigurable distributed VM in the grid environment, we developed the interface to Globus in M-DSA for efficient and secure resource management.

17.4.1 Resource Management and Migration Decision

Dynamics is a property of the grid environments. Compute nodes may join and leave ongoing computations, and some may also become overloaded when their workload is too huge. At the same time, the M-DSA run-time system is a distributed VM spreading across multiple physical servers. Resource management is an important component for the system start-up and service migration, because we need to select appropriate physical servers to accommodate the DSA services.

When the M-DSA system receives an application program, it must first determine what resources are available and secure an appropriate subset for the application. The resource management component will negotiate with a *grid runtime system* (e.g., Globus in our implementation) to obtain a list of available physical servers that have sufficient resources. Then it will broker the allocation and scheduling of application programs on grid resources, or start up the *Bootstrap* daemon on the new destination server when a service migration is necessary. This step will also insert sensors to help the *LoadMonitor* control application execution. Next, the resource management component will generate a script file, which specifies the list of servers on which the application programs will be executed; see the following for a sample script. The Globus Toolkit will start-up the application execution by running the script file.

```
( &(resourceManagerContact="node2.grid.wayne.edu")
      (count=1)
      (label="subjob 1")
      (environment=(GLOBUS_DUROC_SUBJOB_INDEX 0))
      (directory=/wsu/home/song/mdsa/app)
      (executable=/usr/java/j2sdk1.4.2/bin/java)
      (arguments=-classpath .:../classes:../mdsa.jar
              '-Djava.security.policy=mdsa.policy' app.lu.Lu lu/lu.cfg)
)

( &(resourceManagerContact="node8.grid.wayne.edu")
      (count=1)
      (label="subjob 2")
      (environment=(GLOBUS_DUROC_SUBJOB_INDEX 1))
      (directory=/wsu/home/song/mdsa/app)
      (executable=/usr/java/j2sdk1.4.2/bin/java)
      (arguments=-classpath .:../classes:../mdsa.jar
              '-Djava.security.policy=mdsa.policy' app.lu.Lu lu/lu.cfg)
)
```

During program execution, the sensor on each physical server monitors the application performance and reports it to the LoadMonitor. After LoadMonitor collects the performance data from all the participant sensors, it checks whether there are overloaded servers in the system. If that is the case, service migrations will be triggered, and then the resource manager will notify the overloaded servers to transfer the application programs and DSA run-time service to new servers. During this process, the resource manager needs to negotiate with the grid run-time system again to find the new available resources. A rigorous model for migration decision is presented in Chapter 18. It derives the optimal migration timing and a lower bound of the prospective destination server's capacity, based on stochastic optimization and renew process techniques.

17.4.2 Security Protection

Grid computing is concerned with the sharing and coordinated use of diverse resources in distributed "virtual organizations." The dynamic and multidomain nature of the grid environments introduces significant security issues [311]. For instance, in a campus grid, clusters of different departments are connected via an insecure campus network. Unauthorized users may dispatch their programs to grid computers, and even try to exhaust grid resources via denial of service attacks. Therefore, security protection is a vital issue affecting the feasibility of a system in practice.

As discussed in the previous sections, the M-DSA system provides an open, reconfigurable VM environment for parallel computation in grids. The Bootstrap mechanism allows different run-time services to multiplex a physical server. However, it is also possible that distrusted users attempt to run their private services on top of the Bootstraps. To tackle this problem, we devised *service tags* to distinguish registered service programs from others. Each Bootstrap daemon on a physical server maintains a list of tags, which map one-to-one to the set of registered services. The tag list is obtained from a trusted server in the system. When a service pack is transmitted to a Bootstrap daemon from a remote server, the daemon checks the tag field of the pack. If the tag matches one in the tag list, the corresponding service program is permitted to be restored and initiated on the server. Otherwise, the service pack will be discarded. To prevent malicious users on a network to extract the tag information from a valid pack, we encrypted its tag field. A digital signature is also included to ensure the packet integrity. The body of service pack is not encrypted, due to performance concerns and the observation that a service pack for parallel computation usually does not contain sensitive information. Currently, we apply symmetric keys to encrypt the tag field, and sign and verify the digest of a pack. Other sophisticated key management techniques can be incorporated in later versions.

During the execution of a parallel application, the DsaAgents of different servers will contact one another for array accesses. To protect the communication, among virtual servers, from being tampered by network users, we introduced a *selective signing* approach. That is, data, traveling cross two clusters, are appended with a digital signature and the receivers check the digest. However, we assume a cluster is a trusted environment. So intracluster communication is free from digital signature to reduce delay. When a service migration occurs within a cluster, the digital signature of a service pack is also avoided. Our security mechanisms protect system resources from malicious usage, and ensure safe and efficient execution of parallel applications on M-DSA VMs.

17.5 Experiment Results

Our experiment platform is a campus grid in the Wayne State University. The cluster in the Institute of Scientific Computing is composed of eight SUN Enterprise SMP Servers. Five of them are 4-way E3500 with 1024 MB of memory and the other three are 4-way E3000 with 512 MB of memory. Each processor module has one UltraSPARC II with 400 MHz (E3500) or 336 MHz (E3000), and 4 MB of cache. The cluster in the department of Electrical and Computer Engineering has four DELL SMP Servers. Each of them possesses two 774MHz Pentium III processor and 256 KB of cache. They use the Red Hat Linux 8.0 with 2.4.18-14smp kernel. The com-

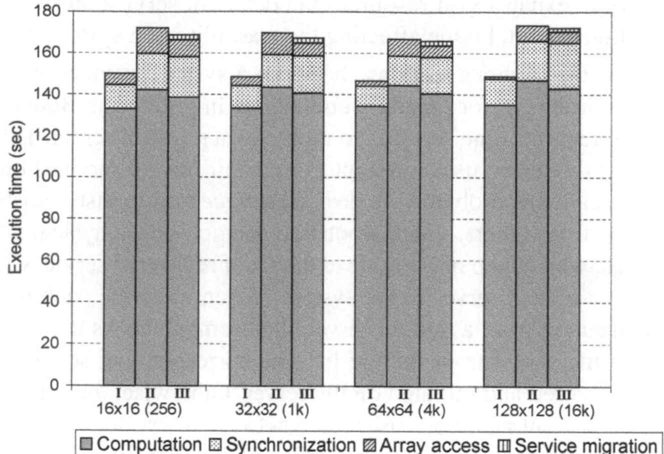

Figure 17.5: LU factorization of 2048 x 2048 double matrix. I: execution time in the Sun Enterprise cluster with no service migration. II: execution time in the campus grid with no service migration. III: execution time in the campus grid with an inter-cluster service migration.

puters within each cluster are connected through a 100 Mbps Fast Ethernet switch. The two clusters are located in different buildings and the inter-cluster connection has a bandwidth of 60 Mbps.

We ported the applications of LU factorization and FFT from the SPLASH-2 [331] benchmark suite to evaluate the performance of service migration on the M-DSA run-time system. All codes were written in Java, compiled in JDK 1.4.1.

17.5.1 Execution Time of LU and FFT on M-DSA

To evaluate the performance of the M-DSA run-time system in a grid environment and impact of service migration on the overall performance of a parallel application, we tested the execution time of LU and FFT running within a cluster, and on the campus grid (without and with service migration). A distributed VM was constructed out of eight computers and each of them was assigned eight computational agents.

Figure 17.5 depicts the execution time breakdown of LU, factorizing $2048x2048$ double matrix. We changed the block size from $16x16$ to $128x128$, and measured the costs of computation, synchronization, shared array access, and service migration in three different scenarios. In Scenario I, the LU application was executed within the SUN Enterprise cluster and there was no service migration. It is shown that the overall execution time decreases as block size increases, and it increases after the block size is more than $64x64$. It is because the data locality can be better exploited with a larger block, but this benefit is compromised by the false sharing when a threshold of block size is exceeded. We also notice the synchronization cost is significant,

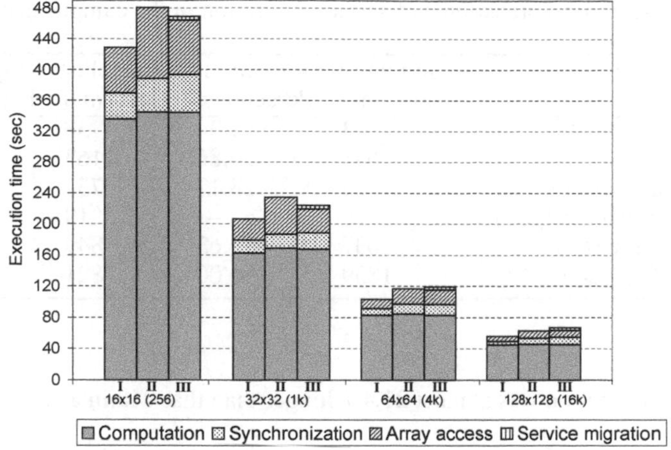

Figure 17.6: FFT for 2048 x 2048 double arrays with complex elements. I: execution time in the Sun Enterprise cluster with no service migration. II: execution time in the campus grid with no service migration. III: execution time in the campus grid with an intercluster service migration.

when the number of agents and computers is large. In Scenario II, we used four SUN E3500 servers and four DELL servers to execute the LU application. To focus on the performance evaluation of extending M-DSA from a cluster to campus grid, we do not consider the cost of security protections, which will be discussed in the following subsections. In Figure 17.5, the computation cost was a little greater than that in Scenario I, due to the relatively slow processing speed of the DELL servers. The intercluster communication leads to more remote access time of the shared array. As a result, the synchronization cost increased by 43.6% to 52.8%. The overall execution time is less than 17% more of that within a cluster. A service migration occurred in Scenario III, which moved the DSA service from a DELL server to a SUN Enterprise E3500 server. According to Figure 17.5, the computation cost was reduced slightly because of the increased processing speed. However, the synchronization cost was more because the average time waiting for agents to complete a computation phase was increased. The service migration overhead is less than 2.5 seconds, i.e., less than 1.5% of the overall execution time.

The execution time of FFT is shown in Figure 17.6. The execution environment was the same as that for the LU factorization. The inputs were 2048 x 2048 double arrays and each element was a complex number. FFT is a single-writer application with fine-grained access. The intensive shared array access and longer synchronization operations contributed to the slowdown of this application. Its performance in the campus grid degraded, primarily due to the large number of intercluster shared array accesses. The service migration improved the data locality, but introduced more synchronization overheads. The overall execution time of our M-DSA system

Table 17.1: Overhead breakdown of service migration in the campus grid M-DSA.

	Intracluster migration		Intercluster migration	
	Cost (ms)	Percentage	Cost (ms)	Percentage
Initialization (t_I)	43	2.34	58	2.06
Packing (t_P)	641	34.86	1163	41.23
Service transfer (t_{Tserv})	518	28.17	751	26.61
Code transfer (t_{Tcode})	19	—	19	—
Restoration (t_R)	637	34.63	849	30.10
Total ($C_{service}$)	1839	100.00	2821	100.00

with service migration was at most 21.4% longer than that within a cluster.

17.5.2 Service Migration Overhead Breakdown

A service migration moves the service data and states of a virtual server along with the residing computational agents from a physical server to another one. Our approach tries to reduce the overheads of a migration. The service migration cost, $C_{service}$, comes from the following:

- *Initialization cost* t_I: The coordinator checks migration requests; the source and destination Bootstrap daemons are notified of a service migration.
- *Packing cost* t_P: The source server wraps up the DSA service and computational agents into a service pack.
- *Service transfer cost* t_{Tserv}: The source server transfers the service pack to the destination.
- *Code transfer cost* t_{Tcode}: The destination prefetches the service and agent codes from a nearby server.
- *Restoration cost* t_R: Service pack is restored and a virtual server is built on the destination. Computational agents resume execution.

To investigate the impact of these components on a service migration and provide a foundation to further improve the system performance, we measured the overhead breakdown of intracluster and intercluster service migrations with security protections. Table 17.1 lists the experiment results in the LU factorization of 2048 x 2048 double matrix with 32 x 32 block size. The distributed VM resided on four SUN E3500 servers and four DELL servers. The intracluster service migration was within the SUN cluster. The intercluster migration was from the DELL cluster to the SUN cluster, with security protection.

Given the cost of each category, we can derive the overall service migration cost as,

$$C_{service} = t_I + t_P + max\{t_{Tserv}, t_{Tcode}\} + t_R,$$

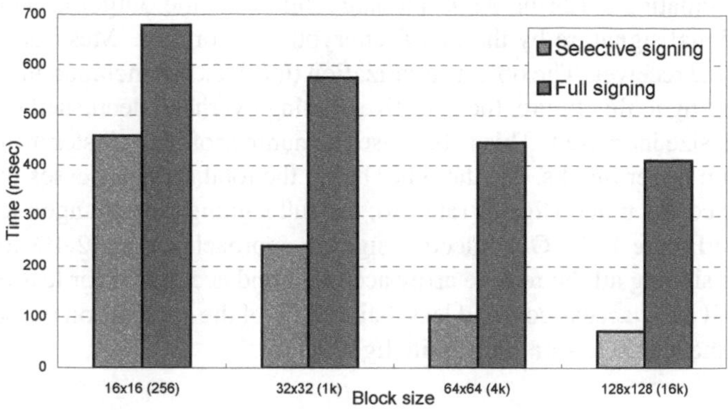

Figure 17.7: Security protection cost in the campus grid M-DSA.

where we take the maximum of t_{Tserv} and t_{Tcode} because service pack and codes are transferred in parallel with the prefetch technique. The t_{Tserv} and t_{Tcode} are determined by the sizes of service pack and codes, respectively. According to Table 17.1, the service packing and restoration are the most costly operations in a service migration. This is partially caused by inefficiency of the Java object serialization procedure. The checkpointing method generated a service pack with 4.2 MB of size, including the data and state of the DSA service and the computational agents. The security protections contributed 297 ms (25.5%) and 210 ms (24.7%) to the packing and restoration costs in the intercluster service migration. The different processing speed of source and destination computers in the intercluster migration accounts for the difference in packing and restoration time. The measured application performance information is piggybacked in the synchronization messages. So, the service initialization time is quite small. From Table 17.1, we can see both types of service migration have a total cost less than 1.8% of the overall execution time of the LU application. Compared with stopping a parallel computation when overload and/or failure occurs in a distributed environment, service migration is an affordable approach to system adaptation.

17.5.3 Security Protection for Intercluster Communication

Security protection is a necessary but time-consuming component in distributed systems. It is an important factor to the system scalability. To evaluate the cost of our security approaches, *selective signing*, for intercluster communication, we measured the execution time for security operations and compared it with a full signing approach, that is, to provide a digital signature for all intercomputer communication even within a cluster.

Figure 17.7 presents the security protection cost in the LU application for 2048 x 2048 double matrix, using four servers in each cluster. The security operation includes calculating a 128-bit MD5 message digest for the outgoing data and generating a digital signature by the 3DSE encryption algorithm. Message integrity is verified by the receiver. The object serialization time was not included in the security cost. According to this figure, the selective signing overhead decreased significantly as the block size increased. This is because the number of intercluster array accesses declines with larger blocks. On the other hand, the total array accesses in the entire system reduce at a much slighter rate. So, the full signing cost changed less rapidly, as shown in Figure 17.7. Our selective signing approach saved 32.4% to 82.1% of the time for signing all the remote array accesses, and accounted for less than 3.72% of the overall array access cost and less than 0.27% of the application execution time. Therefore, the selective signing is quite lightweight.

17.6 Related Work

Our service migration mechanism was inspired by the *capsule* migration in Stanford's Collective project [272]. The Collective capsule encapsulates the complete state of a machine, including the entire operating system as well as applications and running processes. VM monitors are utilized to capture system state. Capsules can be suspended from execution and serialized. By transferring a capsule across a network and binding it to a destination computer architecture, a user can resume his/her work after a commute between home and office. The major contribution of capsule migration to our work is that it presented the possibility of moving the underlying service components between computers for computation continuity. However, as designed at the operating system layer, this migration approach incurs considerable overhead.

A VM presents an abstract view of the underlying physical machine or system software to programs that run with it. It allows multiple applications and even different operating systems to run concurrently and multiplex computer resources. It also decouples the execution of applications from each other. The VM abstraction and isolation greatly simplify the development of applications and improve program portability. Since the classical VMs [221] were first designed as a solution to multiplexing shared mainframe resources in the 1970s, it has been an active research area; see [292] for a recent overview of this topic. With the explosive growth of network services, the Internet is evolving to enable the sharing and coordinated use of geographically distributed resources. To support network centric applications, research interests in VM have been recently renewed. Systems like Denali [323], DVM [290], Terra [130], and Collective [272] develop VM techniques for multiple network applications hosting on a single computer, for inter-VM communication by standard network protocols, or for state transfer. However, these systems are not easily reconfigurable, due to the enormous execution contexts. In contrast, application-level VMs can be tuned to become more efficient for specific applications, and it is much easier to migrate these lightweight services in a network. The Global Object Space in JESSICA2 [362], and the DSA [28] in Traveler [340] are such examples.

Figueiredo *et al.* [109] proposed an architecture to provide network-based services in computational grids, based on classical VMs. To efficiently manage grid resources, it allows a VM to be created on any computer that has sufficient capacity to support it. Virtual server migration is realized by moving an entire computing environment to a remote virtualized compute servers. However, the migration details and associated With the OS and middleware level support, SODA [159] constructs a distributed VM for an application service on demand. Each virtual service node provides stationary runtime support for Grid computation. Overheads were not discussed by the authors.

The VM technology and virtual server migration technique have also been applied to peer-to-peer (P2P) systems. Chord [301] was one of the first that used the notion of virtual servers to improve load balance. By allocating log$N$ virtual servers on each real homogeneous node, Chord ensures with high probability the number of objects per node is within a constant factor from the optimum. CFS [77] accounts for node heterogeneity by ensuring the number of allocated virtual servers at each node is proportional to the node capacity. System adaptation is realized by deleting and creating virtual servers according to the load of each physical server. In [255], Rao *et al.* presented three heuristic algorithms to achieve virtual server migration in structured P2P systems. Their simulation results indicated that virtual server migration was able to balance the system load to some extent even with a simple migration scheme. Their migration approach only tries to redistribute the identifier space among peers and does not involve transferring any run-time states. So, it is quite simplified compared with our service migration mechanism, which achieves computation continuity by moving the run-time supporting services.

Besides, there is a large collection of process/thread migration approaches [86, 87, 158, 163, 199] for the purpose of dynamic load distribution, fault resilience, and data access locality. However, it is difficult to extend them into the heterogeneous grids, because of the portability problem of the system software and programming languages that they relied on. JESSICA2 [362] is a distributed JVM supporting multithreaded Java applications on a set of heterogeneous computers. It realized thread migration by transferring the thread states between networked nodes. However, the just-in-time (JIT) compilation mode makes the state capture procedure quite complicated and machine dependent. Besides, the success of its thread migration is based on the existence of a Global Object Space, which provides a shared memory for the distributed nodes. HPCM [91] is a middleware supporting process migration in heterogenous environments. However, this is achieved by the scheduler moving pre-processed code to the new destination with a pre-run and machine-specific runtime system. As a result, when a node failure occurs, the entire system has to be stopped. So, process/thread migration alone cannot achieve service reconfiguration and cannot provide adequate adaptation to the changing environment of wide area computing.

Finally, we note that for adaptive grid computing, Berman *et al.* [101] proposed the Application Level Scheduling (AppLes) to effectively manage the distributed and heterogeneous grid resources; AppLes allocates system resources with a feedback-like control scheme. The scheduling approaches are selected according to the run-time behaviors of different applications. The AppLes methodology has been incorporated into the GrADS [100] project. In addition, the GrADS system uses application migration [165, 312] for adaptivity to the dynamic grid environments. Their migration schemes do not move data and states of a run-time support service. Instead, the application's execution environment on a destination node is constructed by a centralized *Rescheduler*. As a result, much global information has to be maintained by the system, which compromises its scalability.

17.7 Concluding Remarks

In this chapter, we have presented an M-DSA service migration mechanism to support reconfigurable distributed VMs in grid computing. Experiment results from parallel LU factorization and FFT in a campus grid environment show that service migration can achieve system adaptivity with marginal performance degradation.

Service migration complements agent migration in response to the change of the network computing environment. The two forms of migration define a hybrid mobility model for adaptive grid computing. Chapter 18 addresses the decision problem of hybrid mobility for load balancing in reconfigurable DVMs.

Chapter 18

Hybrid Migration Decision in Reconfigurable Distributed Virtual Machines

Service migration is a vital technique to construct reconfigurable virtual machines (VMs). By incorporating mobile agent technology, VM systems can improve their resource utilization significantly via the hybrid mobility of services and computational agents. This chapter addresses the decision problem of hybrid mobility for load balancing in reconfigurable distributed VMs. Chapter 2 defines the load balancing problem in a W5 model. This chapter presents a rigorous treatment of the problem, focusing on migration candidate determination (which), migration timing (when), and destination server selection (where).

18.1 Introduction

VMs provide a powerful layer of abstraction in distributed computing environments. It is certainly possible to deploy them as static computing units. However, this abstraction layer would not be fully exploited unless VMs are instantiated and managed dynamically. Service migration is one approach to their dynamic reconfiguration. Since VMs are usually created to provide services for programs atop them, we use the terms of VM migration and service migration interchangeably. Many recently proposed VM systems incorporate service migration techniques for adaptation in grid and peer-to-peer computing [109, 124, 125, 255, 272].

In addition to service migration, process/thread migration is another way to construct reconfigurable distributed systems. It concerns the transfer of user-level processes or threads across different address spaces. JESSICA2 [362] and MigThread [158] are examples that use thread migration to utilize idle cycles on other computers and to balance workload. In comparison with process/thread migration, agent migration is a more general approach to load balancing, fault tolerance, and data locality. An agent represents an autonomous computation that can migrate in a network to perform tasks on behalf of its creator. The autonomy of the agent makes high-performance grid computing different from cluster-wide process/thread migrations. Multihop mobility allows the agent to move computations across a wide area network and negotiate with others. Agent technology has been applied to grid computing for load scheduling in [46]. Load balancing can be achieved by agent migration in TRAVELER [340] and MALD [47], which are able to react efficiently in response to the change of background load.

In a reconfigurable VM system with hybrid mobility of services and agents, migration decision becomes a crucial performance issue. It deals with three aspects of the problem. First is migration candidate determination, concerning which server should transfer a service and/or which agent should exercise migration. Another one is migration timing, determining when a migration should be performed. Finally, a destination server should be selected so that the performance gain from a migration won't be outweighed by the migration overhead. Literature is littered with heuristic migration algorithms for either service or agent migrations. The objective of this chapter is to provide a rigid stochastic optimization model for this hybrid mobility decision problem. Decision policies gained as a result of this study should be complementary to the existing migration mechanisms.

The hybrid migration decision problem is nontrivial. A VM provides a dynamic environment for the execution of agents. Each agent will carry out its tasks during its life span (corresponding to a *time domain*) and its execution may be performed on different virtual servers due to migrations (referring to this as a *space domain*). Therefore, we have to consider both domains when deriving an optimal migration decision. Their interaction makes the decision process more complicated. Second, the capacities of physical servers in a distributed VM may be different, which is typical in Internet-based computing. This capacity heterogeneity results in different distributions of agent workload changes. Their distributions can even alter from one server to another during service migration. Besides, agent migration is hard to schedule in such a dynamic environment with server reconfiguration. The impact of agent migration on service migration also has to be carefully handled.

In [121], we investigated the hybrid mobility problem with a focus on service/agent migration decisions for load balancing in heterogeneous distributed systems. We related the service/agent migration decision in reconfigurable VM to a dynamic remapping scheduling problem in parallel computing [113, 340, 345]. We formulated the migration issues as two stochastic optimization models, and derived optimal migration policies in both time and space domains, by assuming the workload change of an agent is a random process with arbitrary probabilistic distributions. The rest of this chapter draws substantially on the hybrid mobility paper [121].

18.2 Reconfigurable Virtual Machine Model

An agent-aware VM is composed of a set of virtual servers spreading among distributed physical servers. Each virtual server can accommodate multiple agents to share resource and perform computation and communication. Each physical server

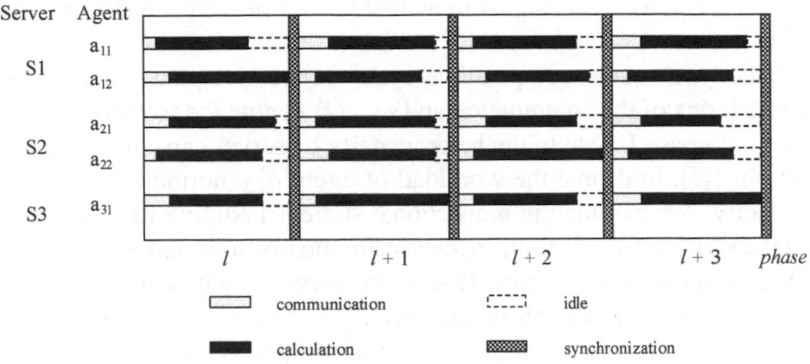

Figure 18.1: Bulk synchronous computations of virtual machine.

can be run in two modes: *dedicated* or *multiprogrammed*. On a dedicated server, only virtual servers of the same application can reside, while a multiprogrammed server allows virtual servers of different applications to share its resource. On the other hand, a VM can be run on a single physical server or a cluster of servers. Correspondingly, on *dedicated* and *multiprogrammed single-server* systems, only service migration can be applied for their reconfigurability. In a *dedicated multiserver* VM, both service and agent migrations are useful to balance workload among multiple physical servers. The most general type is *multiprogrammed mulitserver* VMs. But the asynchrony among different applications makes multiprogrammed servers hard to analyze. In this study, we confine our discussion to the dedicated multiserver reconfigurable VMs.

We consider the bulk synchronous computation as our execution model because of its popularity in scientific and engineering applications. Its computations proceed in phases that are separated by global synchronization. During each phase, they perform calculations independently and then communicate new results with their data-dependent peers. Due to the need of synchronization between steps, the duration or execution time of a step is determined by the most heavily loaded server, on which the agent having the longest finish time resides. Phasewise computation may exhibit varying requirements as the computation proceeds and even a static bulk synchronous computation may have nondeterministic computational characteristics. Figure 18.1 illustrates an example of bulk synchronous computations over a VM with multiple agents on each server.

Consider a distributed application in a heterogeneous environment. A VM $V = \langle s_1, s_2, \ldots, s_N \rangle$, formed by N servers out of available physical servers, is allocated to the application. Let a sequence $C = \langle c_1, c_2, \ldots, c_N \rangle$ denote their computational capacities and each c_i, $i = 1, \ldots, N$, is a constant. The computation on the VM comprises multiple agents, each of which executes a number of tasks that collectively determine its workload. We use a dynamic agent distribution sequence $M = \langle m_1, m_2, \ldots, m_N \rangle$ to represent the number of agents assigned to different servers. Due to a many-to-one mapping between agents and servers, we use a sequence $A_i = \langle a_{i,1}, a_{i,2}, \ldots, a_{i,m_i} \rangle$ to denote the set of agents residing on server s_i in the VM.

In the bulk synchronous computation, agents proceed in phases. Let $l = 0, 1, \ldots$, be the phase index of the computation and $w_{i,k}(l)$ denote the workload of agent $a_{i,k}$ on server s_i at phase l. Due to the heterogeneity of server capacities, we use scaled workload, $\tilde{w}_{i,k}(l)$, to denote the workload of agent $a_{i,k}$ normalized by its quota of server capacity. We assume the proportional share scheduling [317] among mobile agents of the same server. Multiple agents share the resource and each of them has an equal quota of the server capacity. That is, for server s_i with capacity c_i, a residing agent $a_{i,k}$ is assigned a portion of capacity as $c_{i,k} = c_i/m_i$. Therefore, we have $\tilde{w}_{i,k}(l) = w_{i,k}(l)/c_{i,k} = m_i w_{i,k}(l)/c_i$. For simplicity in notation, we will use $w_{i,k}(l)$ to denote the scaled workload henceforth.

As the computation proceeds, the scaled workload of each agent may evolve dynamically. Let $\delta_{i,k}(l-1)$ denote the net change of scaled workload $w_{i,k}(l)$ from phase $l-1$ to l due to the workload generation and/or consumption by agent $a_{i,k}$ during the period. The overall scaled workload of server s_i equals to the summation of its agents' scaled workload, i.e., $w_i(l) = \sum_{k=1}^{m_i} w_{i,k}(l)$. Similarly, its scaled workload change is defined as $\delta_i(l-1) = \sum_{k=1}^{m_i} \delta_{i,k}(l-1)$. Let sequences $w(l) = \langle w_1(l), w_2(l), \ldots, w_N(l) \rangle$ and $\delta(l-1) = \langle \delta_1(l-1), \delta_2(l-1), \ldots, \delta_N(l-1) \rangle$ denote the scaled workload distribution at phase l and the scaled workload change distribution from phase $l-1$ to l of the VM, respectively. Then the bulk synchronous computation can be characterized by an additive dynamic system:

$$w(l) = w(l-1) + \delta(l-1), \tag{18.1}$$

where the scaled workload change is independent of the server workload. The scaled workload is described by a Markov chain, which is assumed in many other studies on bulk synchronous computations [113, 345]. As a result of scaled workload change between phases, the computation of each server becomes nondeterministic. Although this study focuses on the additive scaled workload evolution model, our analysis could be extended to other dynamic systems as discussed in [339].

During a service migration, the residing virtual servers on a physical server can be transferred as a whole. We can treat the virtual servers on each physical server as a set and the two terms; physical and virtual servers will be used interchangeably. When a server decides to perform a service migration to another available server during the computation, the correspondent entry in the capacity sequence *C* will be replaced by the destination server's capacity. Similarly, agent migration leads to a modification of the agent distribution sequence *M*, as agents land on or leave from servers of the VM. So the migration decision problem can be tackled based on renewal processes. We use sequences $C^0 = \langle c_1^0, c_2^0, \ldots, c_N^0 \rangle$ and $M^0 = \langle m_1^0, m_2^0, \ldots, m_N^0 \rangle$ to represent the capacities and agent distribution on the original servers, i.e., the servers allocated to the application at its initiation. The scaled workload of an agent a_i at phase l and its scaled workload change from phase $l-1$ to l on its original server are denoted by $w_{i,k}^0(l)$ and $\delta_{i,k}^0(l-1)$, respectively. We assume initially agents have equal scaled workload at phase $l = 0$.

To make the migration decision problem tractable, we make some simplifying assumptions.

Assumption 1: The initial scaled workload changes of different agents on a server, $\delta^0_{i,k}(\cdot)$, are i.i.d. random variables and they are distribution free. Agent autonomy makes the construction of independent agents easier by utilizing data locality and migration.

Assumption 2: Service/agent migration keeps the probabilistic distribution of the scaled workload change of an agent unchanged. But it may have a different mean after migration. This is reasonable because the scaled workload change equals to the amount of agent workload change relative to its capacity quota. This quota is a constant during the agent execution on a server with a fair-share scheduling scheme.

18.3 Service Migration Decision

A VM is composed of a set of servers executing tasks of their residing agents and synchronizing with each other by performing barrier operations between phases. By the dynamic system in (18.1), it is expected that the servers' scaled workload will change with time and cause overloaded states. Since the duration of a phase is determined by the heavily loaded servers, the overall system performance may deteriorate in phase. Service migrations aim to eliminate the performance bottleneck of a VM by increasing capacities of the overloaded components. A service migration involves the tasks of transmitting service states and data along with the residing agents to a destination server with a higher capacity, continuing services, and resuming agent execution there. Although we do not focus on the details of how to perform service migrations, it is clear that a migration operation would incur significant run-time overhead and the adaptive computation cannot afford frequent migrations.

The objective of the service migration decision is to obtain the optimal timing in phase to minimize the migration frequency, and the low bound of the destination server capacity to guide the server selection. In the following, we suppose there are only service migrations during the execution of a VM. The agent migration decision problem and its interplay with service migration will be tackled in Section 18.4.1.

18.3.1 Migration Candidate Determination and Service Migration Timing

Although servers of a VM need to perform global barrier synchronization between phases, their executions within a phase can be assumed to be independent. This is particularly the case in agent-based systems, where autonomous agents perform their own tasks with local services and data sets. A service migration decision is made by a server based on its overall workload relative to its capacity. In the following, we use a set of simplified notations, in which s denotes any server in a VM, c and m

represent its capacity and number of agents, $\langle a_1, a_2, \ldots, a_m \rangle$ is the sequence of its agents, and $w_i(l)$ and $\delta_i(l)$ denote the scaled workload and workload change of a residing agent a_i at phase l.

For server s, its overall workload at a phase is the summation of all its agents' workloads, if we do not consider their overlapping. We define *expected relative workload* of server s at phase l as

$$r(l) \;=\; E[\,\frac{\sum_{i=1}^{m} w_i(l)}{m}\,]. \tag{18.2}$$

Expected relative workload represents a server's overall workload with regard to its capacity. We use the scaled workload of each agent in order to ensure workload of different servers in the VM are comparable with each other.

Most of previous work designs migration strategy to minimize the degree of overload. Due to the nonnegligible run-time overhead in migration, a certain degree of overload must be tolerated. We reformulate the decision problem as "Given a bound of overload, find the minimal migration frequency." Therefore, the optimal migration timing is obtained by finding the maximal phase l^* for a given overload bound R, $0 < R < 1$, i.e.,

$$r(l) \;\le\; R. \tag{18.3}$$

R reflects the degree to which a server is considered to be overloaded. To find the maximal phase in (18.3), we need to calculate the expected value of scaled workload $w_i(l)$ for each residing agent. But the heterogeneity in terms of different server capacities and alteration of an agent's scaled workload due to service migration, add complexity to its calculation. Next, we try to derive the expression of scaled workload $w_i(l)$ with reference to the value $w_i^0(l)$ on the original server.

Consider server s makes a service migration at certain phase k. Let s' denote the server after migration, in the *space domain*. We use (c', m') to represent its capacity and number of agents, respectively. For a residing agent, say a_i, its scaled workload and workload change after migration are denoted by $(w_i'(\cdot), \delta_i'(\cdot))$. The *workload conservation property* states the workload of an agent remains unchanged at the moment of migration. That is, $w_i'(k)c'/m' = w_i(k)c/m$. In the *time domain*, with the additive scaled workload evolution model in (18.1), it is clear $w_i'(l) = w_i'(k) + \sum_{t=k}^{l-1} \delta_i'(t)$, for phases $l > k$. If there is no service migration at phase k, the scaled workload becomes $w_i(l) = w_i(k) + \sum_{t=k}^{l-1} \delta_i(t)$, for $l > k$. By Assumption 2, we have $\delta_i'(t)c'/m' = \delta_i(t)c/m$. Thus, we derive $w_i'(l) = w_i(l)cm'/(c'm)$. Together with (c^0, m^0) for the original server and $m = m^0$ due to no agent migration, the scaled workload of agent a_i on server s can be calculated as

$$w_i(l) \;=\; \frac{w_i^0(l)}{\tilde{c}}, \tag{18.4}$$

where $\tilde{c} = c/c^0$ is called the *relative capacity*. We can see that the scaled workload of an agent equals its initial scaled workload relative to the ratio of capacities between the current server and original one. This is caused by service migrations in the space domain.

Let μ_i be the means of scaled workload changes $\delta_i^0(\cdot)$, for $i = 1, \ldots, m$. According to the additive workload evolution model, the scaled workload of agent a_i on its original server at phase l is $w_i^0(l) = w_0 + \sum_{t=0}^{l-1} \delta_i^0(t)$, where $w_0 = w_i^0(0)$ is scaled workload at the phase 0, i.e., the initial scaled workload of an agent on its original server. Since the scaled workload changes $\delta_i^0(\cdot)$ in different phases are i.i.d. random variables, we have the expected value of scaled workload $E[w_i(l)] = E[w_i^0(l)]/\tilde{c} = (w_0 + l\mu_i)/\tilde{c}$. Thus, the expected relative workload of server s becomes

$$r(l) = \frac{\sum_{i=1}^{m}(w_0 + l\mu_i)}{m\tilde{c}} = \frac{w_0 + l\bar{\mu}}{\tilde{c}}, \tag{18.5}$$

where $\bar{\mu} = (\sum_{i=1}^{m} \mu_i)/m$ is the average change rate of the scaled workload. If $\bar{\mu} \leq 0$, then the overall scaled workload of a server tends to remain unchanged or decrease, as the computational phase proceeds. It means there is no need to perform service migration any more. So, we only consider the cases with $\bar{\mu} > 0$ in our discussion. With $r(l) \leq R$, we have the following theorem to determine the optimal timing for service migrations.

Theorem 18.1 *For any server s with capacity c in a virtual machine, suppose the scaled workload changes of different agents on the original server, $\delta_i^0(\cdot)$, $i = 1, \ldots, m$, have means μ_i and their average $\bar{\mu} > 0$. For a given overload bound R, $0 < R < 1$, of the expected relative workload, the optimal service migration timing l^* in phase, under Assumption 1 and Assumption 2, is*

$$l^* = \frac{\tilde{c}R - w_0}{\bar{\mu}}. \tag{18.6}$$

Intuitively, Theorem 18.1 states that the optimal service migration timing of a server equals to its available server capacity, apart from the initial workload, divided by the average rate of workload increase. Equation (18.6) reflects the influence from both the time domain (by the average mean of scaled workload changes) and the space domain (by the relative server capacity $\tilde{c}$). It shows the timing is proportional to the relative server capacity. This is because both w_0 and each μ_i are defined relative to capacity of the original server. To reduce the service migration frequency, we can select the destination server with the highest capacity. But this greedy strategy may allocate more resource to less heavily loaded servers, which leads to resource imbalance in the VM. We will discuss this trade-off in Section 18.3.2.

18.3.2 Destination Server Selection

Servers in a VM independently decide when to perform service migrations based on Theorem 18.1. At phases of the optimal service migration timing, their services and agents will be transmitted to the destination servers. As we have discussed in the previous subsection, the selection of a destination server is a trade-off between the local and global performance optimizations. From an individual server's perspective, it is preferable to obtain an available server with the highest capacity as its destination so that its service migration frequency becomes minimal. But from the perspective of the entire VM, it is better to allocate more powerful servers to more heavily-loaded servers. In this section, we introduce a migration gain function and try to find the minimal capacity, that a prospective destination server must have, for a given target gain value.

For server s in a VM, the execution time of its agent a_i at phase l, $t_i(l)$, is proportional to its workload relative to the capacity quota, i.e., its scaled workload, as $t_i(l) \propto w_i(l)$. The duration of a computational phase on a server is determined by its agent with the longest finish time, *i.e.* $t(l) = \max_{i=1..m} t_i(l)$. Consider server s makes a service migration to an available server, denoted by s', with capacity c' at the optimal migration timing l^* according to Theorem 18.1. We require $c' > c$; otherwise this migration is meaningless w.r.t. the system performance improvement. Thus the execution time of each agent on server s' will be reduced. We define the *expected migration gain* for this service migration as

$$g(c') = E[\sum_{l=l^*}^{k-1}(t(l) - t'(l))] - E[O(w(l^*))], \tag{18.7}$$

where k is the prospective optimal migration timing for server s' with its supposed capacity c' according to Theorem 18.1. The execution time on server s' is $t'(l) \propto \max_{i=1..m} w'_i(l)$, and $O(\cdot)$ returns the migration overhead for a given scaled workload. The expected migration gain function describes the performance improvement, in terms of agents' execution time on the two servers until the next service migration, compared with the migration overhead. We do not use execution time of the entire VM as a metric to express performance gain, due to the unpredictable behaviors of other servers in the VM and the independency of server execution. We expect the performance benefits in terms of a value G from a service migration. The target gain, G, is measured relative to the migration overhead. So, the selection of a destination server in a service migration is guided by finding the low bound of server capacity for a given nonnegative target gain value G of the migration gain function, i.e.,

$$g(c') \geq G. \tag{18.8}$$

A special case is to have $G = 0$, which means no benefit is yielded from a migration except to counteract the migration overhead.

By increasing G, capacities of the destination servers will become higher accordingly. But to avoid resource allocation imbalance in the VM, we need to choose the value of G appropriately. Since the state information of services on a server is usually maintained by a relatively limited number of variables, the main portion of the service migration overhead is caused by transferring codes and data of services. The size of data set is likely to change little since its initial assignment to a server and the code of services are immutable. So, the migration overhead of a server can be represented by a constant, denoted by H. However, our optimization approach is still applicable with the general overhead function, $O(\cdot)$. Since the execution time of an agent, $t_i(l)$, is solely determined by its scaled workload $w_i(l)$, we will use these two terms interchangeably. Next, we transform the migration gain function to find the minimal capacity of the destination server for a given target gain value.

With the expressions of $t(l)$, $t'(l)$ and $w_i(l)$, the expected migration gain (18.7) becomes

$$\begin{aligned} g(c') &= E[\sum_{l=l^*}^{k-1}(\max_{i=1..m}\frac{w_i^0(l)}{\tilde{c}} - \max_{i=1..m}\frac{w_i^0(l)}{\tilde{c}'})\,] - H \\ &= (\frac{1}{\tilde{c}} - \frac{1}{\tilde{c}'})\sum_{l=l^*}^{k-1} E[\max_{i=1..m} w_i^0(l)\,] - H, \end{aligned} \quad (18.9)$$

where $\tilde{c}' = c'/c^0$. Let μ_i and σ_i^2 be the mean and variance of the scaled workload change $\delta_i^0(\cdot)$. Since $w_i^0(l) = w_0 + \sum_{t=0}^{l-1}\delta_i^0(t)$ and the phasewise scaled workload changes of each agent $\delta_i^0(\cdot)$ are i.i.d. random variables w.r.t. phases, the central limit theorem of statistics guarantees that as l gets large, the distribution of $w_i^0(l)$ tends to become normally distributed with mean $w_0 + l\mu_i$ and variance $l\sigma_i^2$. Because $\delta_i^0(\cdot)$, for $i = 1, \ldots, m$, are independent random variables, scaled workloads $w_i^0(\cdot)$ are also independently distributed. So we have

$$Pr\{\max_{i=1,\ldots,m} w_i^0(l) \le x\} = \prod_{i=1}^{m} Pr\{w_i^0(l) \le x\}.$$

Since each $w_i^0(\cdot) \ge 0$, it is clear

$$\begin{aligned} E[\max_{i=1,\ldots,m} w_i^0(l)\,] &= \int_0^\infty Pr\{\max_{i=1,\ldots,m} w_i^0(l) > x\}\,dx \\ &= \int_0^\infty [1 - \prod_{i=1}^{m}\int_{-\infty}^{x} f_i(u)\,du]\,dx, \end{aligned} \quad (18.10)$$

where $f_i(u)$ is the probability density function of normal distribution with parameters $w_0 + l\mu_i$ and $l\sigma_i^2$ for scaled workload $w_i^0(l)$. Therefore, we can derive the low bound of the destination server capacity, as shown in the following theorem.

Theorem 18.2 *For any server s with capacity c in a virtual machine, suppose the scaled workload changes of different agents on the original server, $\delta_i^0(\cdot)$, $i = 1, \ldots, m$, have means μ_i and variances σ_i^2. For a given nonnegative target gain G of the expected migration gain function, the minimal capacity that a prospective destination server must have during a service migration at the optimal phase l^*, under Assumption 1 and Assumption 2, is*

$$c^* = \max\{\, c^+,\; c + \frac{\bar{\mu}}{R}c^0 \,\}, \quad (18.11)$$

where R is the overload bound of the expected relative workload in (18.3) and c^+ is the solution to equation:

$$(\frac{1}{\tilde{c}} - \frac{1}{\tilde{c}'})\sum_{l=l^*}^{k-1}[\int_0^\infty (1 - \prod_{i=1}^{m}\int_{-\infty}^{x} f_i(u)\ du)\ dx\ ] - H = G, \tag{18.12}$$

in which the optimal service migration phases $l^* = (\tilde{c}R - w_0)/\bar{\mu}$ and $k = (\tilde{c}'R - w_0)/\bar{\mu}$ by Theorem 18.1.

Proof By applying $E[\max_{i=1,\ldots,m} w_i^0(l)]$ in (18.10) to the migration gain function, we get the left-hand side part of (18.12). It is clear the resulting gain function monotonously increases as its parameter c' becomes greater. So for a given target gain value G, the minimal capacity satisfying $g(c') \geq G$ is the solution to equation $g(c') = G$, denoted by c^+.

At the same time, the prospective optimal migration phase k for the new server s' should be greater than phase l^* for server s. Otherwise, a new service migration will occur as soon as the previous one completes. We call this phenomena *idle migrations*. So, we have $k \geq l^* + 1$, i.e., $(R\tilde{c}' - w_0)/\bar{\mu} \geq (R\tilde{c} - w_0)/\bar{\mu} + 1$ and obtain $c' \geq c + \bar{\mu}c^0/R$. Therefore, the minimal capacity of a destination server takes the form in the theorem. □

The expected gain function is hard to calculate by (18.10). A special case is when the scaled workload changes $\delta_i^0(\cdot)$, for $i = 1, \ldots, m$ are i.i.d. random variables with the same mean and variance. This is reasonable for the single program multiple data (SPMD) applications, in which agents execute the same program on their own data sets and they have similar workload change behaviors. For these applications, the minimal destination server capacity can be determined by the following corollary.

Corollary 18.1 *For any server s with capacity c in a virtual machine, suppose the scaled workload changes of different agents on the original server, $\delta_i^0(\cdot)$, $i = 1, \ldots, m$, have the same mean μ and variance σ^2. For a given non-negative target gain G of the expected migration gain function, the minimal capacity that a prospective destination server must have during a service migration at the optimal phase l^*, under Assumption 1 and Assumption 2, is determined by (18.11), where c^+ is the solution to equation:*

$$(\frac{1}{\tilde{c}} - \frac{1}{\tilde{c}'})\sum_{l=l^*}^{k-1}[\ w_0 + l\mu + \alpha(m)\sigma\sqrt{l}\ ] - H = G, \tag{18.13}$$

in which $\alpha(m) = (2 \ln m)^{1/2} - (\ln \ln m + \ln 4\pi)/[2(2 \ln m)^{1/2}] + \gamma/(2 \ln m)^{1/2}$ and γ is Euler's constant ($0.5772\cdots$).

Proof Since the scaled workload change $\delta_i^0(\cdot)$, $i = 1, \ldots, m$, have the same mean μ and variance σ^2, the scaled workload $w_i^0(l)$, $i = 1, \ldots, m$ are i.i.d. normally distributed random variables with the same mean $w_0 + l\mu$ and variance $l\sigma^2$. According to the extreme value theory [8], we have

$$E[\max_{i=1,\ldots,m} w_i^0(l)] \approx w_0 + l\mu + \alpha(m)\sigma\sqrt{l}. \tag{18.14}$$

Therefore, the expected migration gain equals to the left-hand side of (18.13). Thus the minimal destination server capacity is determined according to Theorem 18.2. □

We can see the minimal destination server capacity ensures the performance improvement due to a service migration equal to a given target gain in addition to the migration overhead. Since servers make their migration decisions independently, it is possible that two or more servers may decide to migrate to the same destination server according to Theorem 18.2. This phenomenon is called *destination conflict* in service migration. To solve the conflict and reduce migration frequency, we allow the server with the largest prospective next optimal migration timing, determined by Theorem 18.1, to take the destination server. Other servers will try to choose those from the remaining available servers with the least sufficient capacities. After selecting a destination server and completing service migration to it, we update the server sequence of the VM and the capacity distribution sequence *C*. Since then, the renewal process begins a new round of VM computation and each server determines its next migration timing.

18.4 Hybrid Migration Decision

The properties of autonomy and multihop mobility make the mobile agent as a suitable technique to achieve load balancing in distributed computing. Like service migration, there is also a decision problem for agent migration. But the dynamic environment with server reconfiguration and interaction among different agents greatly add to the difficulty of its decision. In this section, we model the agent migration decision problem by dynamic programming and extend the service migration decision to incorporate hybrid mobility.

18.4.1 Agent Migration Decision

To adapt to the change of server capacities, agents may need to migrate. The decision question is when to migrate and which server the agent should travel to. A locally optimal strategy can be derived if we assume individual agents make their migration decisions independently. But unlike service migration, different agents may migrate to the same server. The interaction among them may cause this migration strategy useless. So, certain global information about the agent or workload distribution is needed. If we consider the global optimization of agent migration, the decision problem is equivalent to the task scheduling in distributed computing. In general, it is a NP-problem to find the globally optimal task distribution. However, in the bulk synchronous computation model for SPMD, agents proceed their computation in phases and their carry out the same task at each phase. It is possible to find a feasible optimal solution to agent migration with some global information. We relate the coordinated decision problem of agent migration to the remapping problem in parallel computing and use dynamic programming to derive a globally optimal solution.

We define the state of a VM as the workload distribution of different servers at a certain phase. Assume the choice of agent migration while in state v incurs a cost $C(v, u)$, where u is a binary decision: migrate or not. $C(v, u)$ may be random. The decision process then passes into another state. The probability $p_{vq}(u)$ of passing into state q from v is dependent on the action chosen in state v. The expected total cost of a decision policy is the expected sum of the costs incurred at each decision step. An optimal decision policy minimizes the expected total cost.

Let $\mathcal{J}(v)$ be the expected total cost of the VM, which starts in state v and which is governed by the control decisions. Then

$$\mathcal{J}(v) = \min_u \{C(v, u) + \sum_{q \in I} p_{vq} \mathcal{J}(q)\}, \tag{18.15}$$

where I is the set of states that the VM may enter from state v due to agent migration. For any decision process state v, let $next(v)$ denote the set of states reachable from v in one phase and $travel(v)$ be the entry state of an agent migration. Each decision state $r = \langle r_1, \ldots, r_N, l+1 \rangle \in next(\langle v_1, \ldots, v_N, l \rangle)$ has a transition probability

$$Pr\{r|v\} = \prod_{i=1}^{N} Pr\{\ r_i|v_i\ \},$$

where $Pr\{r_i|v_i\}$ is the probability of chain i passing from state v into state r_i in one phase. The execution time of the VM at state v is determined by the most heavily loaded server, i.e., $t(v) = \max_i\{v_i\}$. Each v_i equals to the scaled workload of the agent with the longest finish time. That is $v_i = \max_k\{w_{i,k}(l)\}$. Equation (18.15) can be rewritten as

$$\mathcal{J}(v) = \min_u \begin{cases} t(v) + \sum_{v' \in next(v)} Pr(v'|v)\mathcal{J}(v') \\ t(v) + C(v, u) + \sum_{v' \in next(travel(v))} Pr(v'|v)\mathcal{J}(v'), \end{cases} \tag{18.16}$$

where the bottom equation on the right-hand side is the cost function associated with agent migration, and the top equation is associated with no migration.

According to the theory of Markov decision processes, the optimal decision to make from state v is the decision which minimizes the right-hand side of (18.16). If the number of phases the agent takes is some random variable with finite mean, then the system of equations given by (18.16) can be solved. The iteration algorithm proposed in [345] can be applied to solve the equation.

18.4.2 Interplay between Service and Agent Migration

With the introduction of agent migration, the set of residing agents on each server of a VM will not be immutable any more. An agent may land on or leave from a server at any phase. So the service migration decision problem is quite different from the discussion in Section 18.3.

An agent may decide to migrate to a server that will perform a service migration to another server at the same phase. So the information, e.g., server capacity, for the agent to make a decision may be incorrect at the moment of its migration. To avoid this problem, we require service migrations to be decided and performed first, and after their completion, agents decide whether to migrate or not with the new system information. Now the agent distribution sequence $M=\langle m_1, m_2, \ldots, m_N\rangle$ of a VM is no longer a constant. Instead, we treat it as a sequence of random variables whose distributions are determined by the dynamic agent migration strategy discussed in Section 18.4.1. At a certain computational phase, the residing agents on server s are denoted by $\langle a_1, a_2, \ldots, a_m\rangle$. They may come from different servers via agent migration. These agents are assigned to their original servers, represented by $\langle s_{(1)}, s_{(2)}, \ldots, s_{(m)}\rangle$, at phase $l = 0$. Entries in this sequence may be identical, which reflects some agents come from the same server. The optimal service migration timing for a server in Theorem 18.1 will be extended by considering the dynamic composition of its agent set. As we can see, it is quite complicated to make optimal decisions with both service and agent migrations for general applications. Here we consider a simplified case where agents on their original servers have the same capacity quota. It can be realized by the task scheduler at the system initiation. The optimal service migration timing with agent migration is described in the following theorem.

Theorem 18.3 *For any server s with capacity c in a virtual machine, its m agents may migrate from different servers. Suppose these residing agents are, at the initial phase, equal to initial capacity quota q_0. Their scaled workload changes, $\delta_i^0(\cdot)$, $i = 1, \ldots, m$, have the same mean μ. For a given overload bound R, $0 < R < 1$, of the expected relative workload, the optimal service migration timing l^* in phase, under Assumption 1 and Assumption 2, is*

$$l^* = \frac{1}{\mu}\left(\frac{cR}{q_0 E[m]} - w_0\right). \tag{18.17}$$

Proof For a residing agent a_i on server s, suppose it originally resides on server $s_{(i)}$, whose capacity and initial number of agents are $c_{(i)}^0$ and $m_{(i)}^0$, respectively. Its initial capacity quota $q_0 = c_{(i)}^0 / m_{(i)}^0$. According to the previous discussion, its scaled workload at phase l is $w_i(l) = w_i^0(l) c_{(i)}^0 m / (c m_{(i)}^0) = w_i^0(l) q_0 m / c$. Since $w_i^0(l)$ tends to have a normal distribution with mean $w_0 + l\mu$, we have $E[w_i^0(l) q_0] = (w_0 + l\mu) q_0$. Similarly, the initial scaled workloads of other agents on server s are also normally distributed. The m random variables $w_1^0(l)q_0, \ldots, w_m^0(l)q_0$ are i.i.d. with the same mean $(w_0+l\mu)q_0$. Since m is also a random variable, we have

$$\begin{aligned} r(l) &= \frac{1}{c} \cdot E[\sum_{i=1}^{m} w_i^0(l)q_0] = \frac{1}{c} \cdot E[E[\sum_{i=1}^{m} w_i^0(l)q_0 \mid m]] \\ &= \frac{1}{c} \cdot E[mE[w_i^0(l)q_0]] = \frac{1}{c} \cdot E[m]E[w_i^0(l)q_0] = \frac{1}{c}q_0(w_0 + l\mu)E[m]. \end{aligned}$$

By applying the above equation to $r(l) \leq R$, the optimal service migration timing l^* of server s can be derived as in (18.17). ☐

According to Theorem 18.3, we can see the optimal service migration timing is reversely proportional to the expected value of agent number m. This can be explained that as the number of agents on a server increases, its overall workload tends to become greater and this leads to a reduced migration interval. Compared with Theorem 18.1, (18.17) is additionally related to the initial capacity quota of each agent due to agent and service migration. The destination server selection policy in Section 18.3.2 can also be extended when incorporating agent migration.

18.5 Simulation Results

To verify and analyze results of the migration decision according to our optimal model, we performed experiments to simulate the random processes of agent execution in a reconfigurable VM with 30 agents on each server. Each data point of the simulation results was an average of 200 runs. The 95% confidence interval results are also presented to demonstrate the robustness of the estimates. The distribution functions of agent workload evolution characterize the workload dynamics in three of various applications.

Case 1. In the first experiment, we were to find and verify the optimal computational phase before a service migration according to Theorem 18.1 with the objective of minimizing the migration frequency. The agents on a server were divided into three groups with ten agents in each one. They might be run in different phases and their initial scaled workload changes obeyed the following distribution functions, respectively:

$$\delta_1^0(l) = \begin{cases} 0.1 & \text{w.p. } 0.50, \\ 0 & \text{w.p. } 0.50. \end{cases} \quad \delta_2^0(l) = \begin{cases} 0.1 & \text{w.p. } 0.50, \\ 0 & \text{w.p. } 0.25, \\ -0.1 & \text{w.p. } 0.25. \end{cases} \quad \delta_3^0(l) = \begin{cases} 0.1 & \text{w.p. } 0.25, \\ 0 & \text{w.p. } 0.50, \\ -0.1 & \text{w.p. } 0.25, \end{cases}$$

where w.p. means "with probability." The distribution of scaled workload change $\delta_1^0(l)$ implies the total workload of an agent keeps increasing. This is valid in the early phase of some search algorithms, such as the branch-and-bound method. The branching procedure recursively expands a (sub)problem into subproblems such that

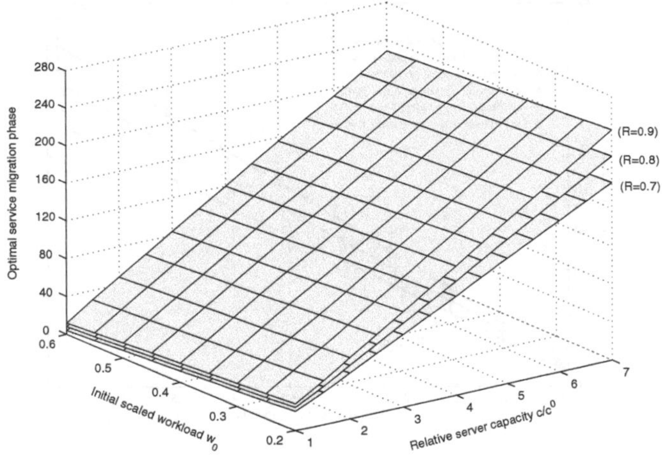

Figure 18.2: Optimal service migration timing (numerical results).

the size of subproblem set is on the steady increase. The second distribution $\delta_2^0(l)$ indicates the bounding procedure computes the expanded subproblems, which completes part of agent's workload. But the agent tends to generate new subproblems with higher probability as the computation proceeds. In the third phase $\delta_3^0(l)$, the rates of subproblem generation and completion tend to be equal so that the agent's expected workload remains unchanged for a certain period.

We varied the initial scaled workload w_0 from 0.2 to 0.6 and the capacity of each server relative to the original server of its residing agents, $\tilde{c}$, from 1 to 7. The corresponding optimal migration timing changed from the 4th to 244th phase as shown in Figure 18.2. The bound of overload, R, indicates the degree to which certain overload must be tolerated by a server to avoid frequent migration. Its value should be set according to the run-time overhead of service migration. By varying the overload bound R from 0.7 to 0.9, we can see the migration timing increases. Even with $w_0 = 0.5$, $R = 0.7$, and $\tilde{c} = 4$, the optimal migration phase is the 100th, at which much computation has been performed. So service migration is a feasible approach to realize VM reconfiguration. Figure 18.3 shows the migration phases simulated by random processes when the initial scaled workload w_0 is 0.5. This does not affect the experiment results. We can see our migration decision provides an accurate low bound in determining the service migration timing. The difference between simulated and predicted results tends to decrease as the overload bound R increases. The small confidence interval indicates the robustness of the estimates.

Case 2. The purpose of the second experiment is to verify the accuracy of destination server selection by Theorem 18.2 for a service migration. The agent composition and distribution functions of scaled workload change were the same as those in Case 1. The initial scaled workload of agents and overload bound were set as $w_0 = 0.5$ and $R = 0.8$, respectively. The service migration overhead was $H = 50$, as 100

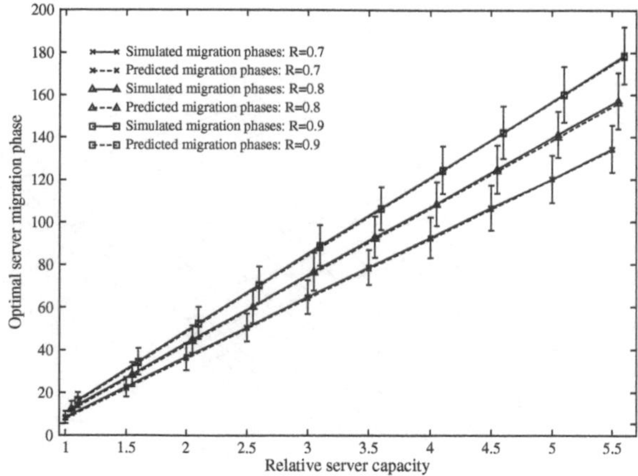

Figure 18.3: Optimal service migration timing with $w_0 = 0.5$ (simulation results).

times greater than the initial scaled workload. It confirms to our measurements of the run-time overhead of service migration in M-DSA shown in Chapter 17.

Figure 18.4 plots the minimal capacity that a prospective destination server must have. We varied the capacity of a server from 1 to 5.5 relative to the original server and measured the minimal relative capacity of destination server. We can see the predicted value is either equal to or slightly smaller than the simulated result. This is because our migration decision finds the low bound of destination server capacity. The difference is caused by approximating the scaled workload of an agent by a normal distribution according to the central limit theorem. From Figure 18.4, we can also see that the minimum value of destination capacity increases at a sub-linear rate. The target gain value G represents the expected benefits from service migration and it is measured relative to the migration overhead. As shown in this figure, if the target gain G is set to 20% of the migration overhead and the relative capacity of a server is 1, the minimal relative capacity of a destination server is about 2.3. Then during the next optimal migration timing, this service will be further migrated to an available server whose capacity should be at least 4.0 times greater than the original server of its residing agents. After migration, the process will be renewed and agent tasks will be executed until the next migration.

Case 3. In this experiment, we repeated the second experiment, but assuming all agents of a server had the same distribution function of scaled workload change as

$$\delta^0(l) = \begin{cases} 0.1 & \text{w.p. } 0.50, \\ 0 & \text{w.p. } 0.25, \\ -0.1 & \text{w.p. } 0.25. \end{cases}$$

According to Corollary 18.1, the problem of finding the lower bound of destination server capacity can be approximated by using the extreme value theory.

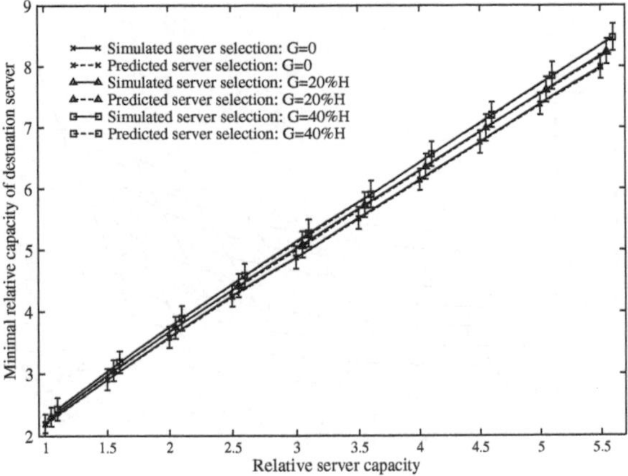

Figure 18.4: Destination server selection.

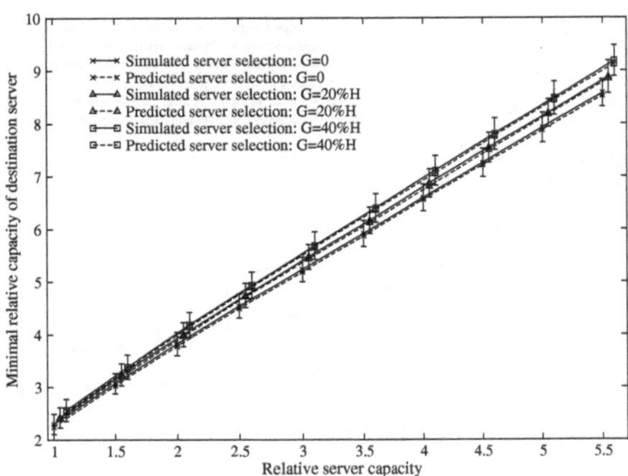

Figure 18.5: Destination server selection with extreme value theory.

Figure 18.5 shows the accuracy of the prediction approximation. The results in solid line are due to simulation measurements. Compared with the predicted values in Case 2, Corollary 18.1 is a little less accurate in destination server selection. It is because the extreme value theory provides an upper bound approximation of the expected value for maximal jointly distributed random variables. But, in Figure 18.5, we can see the predicted results are still very close to the simulated ones. The advantage of applying Corollary 18.1 is to simplify calculation of the expected migration gain function.

Case 4. The decision problem of agent migration has been modeled by dynamic programming in Section 18.4.1. To analyze the interplay between agent and service migration, we conducted the fourth experiment. We applied Theorem 18.3 to extend

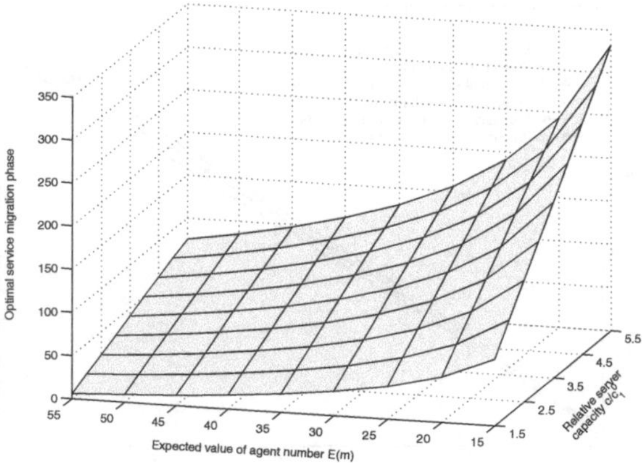

Figure 18.6: Optimal service migration timing for hybrid mobility.

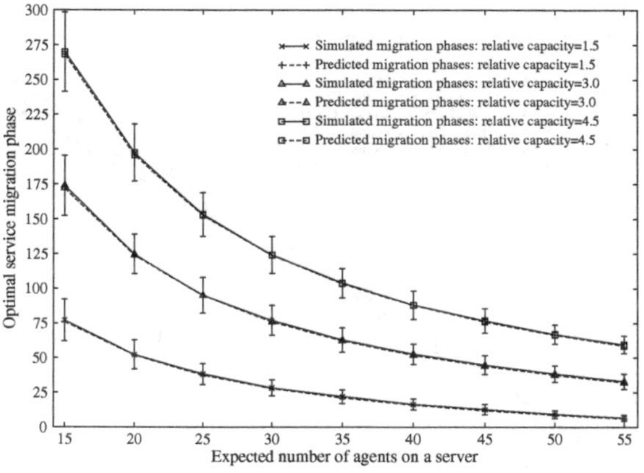

Figure 18.7: Optimal service migration timing for hybrid mobility with $w_0 = 0.5$ and $R = 0.8$.

the optimal service migration decisions in Case 1 and conducted simulations to verify the results. To meet the premise of the theorem, we assigned 30 agents to each server and ensured each agent was allocated the same initial capacity quota. All other settings were the same as in Case 3.

Figure 18.6 plots the optimal service migration timing by incorporating agent migration. We varied the capacity ratio between a server and the original one of the residing agents. For each ratio value, we measured the optimal migration timing corresponding to different expectations of the agent number, $E[m]$. This expected value represents the dynamics of agent migration. From Figure 18.6, we can see the migration timing decreases as the expectation of agent number increases. An extreme was reached when $E[m] = 55$ and server capacity ratio was 1.5. Its migration phase was about 6. This resulted from many agents landing on this server and the overall workload rapidly overwhelming its capacity. As computation proceeded, agents might decide to migrate to other light-loaded servers. Thus, the server's next migration interval became expanded accordingly. The expected value of agent number for each server can be determined dynamically by the history information of agent membership during the computation. Figure 18.7 presents the simulation results. We measured the migration timing for the capacity ratio equal to 1.5, 3.0, and 4.5. We can see the difference between simulated and predicted phases tends to be small around $E[m] = 30$, which is the initial number of agents on each server.

18.6 Concluding Remarks

In this chapter, we have studied the decision problem of hybrid mobility for load balancing in reconfigurable distributed VM. Chapter 2 defines the load balancing problem in a W5 model: who makes a load balancing decision based on what information; which task should be migrated to which server; when or why a migration should be invoked. This chapter presents a rigorous treatment of the problem, focusing on the questions of which, when, and where. By modeling the migration decision problem as stochastic optimization models, we have derived optimal migration policy in both time and space domains for given overload bound and target gain value. We have also analyzed the interplay of the agents and VM services in dynamic environments.

Hybrid mobility is not limited to load balancing for performance. It opens up a new path to fault tolerance in distributed VMs. Hybrid mobility complements other scaling techniques discussed in this book toward highly available Internet services.

References

[1] A. Takefusa et al. Multi-client LAN/WAN peformance analysis of Ninf: A high performance global computing system. In *Proc. of Supercomputing'97*, 1997. pages 302

[2] T. F. Abdelzaher, K. G. Shin, and N. Bhatti. Performance guarantees for Web server end-systems: a control-theoretical approach. *IEEE Transactions on Parallel and Distributed Systems*, 13(1):80–96, 2002. pages 60, 131

[3] M. Accetta, R. Baron, W. Bolosky, D. Golub, R. Rashid, A. Tevanian, and M. Young. Mach: A new kernel foundation for unix development. In *Proc. of the USENIX 1986 Summer Conf.*, 1986. pages 198

[4] C. C. Aggarwal, J. L. Wolf, and P. S. Yu. The maximum factor queue length batching scheme for video-on-demand systems. *IEEE Transactions on Computers*, 50(2):97–110, 2001. pages 40, 78, 81

[5] J. Almeida, M. Dabu, A. Manikutty, and P. Cao. Providing differentiated levels of services in Web content hosting. In *Proc. ACM SIGMETRICS Workshop on Internet Server Performance*, pages 91–102, 1998. pages 62, 63, 64, 72, 103

[6] V. Almeida, A. Bestavros, M. Crovella, and A. de Oliveira. Characterizing reference locality in the www. In *Proc. of IEEE Conf. on Parallel and Distributed Information Systems*, pages 92–103, 1996. pages 175

[7] E. Amir, S. McCanne, and H. Zhang. An application level video gateway. In *Proc. 3rd ACM Multimedia Conf.*, 1995. pages 68, 72

[8] A. H.-S. Ang and W. H. Tang. *Probability Concepts in Engineering Planning and Design, Volume II.* Rainbow Bridge, 1984. pages 353

[9] F. Annexstein, M. Baumslag, and A. L. Rosenberg. Group action graphs and parallel architecture. *SIAM Journal of Computing*, 19:544–569, 1990. pages 172

[10] A. W. Appel. Foundational proof-carrying code. In *Proc. of the 16th Annual IEEE Symp. on Logic in Computer Science*, page 247. IEEE Computer Society, 2001. pages 264

[11] M. Arlitt and T. Jin. A workload characterization study of the 1998 world cup web site. *IEEE Network*, 14(3):30–37, May–June 2000. pages 113

[12] M. Arlitt, D. Krishnamurthy, and J. Rolia. Characterizing the scalability of a large Web-based shopping system. *ACM Transactions on Internet Technology*, 1(1):44–69, 2001. pages 7, 98

[13] M. F. Arlitt and C. L. Williamson. Internet web servers: Workload characterization and performance implications. *IEEE/ACM Transactions on Networking*, 5(5):631–645, Oct. 1997. pages 7, 113, 130

[14] M. Aron, P. Druschel, and W. Zwaenepoel. Cluster reserves: A mechanism for resource management in cluster-based network servers. In *Proc. ACM SIGMETRICS*, pages 90–101, 2000. pages 67, 72

[15] M. Aron, D. Sanders, and P. Druschel. Scalable content-aware request distribution in cluster-based network servers. In *Proc. of 2000 USENIX Annual Technical Conf.*, pages 323–326, 2000. pages 22, 24, 31

[16] A. Awadallah and M. Rosenblum. The vMatrix: A network of virtual mahine monitors for dynamic content distribution. In *Proc. of the 7th Int. Workshop on Web Caching and Content Distribution (WCW'02)*, 2002. pages 195

[17] C. Baeumer, M. Breugst, S. Choy, and T. Magedanz. Grasshopper — a universal agent platform based on OMG MASIF and FIPA standards. `www.grasshopper.de`. pages 233

[18] A. Baiocchi and N. Blefari-Melazzi. Steady-state analysis of the MMPP/G/1/K queue. *IEEE Transactions on Communications*, 41:531–534, 1993. pages 130

[19] J. Baldeschwidler, R. Blumofe, and E. Brewer. Atlas: An infrastructure for global computing. In *Proc. of the 7th ACM SIGOPS European Workshop: Systems Support for Worldwide Applications*, 1996. pages 302

[20] G. Banga, P. Druschel, and J. Mogul. Resource containers: A new facility for resource management in server systems. In *Proc. USENIX Symp. on Operating System Design and Implementation*, 1999. pages 65

[21] J. Banino. Parallelism and fault tolerance in chorus. *Journal of Systems and Software*, pages 205–211, 1986. pages 199

[22] A. Barak and A. Litman. MOS: A multicomputer distributed operating system. *Software — Practice and Experience*, 15(8):725–737, 1985. pages 198

[23] A. Baratloo, M. Karaul, Z. Kedem, and P. Wyckoff. Charlotte: Metacomputing on the Web. *Future Generation Computer Systems*, 15:559–570, 1999. pages 302

[24] P. Barford, A. Bestavros, A. Bradley, and M. Crovella. Changes in web client access patterns: Characteristics and caching implications. *Journal of World Wide Web*, 2:15–28, 1999. pages 175, 182, 185, 186, 190

[25] P. Barford and M. Croella. Critical path analysis of tcp transactions. *IEEE/ACM Transactions on Networking*, 9(3):238–248, June 2001. pages 57

[26] P. Barford and M. Crovella. Generating representative web workloads for network and server performance evaluation. In *Proc. of ACM Sigmetrics'98*, pages 151–160, June 1998. pages 117

[27] P. Barford and M. Crovella. Generating representative web workloads for network and server performance evaluation. In *Proc. of ACM SIGMETRICS*, pages 151–160, 1998. pages 124

[28] R. Basharahil, B. Wims, C.-Z. Xu, and S. Fu. Distributed shared arrays: An integration of message passing and multithreading on SMP clusters. *Journal of Supercomputing*, 31(2):161–184, February 2005. pages 341

[29] J. Baumann, F. Hohl, K. Rothermel, and M. Straber. Mole — concepts of a mobile agent system. *World Wide Web*, 1(3):123–137, 1998. pages 233

[30] M. A. Bender, S. Chakrabarti, and S. Muthukrishnan. Flow and stretch metrics for scheduling continuous job streams. In *Proc. of ACM/SIAM Symp. on Discrete Algorithms (SODA)*, 1998. pages 95

[31] A. W. Berger and W. Whitt. Extending the effective bandwidth concept to networks with priority classes. *IEEE Communications Magazine*, Aug 1998. pages 130

[32] S. Berkovits, J. Guttman, and V. Swarup. Authentication for mobile agents. In G. Vigna, editor, *Mobile Agents and Security, LNCS 1419*, pages 114–136. Springer-Verlag, London, 1998. pages 204, 264

[33] F. Berman. High-performance schedulers. In I. Foster and C. Kesselman, editors, *The Grid: Blueprint for a New Computing Infrastructure*. Morgan Kaufmann, San Francisco, 1998. pages 301

[34] E. Bertino, C. Bettini, E. Ferrari, and P. Samarati. An access control model supporting periodicity constraints and temporal reasoning. *ACM Transactions on Database Systems*, 23(3):231–285, 1998. pages 273

[35] E. Bertino, P. A. Bonatti, and E. Ferrari. TRBAC: A temporal role-based access control model. *ACM Transactions on Information and System Security*, 4(3):191–233, August 2001. pages 273, 274

[36] P. Bieber, J. Cazin, P. Girard, J. Lanet, V. Wiels, and G. Zanon. Checking secure interactions of smart card applets. In *Proc. of the 6th European Symp. on Research in Computer Security (ESORICS'00)*, pages 1–16, 2000. pages 212

[37] A. Bieszczad, T. White, and B. Pagurek. Mobile agents for network management. *IEEE Communications Surveys*, 1(1), 1998. pages 242

[38] S. Blake, D. Black, M. Carlson, E. Davies, W. Z., and W. Weiss. An architecture for differentiated services. *IETF RFC 2475*, 1998. pages 56

[39] A. Bonner and M. Kifer. Concurrency and communication in transaction logic. In *Joint Int. Conf. and Symp. on Logic Programming*, pages 142–156, 1996. pages 255

[40] T. Bourke. *Server Load Balancing.* O'Reilly & Associates, 2001. pages 17, 18, 19

[41] D. P. Bovet and M. Cesati. *Understanding the Linux Kernel.* O'Reilly & Associates, 2000. pages 34

[42] R. Braden, D. Clark, and S. Shenker. Integrated services in the Internet architecture: An overview. *IETF RFC 1633*, 1994. pages 56

[43] B. Brewington, R. Gray, K. Moizumi, D. Kotz, G. Cybenko, and R. Rus. Mobile agents in distributed information retrieval. In M. Klusch, editor, *Intelligent Information Agents*. Springer-Verlag, Heidelberg, 1999. pages 242

[44] S. Brin and L. Page. The anatomy of a large-scale hypertextual web search engine. In *Proc. of WWW7*, 1998. pages 7

[45] O. Buyukkokten, H. Garcia-Molina, and A. Paepcke. Seeing the whole in parts: Text summarization for Web browsing on handheld devices. In *Proc. ACM WWW Conf.*, 2001. pages 68

[46] J. Cao, D. P. Spooner, S. A. Jarvis, S. Saini, and G. R. Nudd. Agent-based grid load balancing using performance-driven task scheduling. In *Proc. of the Parallel and Distributed Processing Symposium (IPDPS)*, Nice, France, April 2003. pages 344

[47] J. Cao, Y. Sun, X. Wang, and S. K. Das. Scalable load balancing on distributed Web servers using mobile agents. *Journal of Parallel and Distributed Computing*, 63(10):996–1005, 2003. pages 344

[48] V. Cardellini, E. Casalicchio, M. Colajanni, and M. Mambelli. Web switch support for differentiated services. *ACM Performance Evaluation*, 29(2):14–19, 2001. pages 66

[49] H. Casanova and J. Dongarra. Using agent-based software for scientific computing in the netsolve system. *Parallel Computing*, 24:1777–1790, 1998. pages 302

[50] S. Chakrabarti, M. van der Berg, and B. Dom. Focused crawling: A new approach to topic-specific web resource discovery. In *Proc. of the 8th Int. World Wide Web Conf. (WWW8)*, pages 545–562, 1999. pages 195

[51] S. Chandra, C. S. Ellis, and A. Vahdat. Application-level differentiated multimedia Web services using quality aware transcoding. *IEEE Journal on Selected Areas in Communications*, 18(12):2544–2265, 2000. pages 68, 72

[52] K. M. Chandy and J. Misra. *Parallel Program Design: A Foundation.* Addison-Wesley, Reading, MA, 1988. pages 248

[53] G. Chen, C.-Z. Xu, H. Shen, and D. Chen. P2P overlay networks of constant degree. In *Proc. of Int. Conf. on Grid and Collaborative Computing (GCC), LNCS 3202*, pages 412–419. Springer-Verlag, 2003. pages 152, 161

[54] H. Chen and P. Mohapatra. Session-based overload control in QoS-ware Web servers. In *Proc. IEEE INFOCOM*, 2002. pages 102

[55] S. Chen, B. Shen, S. Wee, and X. Zhang. Designs of high quality streaming systems. In *Proc. of IEEE INFOCOM'04*, 2004. pages 194

[56] X. Chen and P. Mohapatra. Providing differentiated services from an Internet server. In *Proc. 8th IEEE Int. Conf. on Computer Communications and Networks (ICCCN)*, pages 214–217, 1999. pages 60

[57] X. Chen and P. Mohapatra. Performance evaluation of service differentiating internet servers. *IEEE Transactions on Computers*, 51(11):1,368–1,375, 2002. pages 72

[58] X. Chen, P. Mohapatra, and H. Chen. An admission control scheme for predictable server response time for web accesses. In *Proc. of the 10th Int. World Wide Web Conf. (WWW10)*, 2001. pages 60

[59] X. Chen and X. Zhang. A popularity-based prediction model for web prefetching. *IEEE Computer*, 36(3):63–70, March 2003. pages 193

[60] L. Cherkasova and P. Phaal. Session-based admission control: A mechanism for peak load management of commercial web sites. *IEEE Transactions on Computers*, 51(6):669–685, 2002. pages 61, 75, 102, 103

[61] Chesapeake Computer Consultants. Test TCP (TTCP). `www.ccci.com/tools/ttcp`. pages 294

[62] D. M. Chess. Security issues in mobile code systems. In G. Vigna, editor, *Mobile Agents and Security, LNCS 1419*, pages 1–14. Springer-Verlag, London, 1998. pages 263

[63] B. Christiansen, P. Cappello, M. Ionescu, M. Neary, and K. Schauser. Javelin: Internet-based parallel computing using Java. In *Proc. of 1997 ACM Workshop on Java for Science and Engineering Computation*, June 1997. pages 302

[64] E. Clarke, O. Grumberg, and D. Peled. *Model Checking*. MIT Press, Cambridge, MA, 1999. pages 248, 258, 259

[65] I. Clarke, O. Sandberg, B. Wiley, and T. W. Hong. Freenet: A distributed anonymous information storage and retrieval system. In *Proc. Int. Workshop on Design Issues in Anonymity and Unobservability*, pages 46–66, 2001. pages 10, 151, 152

[66] A. Cohen. In search of Napster II. *TIME Magazine*, February 2001. pages 151

[67] A. Cohen, S. Rangarajan, and H. Slye. On the performance of TCP slicing for URL-Aware rediection. In *Proc. 2nd USENIX Symp. on Internet Technologies and Systems*, 1999. pages 20

[68] C. Collberg and C. Thomborson. Watermarking, tamper-proofing, and obfuscation tools for software protection. *IEEE Transactions on Software Engineering*, 28(8):735–746, 2002. pages 214

[69] CompTIA. Browser-based attacks may pose next big security nightmare, April 12, 2004. `comptia.org`. pages 6

[70] M. Crovella and P. Barford. The network effects of prefetching. In *Proc. of IEEE INFOCOM'98*, pages 1232–1240, 1998. pages 194

[71] M. E. Crovella and A. Bestavros. Self-similarity in world wide web traffic — evidence and possible causes. In *Proc. of SIGMETRICS*, 1996. pages 130

[72] M. E. Crovella, M. Harchol-Balter, and C. D. Murta. Task assignment in a distributed system: Improving performance by unbalancing load. In *Proc. of ACM SIGMETRICS'98*, pages 168–169, 1998. pages 24, 111

[73] C. R. Cunha and C. F. B. Jaccoud. Determining WWW user's next access and its application to prefetching. In *Proc. of the Int. Symp. on Computers and Communication*, pages 6–11, 1997. pages 176, 194

[74] K. M. Curewitz, P. Krishnan, and J. Vitter. Practical prefetching via data compression. In *Proc. of SIGMOD'93*, pages 257–266, May 1993. pages 193

[75] D. Milojicic et al. MASIF: The OMG mobile agent system interoperability facility. In *Proc. of the Int. Workshop on Mobile Agents (MA'98)*, 1998. pages 233

[76] D. Wang et al. Concordia: An infrastructure for collaborating mobile agents. In *Proc. of the 1st Int. Workshop on Mobile Agents (MA'98)*, pages 86–97, 1997. pages 217, 219

[77] F. Dabek, M. F. Kaashoek, D. Karger, R. Morris, and I. Stoica. Wide-area cooperative storage with CFS. In *Proc. of the 18th ACM Symp. on Operating Systems Principles (SOSP)*, October 2001. pages 341

[78] T. D. Dang, S. Molnar, and I. Maricza. Queueing performance estimation for general multifractal traffic. *International Journal of Communication Systems*, 16(2):117–136, 2003. pages 130

[79] S. B. Davidson, H. Garcia-Molina, and D. Skeen. Consistency in a partitioned network: A survey. *ACM Computing Surveys*, 17(3):341–370, 1985. pages 328

[80] B. Davison. Simultaneous proxy evaluation. In *Proc. of 4th Int. Web Caching Workshop*, pages 170–178, March 1999. pages 185

[81] B. Davison. A survey of proxy cache evaluation techniques. In *Proc. of 4th Int. Web Caching Workshop*, pages 67–77, March 1999. pages 185

[82] H. Davulcu, M. Kifer, C. R. Ramakrishnan, and I. V. Ramakrishnan. Logic based modeling and analysis of workflows. In *Proc. of the ACM Symp. on Principles of Database Systems*, pages 25–33, June 1998. pages 259

[83] P. Degano and C. Priami. Enhanced operational semantics: a tool for describing and analyzing concurrent systems. *ACM Computing Surveys*, 33(2):135–176, June 2001. pages 248

[84] DESCHALL. A brute force search of des keyspace. `www.interhack.net/projects/deschall`. pages 302

[85] J. Dilley, B. Maggs, J. Parikh, H. Prokop, R. Sitaraman, and B. Weihl. Globally distributed content delivery. *IEEE Internet Computing*, 6(5):50–58, 2002. pages 28, 33

[86] B. Dimitrov and V. Rego. Arachne: A portable threads system supporting migrant threads on heterogeneous network farms. *IEEE Transactions on Parallel and Distributed Systems*, 9(5):459–469, 1998. pages 341

[87] F. Douglis and J. K. Ousterhout. Transparent process migration: Design alternatives and the sprite implementation. *Software — Practice and Experience*, 21(8):757–785, 1991. pages 341

[88] C. Dovrolis, D. Stiliadis, and P. Ramanathan. Proportional differentiated services: Delay differentiation and packet scheduling. *IEEE/ACM Transactions on Networking*, 10(1):12–26, 2002. pages 62, 77, 89, 117

[89] L. W. Dowdy and D. Foster. Comparative models of the file assignment problem. *ACM Computing Surveys*, 14(2):287–313, 1982. pages 39

[90] P. Druschel and G. Banga. Lazy receiver processing (lrp): A network subsysgtem architecture for server systems. In *Proc. of Operating Systems Design and Implementation (OSDI'96)*, 1996. pages 64

[91] C. Du, X.-H. Sun, and K. Chanchio. HPCM: A pre-compiler aided middleware for the mobility of legacy code. In *Proc. of Int. Conf. on Cluster Computing*, 2003. pages 341

[92] G. Edjlali, A. Acharya, and V. Chaudhary. History-based access control for mobile code. In *Proc. of ACM Conf. on Computer and Communications Security*, pages 38–48, 1998. pages 207, 273

[93] L. Eggert and J. Heidemann. Application-level differentiated services for Web servers. *World Wide Web Journal*, 3(2):133–142, 1999. pages 62, 63

[94] R. Ekwall, P. Urb'an, and A. Schiper. Robust TCP connections for fault tolerant computing. In *Proc. Int. Conf. on Parallel and Distributed Systems*, 2002. pages 281

[95] S. Elnikety, E. Nahum, J. Tracey, and W. Zwaenepoel. A method for transparent admission control and request scheduling in e-commerce web sites. In *Proc. of Int. World Wide Web Conf.*, 2004. pages 102

[96] E. A. Emerson. Temporal and modal logic. In J. van Leeuwen, editor, *Handbook of Theoretical Computer Science*, pages 997–1067. Elsevier Science, 1990. pages 277

[97] W. Emmerich and N. Kaveh. Model checking distributed objects. In *Proc. of the 4th Int. Software Architecture Workshop*, June 2000. pages 258

[98] D. H. J. Epema. Decay-usage scheduling in multiprocessors. *ACM Transactions on Computer Systems (TOCS)*, 16, 1998. pages 131

[99] A. Erramilli, O. Narayan, and W. Willinger. Experimental queueing analysis with long-range dependent packet traffic. *IEEE/ACM Transactions on Networking*, 4(2):209–223, 1996. pages 130

[100] F. Berman et al. The GrADS project: Software support for high-level grid application development. *The International Journal of High Performance Computing Applications*, 15(4):327–344, 2001. pages 342

[101] F. Berman et al. Adaptive computing on the grid using AppLeS. *IEEE Transactions on Parallel and Distributed Systems*, 14(4):369–382, April 2003. pages 324, 341

[102] L. Fan, P. Cao, W. Lin, and Q. Jacobson. Web prefetching between low-bandwidth clients and proxies: Potential and performance. In *Proc. of SIGMETRICS'99*, pages 178–187, May 1999. pages 176, 193

[103] W. Farmer, J. Guttman, and V. Swarup. Security for mobile agents: Authentication and state appraisal. In *Proc. of the 4th European Symp. on Research in Computer Security (ESORICS '96)*, pages 118–130, September 1996. pages 205, 211, 264

[104] W. Farmer, J. Guttman, and V. Swarup. Security for mobile agents: Issues and requirements. In *Proc. of the 19th National Information Systems Security Conf.*, pages 591–597, 1996. pages 203, 211

[105] D. G. Feitelson and L. Rudolph. Metrics and benchmarking for parallel job scheduling. In *Proc. of the Workshop on Job Scheduling Strategies for Parallel Processing (in conjunction with IPDPS'98), LNCS 1459*, pages 1–24. Springer-Verlag, 1998. pages 111

[106] R. Feldmann and W. Unger. The cube-connected cycles network is a subgraph of the bufferfly network. *Parallel Processing Letters*, 2(1):13–19, 1991. pages 172

[107] W. Feller. *An Introduction to Probability Theory and Its Applications, Volume II*. John Wiley & Sons, New York, 1971. pages 135, 136, 139

[108] D. F. Ferraiolo, J. F. Barkley, and D. R. Kuhn. A role based access control model and reference implementation within a corporate intranet. *ACM Transactions on Information and System Security*, 2(1):34–64, 1999. pages 266, 273, 274, 277

[109] R. Figueiredo, P. Dinda, and J. Fortes. A case for grid computing on virtual machines. In *Proc. of the 23rd Int. Conf. on Distributed Computing Systems (ICDCS)*, May 2003. pages 323, 341, 343

[110] T. Finin, R. Fritzson, D. McKay, and R. McEntire. KQML as an agent communication language. In *Proc. of the 3rd ACM Int. Conf. on Information and Knowledge Management (CIKM'94)*, pages 456–463, 1994. pages 279, 305

[111] FIPA. Agent communication language, 1999. `http://www.fipa.org`. pages 279, 305

[112] S. Floyd and V. Jacobson. Random early detection for congestion avoidance. *IEEE/ACM Transactions on Networking*, Aug. 1993. pages 60

[113] N.-T. Fong, C.-Z. Xu, and L. Wang. Optimal periodic remapping of bulk synchronous computations on multiprogrammed distributed systems. *Journal of Parallel and Distributed Computing*, 63(11):1036–1049, Nov. 2003. pages 344, 346

[114] W. Ford. *Computer Communications Security: Principles, Standard Protocols and Techniques*. Prentice-Hall, New York, 1994. pages 209

[115] I. Foster and C. Kesselman. Globus: A metacomputing infrastructure toolkit. `www.globus.org`. pages 301, 333

[116] I. Foster and C. Kesselman. *The Grid: Blueprint for a New Computing Infrastructure*. Morgan Kaufmann, San Francisco, 1998. pages 301, 323

[117] C. Fournet, G. Gonthier, J.-J. Levy, L. Maranget, and D. Remy. A calculus of mobile agents. In *Proc. of Int. Conf. on Concurrency Theory (CONCUR'96)*, pages 406–421, 1996. pages 248

[118] A. Fox, S. D. Gribble, Y. Chawathe, and E. A. Brewer. Adapting to network and client variation using infrastructural proxies: Lessons and perspectives. *IEEE Personal Communications*, 5(4):10–19, 1998. pages 68, 72

[119] G. F. Franklin, J. D. Powell, and A. Emami-naeini. *Feedback Control of Dynamic Systems (4th ed.)*. Prentice-Hall, New York, 2002. pages 119

[120] Free Software Foundation. GSL – GNU Scientific Library. `www.gnu.org/software/gsl`. pages 116

[121] S. Fu and C.-Z. Xu. Migration decision for hybrid mobility in reconfigurable virtual machines. In *Proc. of Int. Conf. on Parallel Processing (ICPP)*, pages 335–342, 2004. pages 344

[122] S. Fu and C.-Z. Xu. A coordinated spatio-temporal access control model for mobile computing in coalition environments. In *Proc. of IEEE IPDPS Workshop on Security in Systems and Networks (SSN)*, 2005. pages 207, 265

[123] S. Fu and C.-Z. Xu. Mobile codes and security. In *Handbooks of Information Security*. John Wiley & Sons, New York, 2005. pages 199

[124] S. Fu and C.-Z. Xu. Service migration in distributed virtual machines for adaptive grid computing. In *Proc. of Int. Conf. on Parallel Processing (ICPP)*, 2005. pages 324, 343

[125] S. Fu, C.-Z. Xu, B. Wims, and R. Basharahil. Distributed shared arrays: A distributed virtual machine with mobility support for reconfiguration. *Journal of Cluster Computing*, 8(4), October 2005. pages 324, 343

[126] A. Fuggetta, G. P. Picco, and G. Vigna. Understanding code mobility. *IEEE Transactions on Software Engineering*, 24(5):342–361, 1998. pages 324

[127] D. Funato, K. Yasuda, and H. Tokuda. TCP-R: TCP mobility support for continuous operation. In *Proc. IEEE Int. Conf. on Network Protocols*, pages 229–236, 1997. pages 280

[128] S. Gallagher. Amazon.com at Linux world: All Linux, all the time, January 2004. `eWeek.com`. pages 8

[129] G. Ganger, M. Engler, M. Kaashoek, H. Briceno, and R. Hunt. Fast and flexible application-level networking on exokernel. *ACM Transactions on Computer Systems*, 20(1):49–83, 2002. pages 65

[130] T. Garfinkel, B. Pfaff, J. Chow, M. Rosenblum, and D. Boneh. Terra: A virtual machine-based platform for trusted computing. In *Proc. of the 19th ACM Symp. on Operating Systems Principles (SOSP)*, October 2003. pages 340

[131] D. J. Gemmell, H. M. Vin, D. D. Kandlur, P. V. Rangan, and L. A. Rowe. Multimedia storage servers: A tutorial. *IEEE Computer*, 28(5):40–49, May 1995. pages 37

[132] E. J. Glover, G. W. Falke, S. Lawrence, W. Birmingham, A. Kruger, C. Giles, and D. Pennock. Improving category specific web search by learning query modifications. In *Proc. of Symp. on Applications and the Internet (SAINT)*, January 2001. pages 195

[133] Gnutella home page. `www.gnutella.com`. pages 10, 151, 152

[134] L. Golubchik, R. R. Muntz, C. Chou, and S. Berson. Design of fault-tolerant large-scale VoD servers: with emphasis on high-performance and low-cost. *IEEE Transactions on Parallel and Distributed Systems*, 12(4):363–386, 2001. pages 38

[135] L. Gong. *Inside Java 2 Platform Security: Architecture, API Design, and Implementation*. Addison-Wesley, Reading, MA, 1999. pages 263

[136] P. Goyal, X. Guo, and H. M. Vin. A hierarchical CPU scheduler for multimedia operating systems. In *Proc. 2nd Usenix Symp. Operating System Design and Implementation*, 1996. pages 64

[137] R. Gray, D. Kotz, G. Cybenko, and D. Rus. D'Agents: Security in a multiple-language, mobile-agent system. In G. Vigna, editor, *Mobile Agents and Security, LNCS 1419*. Springer-Verlag, London, 1998. pages 219

[138] R. Gray, D. Kotz, G. Cybenko, and D. Rus. Mobile agents: Motivations and state-of-the-art systems. In *Handbook of Agent Technology*. AAAI/MIT Press, Cambridge, MA, 2001. pages 219

[139] R. S. Gray. Agent Tcl: A flexible and secure mobile-agent system. In M. Diekhans and M. Roseman, editors, *Proc. of the 4th Annual Tcl/Tk Workshop (TCL 96)*, pages 9–23, Monterey, CA, 1996. pages 202, 216

[140] K. T. Greenfeld. Meet the Napster. *TIME Magazine*, October 2000. pages 10, 151

[141] J. Griffioen and R. Appleton. Reducing file system latency using a predictive approach. In *Proc. of the 1994 Summer USENIX Conf.*, pages 197–207, June 1994. pages 194

[142] A. Grimshaw and W. Wulf. Legion: The next logical step toward the world-wide virtual computer. `www.cs.virginia.edu/~legion`. pages 301

[143] R. Guerin, H. Ahmadi, and M. Naghshineh. Equivalent capacity and its application to bandwidth allocation in high-speed networks. *IEEE Journal of Selected Areas of Communications*, 9:968–981, 1991. pages 130

[144] S. Habert, L. Mosseri, and V. Abrossimov. COOL: Kernel support for object-oriented environments. In *Proc. of the Conf. on Object-Oriented Programming Systems, Languages, and Applications (OOPSLA)*, pages 269–277, 1990. pages 198, 199

[145] R. Han, P. Bhagwat, R. LaMaire, T. Mummert, V. Perret, and J. Rudas. Dynamic adaptation in an image transcoding proxy for mobile web browsing. *IEEE Personal Communications*, 5(6):8–17, 1998. pages 68

[146] M. Harchol-Balter. Task assignment with unknown duration. *Journal of ACM*, 49(2):260–288, 2002. pages 22, 34, 35, 95, 103, 111, 112, 113, 117

[147] M. Harchol-Balter, M. E. Crovella, and C. D. Murta. On choosing a task assignment policy for a distributed server system. *Journal of Parallel and Distributed Computing*, 59(2):204–228, 1999. pages 34, 111

[148] M. Harchol-Balter, B. Schroeder, N. Bansal, and M. Agrawal. Size-based scheduling to improve web performance. *ACM Transactions on Computer Systems*, 21(2):207–233, May 2003. pages 22, 111

[149] C. Harrison, D. Chess, and Kershenbaum. Mobile agents: Are they a good idea? Technical report, IBM Watson Research Center, March 1995. pages 199, 242

[150] E. Hashem. Analysis of random drop for gateway congestion control. Technical Report MIT-LCS-TR-465, MIT, Laboratory of Computer Science, 1989. pages 60

[151] H. Hassoun. *Fundamentals of Artificial Neural Networks*. MIT Press, Cambridge, MA, 1995. pages 179, 180

[152] J. L. Hennessy and D. A. Patterson. *Computer Architecture: A Quantitative Approach (3rd ed.)*. Morgan Kaufmann, San Francisco, 2003. pages 7, 15

[153] C. Hoare. *Communication Sequential Processes*. Prentice-Hall, Englewood Cliffs, NJ, 1985. pages 248

[154] F. Hohl. Time limited blackbox security: Protecting mobile agents from malicious hosts. In G. Vigna, editor, *Mobile Agents and Security, LNCS 1419*. Springer-Verlag, London, 1998. pages 215

[155] T. Ibarkai and N. Katoh. *Resource Allocation Problem — Algorithmic Approaches*. MIT Press, Cambridge, MA, 1988. pages 78

[156] J. Ioannidis, D. Duchamp, and G. Q. Maguire. IP-based protocols for mobile internetworking. In *Proc. of ACM SIGCOMM*, April 2002. pages 280

[157] A. Iyengar, J. Challenger, D. Dias, and P. Dantzig. High-performance web site design techniques. *IEEE Internet Computing*, 4(2):17–26, 2000. pages 29

[158] H. Jiang and V. Chaudhary. Compile/run-time support for thread migration. In *Proc. of the 16th Int. Parallel and Distributed Processing Symp. (IPDPS)*, April 2002. pages 341, 343

[159] X. Jiang and D.-Y. Xu. SODA: A service-on-demand architecture for application service hosting utility platforms. In *Proc. of Int. Symp. on High Performance Distributed Computing*, 2003. pages 341

[160] S. Jin and Z. Bestavros. Temporal locality in web request streams — sources, characteristics, and chaching implications. In *Proc. of SIGMETRICS*, 2000. pages 130

[161] D. Johansen, R. van Renesse, and F. Schneider. Operating system support for mobile agents. In *Proc. of the 5th IEEE Workshop on Hot Topics in Operating Systems*, 1995. pages 202, 216

[162] J. B. D. Joshi, E. Bertino, and A. Ghafoor. Temporal hierarchies and inheritance semantics for GTRBAC. In *Proc. the 7th ACM Symp. on Access Control Models and Technologies (SACMAT'02)*, pages 74–83, June 2002. pages 273

[163] E. Jul, H. Levy, N. Hutchinson, and A. Black. Fine grained mobility in the emerald system. *ACM Transactions on Computer Systems*, 6(1):109–133, 1988. pages 198, 199, 341

[164] K. Kato et al. An approach to mobile software robots for the WWW. *IEEE Transactions on Knowledge and Data Engineering*, 11(4):526–548, July/August 1999. pages 242

[165] K. Kennedy et al. Toward a framework for preparing and executing adaptive grid programs. In *Proc. of the NSF Next Generation Systems Program Workshop (In conjunction with IPDPS'2002)*, pages 322–326, April 2002. pages 342

[166] M. F. Kaashoek and R. Karger. Koorde: A simple degree-optimal distributed hash table. In *Proc. of the 2nd Int. Workshop on Peer-to-Peer Systems (IPTPS)*, 2003. pages 154, 156, 163, 167

[167] N. Kapadia, J. Fortes, and M. Lundstrom. Purdue university network-computing hubs: Running unmodified simulation tools via the WWW. *ACM Transactions on Modeling and Computer Simulation*, Jan. 2000. pages 302

[168] D. Karger, E. Lehman, T. Leighton, M. Levine, D. Lewin, and R. Panigraphy. Consistent hashing and random trees: Distributed cashing protocols for relieving hot spots on the world wide web. In *Proc. of 29th Annual ACM Symp. on Theory of Computing*, 1997. pages 154

[169] N. Karnik and A. Tripathi. A security architecture for mobile agents in Aganta. In *Proc. of the 20th Int. Conf. on Distributed Computing Systems (ICDCS'2000)*, 2000. pages 218

[170] J. Kay and P. Lauder. A fair share scheduler. *Communication of ACM*, 31(1):44–55, 1988. pages 77, 103, 143

[171] F. P. Kelly. Effective bandwidths at multi-class queues. *Queueing Systems*, 9:5–16, 1991. pages 130

[172] G. Kesidis, J. Walrand, and C.-S. Chang. Effective bandwidths for multiclass Markov fluids and other ATM sources. *IEEE/ACM Tranactions on Networking*, 1:424–428, 1993. pages 130

[173] L. Kleinrock. *Queueing Systems, Volume II.* John Wiley & Sons, New York, 1976. pages 97

[174] R. P. Klemm. Webcompanion: A friendly client-side web prefetching agent. *IEEE Transactions on Knowledge and Data Engineering*, 11(4):577–594, July/August 1999. pages 193

[175] M. Kobayashi and K. Takeda. Information retrieval on the Web. *ACM Computing Surveys*, 32(2):144–173, June 2000. pages 177, 181

[176] M. Kona and C.-Z. Xu. A framework for network management using mobile agents. In *Proc. of IEEE IPDPS Workshop on Internet Computing and e-Commerce*, 2001. pages 242, 243

[177] D. Kondo, M. Taufer, C. Brooks, H. Casanova, and A. Chien. Characterizing and evaluating Desktop Grids: An empirical study. In *Proc. of the 18th Int. Parallel and Distributed Processing Symp. (IPDPS)*, April 2004. pages 324

[178] C. Kopparapu. *Load Balancing Servers, Firewalls, and Caches*. John Wiley & Sons, New York, 2002. pages 18, 19

[179] T. Kroeger, D. Long, and J. Mogul. Exploring the bounds of web latency reduction from caching and prefetching. In *Proc. of USENIX Symp. on Internet Technologies and Systems*, pages 13–22, December 1997. pages 175, 176

[180] M. M. Krunz and A. M. Ramasamy. The correlation structure for a class of scene-based video models and its impact on the dimensioning of video buffers. *IEEE Transactions on Multimedia*, 2(1):27–36, 2000. pages 140

[181] G. Kuenning and G. Popek. Automated hoarding for mobile computers. In *Proc. of the ACM Symp. on Operating Systems Principles*, pages 264–275, October 1997. pages 194

[182] J. Kurose and K. Ross. *Computer Networking: A Top-Down Approach*. Pearson Addison-Wesley, Reading, MA, 2002. pages 18, 19

[183] B. Lampson. A note on the confinement problem. *Communications of the ACM*, 16(10):613–615, 1973. pages 207

[184] B. Lampson, M. Abadi, M. Burrows, and E. Wobber. Authentication in distributed systems: Theory and practice. *ACM Transactions on Computer Systems*, 10(4):265–310, 1992. pages 204, 264

[185] D. Lange and M. Oshima. *Programming and Deploying Java Mobile Agents with Aglet*. Addison-Wesley, Reading, MA, 1998. pages 201, 202, 217, 219, 233, 268

[186] D. Lange and M. Oshima. Seven good reasons for mobile agents. *Communications of ACM*, 42(3):88–89, March 1999. pages 199, 242

[187] Le Boudec, J.-Y. and Thiran, P. *Network Calculus: A Theory of Deterministic Queuing Systems for the Internet*, volume LNCS No. 2050. Springer-Verlag, Heidelberg, 2001. pages 130

[188] S. C. M. Lee, J. C. S. Lui, and D. K. Y. Yau. Admission control and dynamic adaptation for a proportional-delay diffserv-enabled web server. In *Proc. of SIGMETRICS'02*, 2002. pages 60

[189] H. Lei and D. Duchamp. An analytical approach to file prefetching. In *Proc. of USENIX 1997 Annual Technical Conf.*, pages 275–288, January 1997. pages 194

[190] W. Leinberger, G. Karypis, and V. Kumar. Job scheduling on the presence of multiple resource requirements. In *Proc. of Supercomputing Conf.*, 1999. pages 82

[191] M. K. H. Leung, J. C. S. Lui, and D. K. Y. Yau. Adaptive proportional delay differentiated services: Characterization and performance evaluation. *IEEE/ACM Transactions on Networking*, 9(6):908–817, 2001. pages 62

[192] S. Levey. All eyes on Google. *Newsweek*, March 2004. pages 7

[193] J. Levy and J. Ousterhout. A safe tcl toolkit for electronic meeting place. In *Proc. of the 1st USENIX Workshop on Electronic Commerce*, pages 133–135, 1995. pages 217

[194] K. Li and S. Jamin. A measurement-based admission-controlled web server. In *Proc. IEEE INFOCOM*, pages 544–551, 2000. pages 72

[195] S. Q. Li and C. L. Hwang. On the convergence of traffic measurement and queueing analysis: A statistical-matching and queueing (SMAQ) tool. *IEEE/ACM Transactions on Networking*, February 1997. pages 130

[196] D. Libes. *Obfuscated C and Other Mysteries*. John Wiley & Sons, New York, 1993. pages 214

[197] D. Lie, C. Thekkath, M. Mitchell, P. Lincoln, D. Boneh, J. Mitchell, and M. Horowitz. Architectural support for copy and tamper resistant software. In *Proc. of 9th Int. Conf. on Architectural Support for Programming Languages and Operating Systems (ASPLOS-IX)*, pages 168–177, 2000. pages 212

[198] T. Lindholm and F. Yellin. *The Java Virtual Machine Specification (2nd ed.)*. Addison-Wesley, Reading, MA, 1999. pages 329

[199] M. Litzkow, M. Livny, and M. Mutka. Condor: A hunter of idle workstations. In *Proc. of the 8th IEEE Int. Conf. on Distributed Computing Systems*, pages 104–111, 1988. pages 198, 301, 341

[200] T. S. Loon and V. Bharghavan. Alleviating the latency and bandwidth problems in WWW browsing. In *Proc. of USENIX Symp. on Internet Technologies and Systems*, pages 219–230, December 1997. pages 176, 193

[201] C. Lu, T. Abdelzaher, J. Stankovic, and S. Son. A feedback control approach for guaranteeing relative delays in web servers. In *Proc. of the IEEE Real-Time Technology and Applications Symp.*, 2001. pages 64, 131

[202] S. Lu and C.-Z. Xu. Mail: A mobile agent itinerary language for correctness and safety reasoning. Technical Report TR-DMSL-2004-02/CIC-04-03, Wayne State University, 2004. pages 249

[203] S. Lu and C.-Z. Xu. A formal framework for agent itinerary specification, security reasoning, and logic analysis. In *Proc. of IEEE ICDCS Workshop on Mobile Distributed Computing*, 2005. pages 249

[204] Y. Lu, T. Abdelzaher, and C. Lu. Feedback control with queueing-theoretic prediction for relative delay guarantees in web servers. In *Proc. of the IEEE Real-Time Technology and Applications Symp.*, 2003. pages 64, 131

[205] W. Lum and F. Lau. A context-aware decision engine for content adaptation. *IEEE Pervasive Computing*, 1(3):41–49, July-Sept 2002. pages 67, 68

[206] W. Lum and F. Lau. On balancing between transcoding overhead and spatial consumption in content adaptation. In *Proc. of Mobicom*, pages 239–250, 2002. pages 67, 68

[207] Q. Lv, P. Cao, E. Cohen, K. Li, and S. Shenker. Search and replication in unstructured peer-to-peer networks. In *Proc. of ACM Int. Conf. on Supercomputing (ICS)*, 2001. pages 152

[208] N. Lynch, D. Malkhi, and D. Ratajczak. Atomic data access in distributed hash tables. In *Proc. of the Int. Peer-to-Peer Symp.*, 2002. pages 170

[209] K. W. R. M. Reisslein and S. Rajagopal. A framework for guaranteeing statistical qos. *IEEE/ACM Transactions Networking*, 10-1:27–42, Feb 2002. pages 130

[210] D. Malkhi, M. Naor, and D. Ratajczak. Viceroy: A scalable and dynamic emulation of the butterfly. In *Proc. of Principles of Distributed Computing (PODC)*, 2002. pages 154, 156, 163

[211] D. A. Maltz and P. Bhagwat. MSOCKS: An architecture for transport layer mobility. In *Proc. of IEEE INFOCOM*, pages 1037–1045, 1998. pages 280

[212] D. G. Manolakis, V. K. Ingle, and S. M. Kogon. *Statistical and Adaptive Signal Processing*. McGraw-Hill, New York, 2000. pages 136, 143

[213] E. P. Markatos and C. E. Chronaki. A top-10 approach to prefetching on the web. In *Proc. of INET'98*, July 1998. pages 193

[214] C. Mascolo, G. Picco, and G.-C. Roman. A fine-grained model for code mobility. In *Joint Proc. of the 7th European Software Engineering Conf. and the 6th ACM Int. Symp. on Foundations of Software Engineering (ESEC/FSC'99)*, 1999. pages 249

[215] P. Maymounkov and D. Mazires. Kademlia: A peer-to-peer information systems based on the XOR metric. *Proc. of the 1st Int. Workshop on Peer-to-Peer Systems (IPTPS)*, 2002. pages 152

[216] A. McCallum, K. Nigam, J. Rennie, and K. Seymore. Building domain-specific search engines with machine learning techniques. In *Proc. of AAAI Spring Symp. on Intelligent Agents in Cyberspace*, 1999. pages 195

[217] P. McCann and G.-C. Roman. Compositional programming abstractions for mobile computing. *IEEE Transactions on Software Engineering*, 24(2):97–110, February 1998. pages 248

[218] D. A. Menascé and V. A. F. Almeida. *Scaling for E-Business Technologies, Models, Performance, and Capacity Planning*. Prentice-Hall, Englewood Cliff, NJ, 2000. pages 8, 129, 130

[219] D. A. Menascé, V. A. F. Almeida, R. Fonseca, and M. A. Mendes. A methodology for workload characterization of e-commerce sites. In *Proc. of the 1st ACM Conf. on Electronic Commerce*, 1999. pages 93, 95, 96, 97, 98, 102, 103

[220] D. A. Menascé, V. A. F. Almeida, R. Fonseca, and M. A. Mendes. Resource management policies for e-commerce servers. In *Proc. of ACM SIGMETRICS Workshop on Internet Server Performance*, 1999. pages 93, 95, 96, 97, 98, 102, 103

[221] R. A. Meyer and L. H. Seawright. A virtual machine time sharing system. *IBM System Journal*, 9(3):199–218, 1970. pages 340

[222] Microsoft. Technical overview of windows server 2003: Cluster services, September 2004. `www.microsoft.com/windowsserver2003`. pages 25, 33

[223] R. Milner. *A Calculus of Communicating Systems, LNCS 92*. Springer-Verlag, Berlin, 1980. pages 248

[224] R. Milner, J. Parrow, and D. Walker. A calculus of mobile processes (Parts I and II). *Information and Computation*, 100:1–77, 1992. pages 248

[225] D. Milojicic, F. Douglis, Y. Paindaveine, R. Wheeler, and S. Zhou. Process migration. *ACM Computing Surveys*, 32(3):241–299, September 2000. pages 198, 302, 324, 328

[226] D. Milojicic, W. LaForge, and D. Chauhan. Mobile objects and agents (MOA). In *Proc. of the 4th USENIX Conf. Object-Oriented Technologies and Systems (COOTS'98)*, 1998. pages 233

[227] NCSA. Biology workbench. `www.workbench.ncsa.edu`. pages 302

[228] G. Necula and P. Lee. Research on proof-carrying code on mobile-code security. In *Proc. of the Workshop on Foundations of Mobile Code Security*, 1997. pages 264

[229] G. C. Necula. Proof-carrying code. In *Proc. of the 24th ACM Symp. on Principles of Programming Languages*, pages 106–119, January 1997. pages 207, 264

[230] J. Nieplocha, R. Harison, and R. Littlefield. Global arrays: A nonuniform memory access programming model for high-performance computers. *The Journal of Supercomputing*, 10:169–189, 1996. pages 307

[231] I. Norros. A storage model with self-similar input. *Queueing Systems*, 16:387–396, 1994. pages 130

[232] T. Okoshi, M. Mochizuki, and Y. Tobe. Mobilesocket: Toward continuous operation for Java applications. In *Proc. of the Int. Conf. on Computer Communications and Networks*, pages 50–57, October 1999. pages 281

[233] D. P. Olshefski, J. Niehl, and E. Nahum. ksniffer: Determining the remote client perceived response time from live packet streams. In *Proc. of Usenix Operating Systems Design and Implementation (OSDI)*, 2004. pages 124

[234] D. Oppenheimer and D. A. Patterson. Architecture and dependability of large-scale Internet services. *IEEE Internet Computing*, 6(5), 2002. pages 15, 29

[235] J. Ousterhout, A. Cherenson, F. Gouglis, M. Nelson, and B. Welch. The sprite network operating system. *IEEE Computer*, 21(2):23–38, 1988. pages 198

[236] P. Dasgupta et al. MAgNET:Mobile agents for networked electronic trading. *IEEE Transactions on Knowledge and Data engineering*, 11(4):509–525, July/August 1999. pages 242

[237] V. Padmanabhan and J. Mogul. Using predictive prefetching to improve world wide web latency. *Computer Communication Review*, 26(3):22–36, July 1996. pages 176, 193, 194

[238] V. N. Padmanabhan and L. Qui. The content and access dynamics of a busy web site: Findings and implications. In *Proc. of the ACM SIGCOMM*, pages 111–123. ACM Press, New York, 2000. pages 7, 195

[239] V. S. Pai, M. Aron, G. Banga, M. Svendsen, P. Drusche, W. Zwaenepoel, and E. Nahum. Locality-aware request distribution in cluster-based network servers. In *Proc. of 8th Int. Conf. on Architectural Support for Programming Languages and Operating Systems (ASPLOS)*, 1998. pages 21, 22, 24, 31, 35

[240] R. Pandey and B. Hashii. Providing fine-grained access control for Java programs via binary editing. *Concurrency: Practice and Experience*, 12(14):1405–1430, 2000. pages 206

[241] A. K. Parehk and R. G. Gallager. A generalized processor sharing approach to flow control in Integrated Sevices networks: the single-node case. *IEEE/ACM Transactions on Networking*, 1(3):344–357, 1993. pages 114, 130

[242] V. Paxson. Fast approximation of self-similar network traffic. Technical Report LBL-36-750, Lawrence Berkeley National Laboratory, 1995. pages 140

[243] V. Paxson and S. Floyd. Wide area traffic: The failure of Possion modeling. *IEEE/ACM Transactions on Networking*, 3(3):226–244, June 1995. pages 113, 117

[244] J. Pelline. MyDoom downs SCO site. *CNET News.com*, February 2004. `news.com.com`. pages 6

[245] G. Picco, G.-C. Roman, and P. McCann. Expressing code mobility in Mobile UNITY. In *Proc. of 6th European Software Engineering Conf. (ESEC/FSC'970)*, 1997. pages 249

[246] C. Plaxton, R. Rajaraman, and A. Richa. Accessing nearby copies of replicated objects in a distributed environment. In *Proc. of ACM Symp. on Parallelism Algorithms and Architectures (SPAA)*, 1997. pages 155

[247] F. P. Preparata and J. Vuillemin. The cube-connected cycles: A versatile network for parallel computation. *Communications of ACM*, 24(5):300–309, 1981. pages 160

[248] A. Prior. *Past, Present and Future*. Clarendon Press, Oxford, 1969. pages 275

[249] R. Puri, K. W. Lee, K. Ramchandran, and V. Bharghavan. An integrated source transcoding and congestion control paradigm for video streaming in the Internet. *IEEE Transactions on Multimedia*, 3(1):18–32, 2001. pages 72

[250] X. Qu, J. X. Yu, and R. P. Brent. A mobile TCP socket. In *Proc. of IASTED Int. Conf. on Software Engineering*, November 1997. pages 281

[251] M. Rabinovich and O. Spatscheck. *Web: Caching and Replication*. Addison-Wesley, Reading, MA, 2002. pages 28

[252] M. Rabinovich, Z. Xiao, F. Douglis, and C. Kamanek. Moving edge side includes to the real edge — the clients. In *Proc. of USENIX Symp. on Internet Technologies and Systems*, 2003. pages 195

[253] R. Rajkumar, C. Lee, J. Lehoczky, and D. Siewiorek. A resource allocation model for QoS management. In *Proc. of 19th IEEE Real-Time Systems Symp. (RTSS)*, pages 298–307, 1997. pages 63, 78

[254] R. Rajkumar, C. Lee, J. Lehoczky, and D. Siewiorek. Practical solutions for QoS-based resource allication problems. In *Proc. of 20th IEEE Real-Time Systems Symp. (RTSS)*, pages 296–306, 1998. pages 63, 132

[255] A. Rao, K. Lakshminarayanan, S. Surana, R. Karp, and I. Stoica. Load balancing in structured p2p systems. In *Proc. of the 2nd Int. Workshop on Peer-to-Peer Systems (IPTPS)*, February 2003. pages 341, 343

[256] S. Ratnasamy, P. Francis, M. Handley, R. Karp, and S. Shenker. A scalable content-addressable network. In *Proc. of ACM SIGCOMM*, pages 329–350, San Diego, CA, 2001. pages 11, 152, 154

[257] J. Ravi, W. Shi, and C.-Z. Xu. Personalized email management at network edges. *IEEE Internet Computing*, 9(2):54–60, March/April 2005. pages 195

[258] Recursion Software, Inc. Voyager and software agents. `www.recursionsw.com/voyager.htm`. pages 202, 233, 268

[259] V. Ribeiro, R. Riedi, M. S. Crouse, and R. G. Baraniuk. Multiscale queueing analysis of long-range-dependent network traffic. In *Proc. of IEEE INFOCOM*, 2000. pages 130

[260] R. L. Ribler, J. S. Vetter, H. Simitci, and D. A. Reed. Autopilot: Adaptive control of distributed applications. In *Proc. of the 7th IEEE Symp. on High-Performance Distributed Computing (HPDC)*, pages 172–179, July 1998. pages 328

[261] A. Riska, W. Sun, E. Smirni, and G. Ciardo. Adaptload: Effective balancing in clustered web servers under transient load conditions. In *Proc. of the 22nd Int. Conf. on Distributed Computing Systems*, July 2002. pages 95, 111

[262] T. G. Robertazzi. *Computer Networks and Systems: Queueing Theory and Performance Evaluation*. Springer-Verlag, New York, 1990. pages 129

[263] G.-C. Roman, P. McCann, and J. Plunn. Mobile UNITY: Reasoning and specification in mobile computing. *ACM Transactions on Software Engineering and Methodology*, 6(3), July 1997. pages 248

[264] A. Roscoe. *The Theory and Practice of Concurrency*. Prentice-Hall, Englewood Cliffs, NJ, 1998. pages 258

[265] O. Rose. Statistical properties of MPEG video traffic and their impact on traffic modeling in ATM systems. In *Proc. of the IEEE 20th Conf. on Local Computer Networks*, pages 397–406, 1995. pages 140

[266] A. Rowstron and P. Druschel. Pastry: Scalable, decentralized object location and routing for large-scale peer-to-peer systems. In *Proc. of the 18th IFIP/ACM Int. Conf. on Distributed Systems Platforms (Middleware)*, 2001. pages 11, 152, 154, 155

[267] S. Roy and B. Shen. Implementation of an algorithm for fast down-scale transcoding of compressed video on the Itanium. In *Proc. 3rd Workshop on Media and Streaming Processors*, pages 119–126, 2001. pages 72

[268] S. Deerwester et al. Indexing by latent semantic analysis. *Journal of the American Society for Information Science*, 41(6):391–407, 1990. pages 181

[269] T. Sander and C. Tschudin. On software protection via function hiding. In *Proc. of Information Hiding, LNCS 1525*, pages 111–123. Springer-Verlag, London, 1998. pages 213

[270] T. Sander and C. Tschudin. Protecting mobile agents against malicious hosts. In G. Vigna, editor, *Mobile Agents and Security, LNCS 1419*, pages 44–60. Springer-Verlag, London, 1998. pages 213

[271] T. Sander and C. Tschudin. Towards mobile cryptography. In *Proc. of the IEEE Symp. on Security and Privacy*, 1998. pages 213, 214

[272] C. P. Sapuntzakis, R. Chandra, B. Pfaff, J. Chow, M. S. Lam, and M. Rosenblum. Optimizing the migration of virtual computers. In *Proc. of the 5th Symp. on Operating Systems Design and Implementation (OSDI)*, pages 377–390, December 2002. pages 340, 343

[273] L. Sarmenta and S. Hirano. Bayanihan: Building and studying web-based volunteer computing systems using Java. *Future Generation Computer Systems*, 15(5/6), 1999. pages 302

[274] R. Sarukkai. Link prediction and path analysis using markov chains. In *Proc. of the 9th Int. World Wide Web Conf.*, 2000. pages 176, 193

[275] S. Schechter, M. Krishnan, and M. Smith. Using path profiles to predict http requests. *Computer Networks and ISDN Systems*, 20:457–467, 1998. pages 176, 193

[276] T. Schroeder, S. Goddard, and B. Ramamurthy. Scalable web server clustering technologies. *IEEE Network*, 14(3):38–45, 2000. pages 18

[277] F. Sebastiani. Machine learning in automated text categorization. *ACM Computing Surveys*, 34(1):1–47, Mar. 2002. pages 181

[278] D. Seeley. A tour of the worm. In *USENIX Winter Conf. Proc.*, pages 287–304, 1989. pages 197

[279] S. Sekiguchi, M. Sato, H. Nakada, S. Matsuoka, and U. Nagashima. Ninf: Network based information library for globally high performance computing. In *Proc. of Parallel Object-Oriented Methods and Applications (POOMA'96)*, 1996. pages 302

[280] D. Serugendo, M. Muhugusa, and C. Tschudin. A survey of theories for mobile agents. *World Wide Web*, 1(3):139–153, 1998. pages 248

[281] P. Server. The great Internet Mersenne prime search. `www.mersenne.org/prime.html`. pages 302

[282] L. Sha, X. Lu, Y. Lu, and T. Abdelzaher. Queueing model based network server performance control. In *Proc. of the 23rd IEEE Real-Time System Symp.*, 2002. pages 64, 131

[283] H. Shen and C.-Z. Xu. Locality-aware randomized load balancing algorithms for structured DHT networks. In *Proc. of Int. Conf. on Parallel Processing (ICPP)*, 2005. pages 173

[284] H. Shen, C.-Z. Xu, and G. Chen. Cycloid: A constant-degree lookup-efficient p2p overlay network. *Performance Evaluation: An Internatinal Journal*, 2005. A preliminary version appeared in *Proc. of IEEE IPDPS'04*. pages 152, 154, 157, 164

[285] K. Shen, H. Tang, T. Yang, and L. Chu. Integrated resource management for cluster-based Internet services. In *Proc. of Usenix Operating Systems Design and Implementation (OSDI)*, pages 225–238, 2002. pages 66

[286] P. J. Shenoy, P. Goyal, S. Rao, and H. M. Vin. Symphony: An integrated multimedia file system. In *Proc. ACM/SPIE Multimedia Computing and Networking*, pages 124–138, 1998. pages 64

[287] P. J. Shenoy and H. M. Vin. Cello: A disk scheduling framework for next generation operating systems. In *Proc. of ACM SIGMETRICS*, pages 44–55, 1998. pages 64

[288] W. Shi and V. Karamcheti. Conca: An architecture for consistent nomadic content access. In *Proc. of the Workshop on Cache, Coherence, and Consistency (WC3'01)*, 2001. pages 195

[289] E. Shriver and C. Small. Why does file system prefetching work? In *Proc. of the 1999 USENIX Annual Technical Conf.*, June 1999. pages 194

[290] E. G. Sirer, R. Grimm, A. J. Gregory, and B. N. Bershad. Design and implementation of a distributed virtual machine for networked computers. In *Proc. of the 17th ACM Symp. on Operating Systems Principles (SOSP)*, pages 202–216, December 1999. pages 323, 340

[291] L. P. Slothouber. A model of Web server performance. In *Proc. of 5th Int. World Wide Web Conf.*, 1996. pages 130

[292] J. Smith and R. Nair. *Virtual Machines: versatile Platforms for Systems and Processes*. Morgan Kaufmann, San Francisco, 2005. pages 340

[293] W. D. Smith. TPC-W: Benchmarking an e-commerce solution. `www.tpc.org/tpcw`. pages 98, 102

[294] A. C. Snoeren and H. Balakrishnan. An end-to-end approach to host mobility. In *Proc. 6th Int. Conf. on Mobile Computing and Networking (MobiCom)*, 2000. pages 280

[295] S. Soman, C. Krintz, and G. Vigna. Detecting malicious Java code using virtual machine auditing. In *Proc. of the 12th USENIX Security Symp.*, 2003. pages 207

[296] E. H. Spafford. The Internet worm: Crisis and aftermath. *Communications of ACM*, 32(6):678–687, 1989. pages 197

[297] M. Squiillante and E. Lazowska. Using processor-cache affinity information in shared-memory multiprocessor. *IEEE Transactions on Parallel and Distributed Systems*, 4(2):131–143, Feb. 1993. pages 33

[298] M. S. Squillante, D. Yao, and L. Zhang. Web traffic modeling and Web server performance analysis. In *Proc. of IEEE Conf. on Decision and Control*, 1999. pages 136

[299] J. Stamos and D. Gifford. Remote evaluation. *ACM Transactions on Programming Languages and Systems*, 12(4):537–565, 1990. pages 198

[300] A. D. Stefano and C. Santoro. Locating mobile agents in a wide distributed environment. *IEEE Transactions on Parallel and Distributed Systems*, 13(8):844–864, Aug. 2002. pages 233

[301] I. Stoica, R. Morris, D. Liben-Nowell, K. M. F. Karger, D. R. Karger, F. Dabek, and H. Balakrishnan. Chord: A scalable peer-to-peer lookup protocol for Internet applications. In *Proc. of the 2001 ACM SIGCOMM Conf.*, 2001. pages 11, 152, 154, 163, 170, 341

[302] M. Strasser and K. Rothermel. Reliability concepts for mobile agents. *International Journal of Cooperative Information Systems*, 7(4):355–382, 1998. pages 226

[303] Z. Su, Q. Yang, and H. Zhang. A prediction system for multimedia prefetching in Internet. In *Proc. of ACM Multimedia*, 2000. pages 193

[304] F. Sultan, K. Srinivasan, D. Iyer, and L. Iftode. Migratory TCP: Highly available Internet services using connection migration. In *Proc. of the 22nd Int. Conf. on Distributed Computing Systems*, July 2002. pages 280

[305] V. Sundaram, A. Chandra, P. Goyal, P. J. Shenoy, J. Sahni, and H. M. Vin. Application performance in the QLinux multimedia operating system. In *Proc. 8th ACM Multimedia Conf.*, pages 127–136, 2000. pages 64

[306] J. Tardo and L. Valente. Mobile agent security and telescript. In *Proc. of IEEE COMPCON*, pages 58–63, 1996. pages 216

[307] A. Tripathi, N. Karnik, M. Vora, T. Ahmed, and R. Singh. Design of the ajanta system for mobile agent programming. *Journal of Systems and Software*, May 2002. pages 202, 218, 219, 226

[308] K. Trivedi. *Probability and Statistics with Reliability, Queueing, and Computer Science Applications*. Prentice-Hall, Englewood Cliffs, NJ, 1982. pages 89

[309] E. Truyen, B. Robben, B. Vanhaute, T. Coninx, W. Joosen, and P. Verbaeten. Portable support for transparent thread migration in Java. In *Proc. of the 4th Int. Symp. on Mobile Agents*, pages 29–43. Springer-Verlag, 2000. pages 332

[310] A. Unypoth and P. Sewell. Nomadic pict: Correct communication infrastructure for mobile computation. In *Proc. of the 28th Annual ACM Symp. on Principles of Programming Languages*, January 2001. pages 248

[311] V. Welch et al. Security for grid services. In *Proc. of the 12th IEEE Int. Symp. on High Performance Distributed Computing (HPDC)*, pages 48–57, June 2003. pages 334

[312] S. Vadhiyar and J. Dongarra. A performance oriented migration framework for the grid. In *Proc. of the 3rd Int. Symp. on Cluster Computing and the Grid (CCGrid)*, pages 130–137, May 2003. pages 342

[313] N. Venkatasubramanian and S. Ramanathan. Load management in distributed video servers. In *Proc. IEEE Int. Conf. on Distributed Computing Systems (ICDCS)*, pages 31–39, 1997. pages 40

[314] G. Vigna. Cryptographic traces for mobile agents. In G. Vigna, editor, *Mobile Agents and Security, LNCS 1419*, pages 137–153. Springer-Verlag, London, 1998. pages 211

[315] A. Vogel, B. Kerherve, G. Bochmann, and J. Gecsei. Distributed multimedia and QoS: A survey. *IEEE Multimedia*, 2(2):10–19, 1995. pages 40

[316] W3C. Compositie capability/preference profile (cc/pp): A user side framework for content negotiation. `www.w3c.org/mobile/ccpp`. pages 59

[317] C. A. Waldspurger and W. E. Weihl. Lottery scheduling: Flexible proportional-share resource manangement. In *Proc. of 1st Symp. on Operating System Design and Implementation (OSDI'94)*, pages 1–11, November 1994. pages 65, 77, 114, 346

[318] T. Walsh, N. Paciorek, and D. Wong. Security and reliability in concordia. In *Proc. of the 31st Hawaii Int. Conf. on Systems Sciences (HICSS)*, pages 44–53, 1998. pages 218

[319] J. Warmer and A. Kleppe, editors. *The Object Constraint Language: Precise Modeling with UML.* Addison-Wesley, Reading, MA, 1998. pages 275

[320] J. Wei, C. Xu, and X. Zhou. A robust packet scheduling algorithm for proportional delay differentiation services. In *Proc. of the IEEE Global Telecommunications Conf. (Globecom)*, Nov. 2004. pages 62, 77, 89

[321] J. Wei and C.-Z. Xu. eQoS: Provisioning of client-perceived end-to-end QoS guarantees in Web servers. In *Proc. of Int. Workshop on Quality of Services (IWQoS)*, 2005. pages 64, 127

[322] J. Wei, X. Zhou, and C.-Z. Xu. Robust processing rate allocation for proportional slowdown differentiation on Internet servers. *IEEE Transactions on Computers*, 2005. (In press). pages 64, 112, 121, 124, 131

[323] A. Whitaker, M. Shaw, and S. Gribble. Denali: Lightweight virtual machines for distributed and networked applications. In *Proc. of the USENIX Technical Conf.*, June 2002. pages 340

[324] J. E. White. Mobile agents make a network an open platform for third-party developers. *IEEE Computer*, 27(11):89–90, November 1994. pages 215, 219

[325] W. Willinger, R. Govindan, S. Jamin, V. Paxson, and S. Shenker. Scaling phenomena in the Internet: Critically examining criticality. *Proc. of the National Academy of Sciences of USA*, 99 (Suppl. 1):2573–2580, February 2002. pages 117

[326] P. Wojciechowski and P. Sewell. Nomadic Pict: Language and infrastructure design for mobile agents. *IEEE Concurrency*, pages 42–52, April/June 2000. pages 248

[327] P. T. Wojciechowski. Algorithms for location-independent communication between mobile agents. In *Proc. of AISB Symp. on Software Mobility and Adaptive Behaviour*, pages 10–19, March 2001. pages 202, 280

[328] J. L. Wolf and P. S. Yu. On balancing the load in a clustered Web farm. *ACM Transactions on Internet Technology*, 1(2):231–261, 2001. pages 78

[329] R. W. Wolff. *Stochastic Modeling and the Theory of Queues*. Prentice-Hall, Englewood Cliffs, NJ, 1999. pages 34

[330] D. Wong, N. Paciorek, and D. Moore. Java-based mobile agents. *Communication of ACM*, 42(3):92–101, March 1999. pages 219

[331] S. Woo, M. Ohara, E. Torrie, J. Singh, and A. Gupta. The SPLASH-2 programs: Characterization and methodological considerations. In *Proc. of the 22nd Annual Int. Symp. on Computer Architecture*, June 1995. pages 336

[332] K. Wu, P. S. Yu, and J. Wolf. Segment-based proxy caching of multimedia streams. In *Proc. of Int. World Wide Web Conf.*, 2001. pages 194

[333] C. Xia and Z. Liu. Queueing systems with long-range dependent input process and subexponential service times. In *Proc. of the ACM SIGMETRICS*, pages 25–36, 2003. pages 130

[334] C.-Z. Xu. Naplet: A flexible mobile agent framework for network-centric applications. In *Proc. of IEEE IPDPS Workshop on Internet Computing and E-Commerce (ICEC)*, pages 219–226, April 2002. pages 219

[335] C.-Z. Xu and S. Fu. Privilege delegation and agent-oriented access control in Naplet. In *Proc. of IEEE ICDCS Workshop on Mobile Distributed Computing (MDC)*, pages 493–497, April 2003. pages 203, 207, 265, 289

[336] C.-Z. Xu and T. Ibrahim. Neural net based prefetching to tolerate web access latency. In *Proc. of 20th IEEE Int. Conf. on Distributed Computing Systems*, 2000. pages 176

[337] C.-Z. Xu and T. Ibrahim. Keyword-based semantic prefetching in Internet news services. *IEEE Transactions on Knowledge and Data Engineering*, 16(5):601–611, May 2004. pages 176

[338] C.-Z. Xu and F. Lau. *Load Balancing in Parallel Computers: Theory and Practice*. Kluwer Academic, Dordrecht, 1997. pages 34, 46, 198

[339] C.-Z. Xu, L. Wang, and N.-T. Fong. Stochastic predication of execution time for dynamic bulk synchronous computation. *Journal of Supercomputing*, 21(1):91–103, January 2002. pages 346

[340] C.-Z. Xu and B. Wims. Mobile agent based push methodology for global parallel computing. *Concurrency: Practice and Experience*, 14(8):705–726, July 2000. pages 242, 280, 303, 341, 344

[341] D. Xu, J. Yin, Y. Deng, and J. Ding. A formal architecture model for logical agent mobility. *IEEE Transactions on Software Engineering*, 29(1):31–45, January 2003. pages 249

[342] M. Xu and C.-Z. Xu. Decay function model for resource configuration and adaptive allocation on Internet servers. In *Proc. of 12th IEEE Int. Workshop on Quality-of Services (IWQoS 2004)*, pages 37–46, 2004. pages 131

[343] Z. Xu, C. Tang, and Z. Zhang. Building topology-aware overlays using global soft-state. In *Proc. of Int. Conf. on Distributed Computing Systems*, 2003. pages 155

[344] B. Yee. *Using Secure Coprocessors*. PhD thesis, School of Computer Science, Carnegie Mellon University, Pittsburg, 1994. pages 212

[345] G. Yin, C.-Z. Xu, and L. Wang. Optimal remapping in dynamic bulk synchronous computations via a stochastic control approach. *IEEE Transactions on Parallel and Distributed Systems*, 14(1):51–62, January 2003. pages 344, 346, 354

[346] J. Youn, M. T. Sun, and C. W. Lin. Motion vector refinement for high-performance transcoding. *IEEE Transactions on Multimedia*, 1(1):30–40, 1999. pages 72

[347] A. Yu and W. Cox. Java/DSM: A platform for heterogeneous computing. In *Proc. of ACM 1997 Workshop on Java for Science and Engineering Computation*, June 1997. pages 324

[348] V. C. Zandy and B. P. Miller. Reliable network connections. In *Proc. 8th Annual ACM/IEEE Int. Conf. on Mobile Computing and Networking*, pages 95–106, September 2002. pages 281

[349] W. Zhang. Linux virutal server for scalable network services. In *Proc. of Ottawa Linux Symp.*, 2000. Also available at `www.linux-vs.org`. pages 19

[350] Y. Zhang and S. Dao. A persistent connection model for mobile and distributed systems. In *Proc. of Int. Conf. on Computer Communications and Networks*, September 1995. pages 280

[351] B. Zhao, J. Kubiatowicz, and A. Joseph. Tapestry: An infrastructure for fault-tolerant wide-area location and routing. Technical Report UCB/CSD-01-1141, Computer Science Division, UC Berkeley, 2001. pages 11, 152

[352] T. Zhao and V. Karamcheti. Enforcing resource sharing agreements among distributed server cluster. In *Proc. of IEEE Int. Parallel and Distributed Processing Symp. (IPDPS)*, 2002. pages 103

[353] X. Zhong and C.-Z. Xu. A reliable connection migration mechanism for synchronous communication in mobile codes. In *Proc. of Int. Conf. on Parallel Processing (ICPP)*, pages 431–438, 2004. pages 281

[354] X. Zhong and C.-Z. Xu. Optimal time-variant resource allocation for Internet servers with delay constraints. In *Proc. of 11th IEEE Real-Time Technology and Applications Symp. (RTAS)*, Mar. 2005. pages 149

[355] S. Zhou, X. Zheng, J. Wang, and P. Delisle. Utopia: A load sharing facility for large, heterogeneous distributed computer systems. *Software — Practice and Experience*, 23(12):1305–1336, 1993. pages 198, 301

[356] X. Zhou, J. Wei, and C.-Z. Xu. Modeling and analysis of 2d service differentiation on e-commerce servers. In *Proc. of the IEEE Conf. on Distributed Computing Systems (ICDCS)*, pages 740–747, 2004. pages 63, 95, 101

[357] X. Zhou, J. Wei, and C.-Z. Xu. Processing rate allocation for slowdown differentiation on Internet servers. In *Proc. of Int. Parallel and Distributed Processing Symp. (IPDPS)*, 2004. pages 112

[358] X. Zhou and C.-Z. Xu. Optimal video replication and placement on a cluster of video-on-demand servers. In *Proc. of IEEE 31st Int. Conf. on Parallel Processing (ICPP)*, pages 547–555, 2002. pages 39, 42, 49, 50

[359] X. Zhou and C.-Z. Xu. Request redirection and data layout for network traffic balancing in cluster-based video-on-demand servers. In *Proc. of IEEE IPDPS Workshop on PDIVM*, 2002. pages 39, 51

[360] X. Zhou and C.-Z. Xu. Harmonic proportional bandwidth allocation and scheduling for service differentiation on streaming servers. *IEEE Transactions on Parallel and Distributed Systems*, 15(9):835–848, September 2004. pages 63, 69, 72, 91, 131

[361] H. Zhu, H. Tang, and T. Yang. Demand-driven service differentiation for cluster-based network servers. In *Proc. of IEEE INFOCOM*, pages 679–688, 2001. pages 66, 72, 80, 83, 95, 111

[362] W. Zhu, C.-L. Wang, and F. C. M. Lau. JESSICA2: A distributed Java virtual machine with transparent thread migration support. In *Proc. of the 4th Int. Conf. on Cluster Computing*, September 2002. pages 324, 331, 341, 343

Index

Copyright Permission

- Portions of Chapter 3 draw, with permission of IEEE, on the article "Optimal video replication and placement on a cluster of video-on-demand servers" in *Proc. of Int. Conf. on Parallel Processing (ICPP)*, IEEE Computer Society, 2002.
- Portions of Chapter 5 draw, with permission of IEEE, on the article "Harmonic proportional bandwidth allocation and scheduling for service differentiation on streaming servers" in *IEEE Transaction on Parallel and Distributed Systems*, September 2004.
- Portions of Chapter 6 draw, with permission of IEEE, on the article "Modeling and analysis of 2D service differentiation on e-commerce servers" in *Proc. of Int Conf. on Distributed Computing Systems (ICDCS)*, IEEE Computer Society, 2004.
- Portions of Chapter 7 draw, with permission of IEEE, on the article "Robust processing rate allocation for proportional slowdown differentiation on Internet servers" in *IEEE Transactions on Computers*, 2005.
- Portions of Chapter 8 draw, with permission of IEEE, on the article "Decay function model for resource configuration and adaptive allocation on Internet servers" in *Proc. of Int. Workshop on Quality of Services (IWQoS)*, IEEE Computer Society, 2004.
- Portions of Chapter 9 draw, with permission of IEEE, on the article "Cycloid: a constant-degree lookup-efficient P2P overlay network" in *Proc. of Int. Parallel and Distributed Processing Symp. (IPDPS)*, IEEE Computer Society, 2004.
- Portions of Chapter 10 draw, with permission of IEEE, on the article "Neural net based prefetching to tolerate WWW latency" in *Proc. of Int. Conf. on Distributed Computing Systems (ICDCS)*, IEEE Computer Society, 2000, and the article "Keyword-based semantic prefetching in Internet news services" in *IEEE Transactions on Knowledge and Data Engineering*, May 2004.
- Portions of Chapter 13 draw, with permission of IEEE, on the article "A formal framework for agent itinerary specification, safety reasoning, and logic analysis" in *Proc. of Int. Workshop on Mobile Distributed Computing (MDC)*, IEEE Computer Society, 2005.
- Portions of Chapter 15 draw, with permission of IEEE, on the article "A reliable connection migration mechanism for synchronous communication in mobile codes" in *Proc. of Int. Conf. on Parallel Processing (ICPP)*, IEEE Computer Society, 2004.
- Portions of Chapter 16 draw, with permission of IEEE, on the article "Traveler: a mobile agent based infrastructure for wide area parallel computing" in *Proc. Joint Int. Symp. on Agent Systems and Applications and on Mobile Agents (ASA/MA)*, IEEE Computer Society, 1999.
- Portions of Chapter 17 draw, with permission of IEEE, on the article "Service migration in distributed virtual machines for adaptive grid computing" in *Proc. of Int. Conf. on Parallel Processing (ICPP)*, 2005.
- Portions of Chapter 18 draw, with permission of IEEE, on the article "Migration decision for hybrid mobility in reconfigurable virtual machines" in *Proc. of Int. Conf. on Parallel Processing (ICPP)*, IEEE Computer Society, 2004.

Milton Keynes UK
Ingram Content Group UK Ltd.
UKHW021827071024
449327UK00021B/1454

9 780367 392666